Randomization Tests
Fourth Edition

STATISTICS: Textbooks and Monographs

Recent Titles

Randomization Tests
Fourth Edition

Eugene S. Edgington
University of Calgary
Alberta, Canada

Patrick Onghena
Katholieke Universiteit Leuven
Belgium

 Chapman & Hall/CRC
Taylor & Francis Group
Boca Raton   London   New York

Chapman & Hall/CRC is an imprint of the
Taylor & Francis Group, an informa business

Chapman & Hall/CRC
Taylor & Francis Group
6000 Broken Sound Parkway NW, Suite 300
Boca Raton, FL 33487-2742

International Standard Book Number-10: 1-58488-589-0 (Hardcover)
International Standard Book Number-13: 978-1-58488-589-4 (Hardcover)

Library of Congress Cataloging-in-Publication Data

Edgington, Eugene S., 1924-
    Randomization tests. -- 4th ed. / Eugene Edgington and Patrick Onghena.
       p. cm. -- (Statistics, textbooks and monographs ; v. 191)
    Includes bibliographical references and index.
    ISBN 978-1-58488-589-4 (alk. paper)
    1. Statistical hypothesis testing. I. Onghena, Patrick. II. Title. III. Series.

QA277.E32 2007
519.5'6--dc22
                                                 2006032352

Visit the Taylor & Francis Web site at
http://www.taylorandfrancis.com

and the CRC Press Web site at
http://www.crcpress.com

Preface to the Fourth Edition

The number of innovative applications of randomization tests in various fields and recent developments in experimental design, significance testing, computing facilities, and randomization test algorithms necessitated a new edition of *Randomization Tests*. Also, the rapidly growing number of books on resampling and computer-intensive procedures that confuse randomization tests with permutation tests that do not require random assignment to treatments indicates that the time has come to distinguish, re-emphasize, and clarify the unique approach offered by randomization tests.

Large parts of the third edition were rewritten and reorganized while retaining the general approach of the previous editions. The book now focuses more on the design than on the test statistics, which is reflected in the revised chapter titles, and there is an expansion and elaboration of issues that often have led to discussion and confusion in the research literature, like the type of hypotheses that are being tested and the role of random assignment.

Like previous editions of *Randomization Tests*, this edition can serve as a practical source book for researchers but it puts more emphasis on randomization test rationale. A sound understanding of this rationale will enable the experimenter to gain large benefits from the randomization test approach. Furthermore, a more didactical approach in the exposition and the inclusion of exercises and many new examples make the book more accessible as a textbook for students. And for the sake of simplification, we used "he" to refer to the experimenter.

The most important changes are in the first three chapters, which have been completely rewritten to emphasize the irrelevance and implausibility of the random sampling assumption for the typical experiment and to drive home the message that randomization tests constitute the most basic and valid statistical tests for randomized experiments. Another major change is the combination of repeated-measures and randomized block designs in Chapter 6 after a discussion on factorial designs and interactions in Chapter 5. References to general randomization and permutation test software and publications on permutation algorithms have been integrated in the separate chapters and an overview of free and commercial computer programs with the essential contact information has been included in Chapter 15. Also, in Chapter 11 on N-of-1 designs, more attention is given to the practicality of the tests and the availability of user-friendly software to perform them.

For each chapter, the Fortran codes have been moved from the text to the CD that accompanies this book to make them immediately accessible in electronic format. Furthermore, on the CD an executable randomization test software package, RT4Win, can be found that implements the main programs that are discussed in the book. The CD also contains SCRT, a program for randomization tests in N-of-1 trials as discussed in Chapter 11, and COMBINE, a program to combine P-values as discussed in Chapter 6.

For this fourth edition of *Randomization Tests*, the following website is made available: *http://ppw.kuleuven.be/cmes/rt.htm*. This website contains solutions to the exercises and updates about new releases of RT4Win and other information that is relevant for the book and randomization tests in general.

We want to thank Daniele De Martini, Cyrus Mehta, Fortunato Pesarin, Luigi Salmaso, and Dean McKenzie for their openhearted and thought-provoking discussion of the many ideas reflected in this fourth edition while they were visitors at the Katholieke Universiteit Leuven. Also, warmest thanks are extended to Vance Berger, Peter Bruce, John Ferron, Phillip Good, Tom Kratochwill, Joel Levin, Norman Marsh, and the other members of the exact-stats discussion list. The many e-mails from Dick May and John Todman and meetings with them are truly appreciated. Special thanks and obeisance go to Cliff Lunneborg for his continued inspiration, encouragement, and generosity, but who sadly died in a diving accident while the fourth edition was getting its final shape. We want to dedicate this book to Cliff.

Four people at the Katholieke Universiteit Leuven have been especially helpful during the preparation of this fourth edition. We want to thank Ria Nartus and Bartel Wilms for the indispensable administrative and technical support that is needed for a project like this; Sigrid Van de Ven, who helped us out with the tables and figures faster than anyone can imagine; and Ming Huo for a wonderful job in developing the Randomization Tests "four" Windows (RT4Win) software package.

Finally and most importantly, we want to thank Vera Corfield, Heidi Kennes, An Onghena, and Bieke Onghena for their love and support.

Eugene S. Edgington

Patrick Onghena

Preface to the Third Edition

In recent years, a growing interest in randomization tests and their applications has been reflected in numerous articles and books. There has been widespread application of tests and programs given in the earlier editions, and many computer programs have been produced in recent years by researchers in various fields, as well as by statisticians and computer scientists. Whereas at one time, access to a mainframe computer installation was necessary to make randomization tests practical for general use, personal computers are now quite adequate. Researchers in many fields have acknowledged that in the modern age of computers, randomization tests frequently will be the tests of choice.

In this edition, additional factorial designs are introduced and completely randomized factorial and randomized block designs are covered in separate chapters, Chapters 6 and 7, to clarify the distinction between two basically different factorial designs.

In parametric statistics, "random models" — in which there is random selection of treatment levels — are sometimes advocated for expanding the scope of the treatments about which inferences can be drawn. However, the test itself and consequently the P-value is usually unaffected by employment of a random model instead of a "fixed model" in which treatment levels are systematically chosen. Alternatively, for randomization tests random selection of treatment levels permits a different type of test in which the sensitivity of a test can be substantially increased. This principle is demonstrated in Chapter 11.

Chapter 12, on single-subject randomization tests, has been extensively reorganized and supplemented. The problems of carryover effects are shown to be essentially the same as for repeated-measures experiments. Special "N-of-1" service facilities have conducted a large number of single-subject randomized clinical trials in recent years, and to facilitate the creation of similar services elsewhere, we describe the manner in which those services evolved from the study and application of single-subject randomization tests. The principles of single-subject randomization tests can be useful for intensive study of other units than subjects, and this is demonstrated by their application to single nerve cells and other single units in neuroscience.

Chapters 13 and 14 are entirely new and their primary function is to show how, with no additional assumptions, randomization tests can test "null" hypotheses of specified amounts or types of treatment effect, not just the traditional null hypothesis of no effect.

Chapter 15, "Theory," has been doubled in length to provide a more detailed analysis of the relationship among the randomization (random assignment procedure), the null hypothesis, and the permuting of data for significance determination.

The Appendix lists sources of free and commercial software for performing permutation tests. Such software can be employed in running randomization tests. New computing algorithms that speed up permutation tests considerably are discussed. (In the body of the book, the earlier Fortran program listings are retained, in part because of their practical value but primarily because their simplicity is an asset for illustrating fundamental features of randomization tests and randomization test software.)

I am grateful to Lyle Jones for varied assistance he has provided over the years. In his 1967 review recommending publication of my book, *Statistical Inference: The Distribution-Free Approach* (McGraw-Hill, 1969), he foresaw the impact of randomization tests, and in 1978 he helped fulfill that prophesy as a member of the Brillinger, Jones, and Tukey Statistical Task Force — which, in a widely cited report to the U.S. Secretary of Commerce, strongly recommended the use of randomization tests for the analysis of weather modification data.

An expression of appreciation is due Rose Baker of the University of Salford who contributed the Appendix on modern permutation test software, a topic on which she is eminently qualified to write.

I thank Patrick Onghena for our years of correspondence about randomization tests and I greatly appreciate his and Luc Delbeke's hospitality during my visit at the University of Leuven in Belgium.

At the University of Calgary, four people have been especially helpful during preparation of this third edition. Anton Colijn and Larry Linton generously took time to explain the technical details of permutation test software; Alison Wiigs exhibited great skill and diligence in formatting the manuscript; and Vera Corfield supplied a judicious mixture of encouragement and objective criticism.

Eugene S. Edgington

Preface to the Second Edition

This edition includes new material distributed throughout the book. The major differences between this and the first edition occur in Chapters 1, 6, 11, and 12 in this edition.

Section 1.13, "Major Advances," was added to Chapter 1 for readers who wanted to learn something about the historical background of randomization tests. Reading the works of R.A. Fisher and E.J.G. Pitman in preparing Section 1.13 led to the realization that Fisher often has been given credit for contributions made by Pitman. For example, statisticians have described the "Fisher randomization test," applied to Darwin's data on plant length, as if it did not require the assumption of random sampling, and that is incorrect. Fisher's test required random sampling as well as random assignment; it was Pitman who pointed out that randomization tests could be performed in the absence of random sampling. One must turn to the works of Pitman — not Fisher — to find a theoretical basis for randomization or other permutation tests. Also outlined in Section 1.13 are major advances in theory and application that have been made since Fisher and Pitman.

In response to a widespread interest in randomization tests for factorial designs, Chapter 6 was completely rewritten and a computer program was added. The discussion of interactions was expanded to show how randomization tests sensitive to interactions may be conducted.

Chapter 11 is a new chapter that presents randomization tests that do not fit well into the other chapters. The tests are of practical value and are also useful for demonstrating the potential of the randomization test approach for increasing the flexibility of experimental design and data analysis.

Chapter 12, on randomization test theory, is the most important addition to *Randomization Tests*. It provides a sound theoretical framework for understanding and developing randomization tests based on a principle given in a 1958 article by J.H. Chung and D.A.S. Fraser.

Some of the randomization tests in this edition were developed with the financial support of the Natural Sciences and Engineering Research Council of Canada (Grant number A5571). Otto Haller wrote the two new computer programs, Program 6.1 and Program 11.1.

Eugene S. Edgington

Preface to the First Edition

This book is intended primarily as a practical guide for experimenters on the use of randomization tests, although it can also be used as a textbook for courses in applied statistics. The book will be useful in all fields where statistical tests are commonly used. However, it is likely to be most relevant in the fields of psychology, education, biology, and medicine. No special mathematical background is required to understand the book but it is assumed that the reader has completed an introductory statistics course.

Experimental design books and others on the application of statistical tests to experimental data perpetuate the long-standing fiction of random sampling in experimental research. Statistical inferences are said to require random sampling and to concern population parameters. However, in experimentation random sampling is very infrequent; consequently, statistical inferences about populations are usually irrelevant. Thus, there is no logical connection between the random sampling model and its application to data from the typical experiment. The artificiality of the random sampling assumption has undoubtedly contributed to the skepticism of some experimenters regarding the value of statistical tests. However, what is a more important consequence of the failure to recognize the prevalence of nonrandom sampling in experimentation is overlooking the need for special statistical procedures that are appropriate for nonrandom samples. As a result, the development and application of randomization tests have suffered.

Randomization tests are statistical tests in which the data are repeatedly divided, a test statistic (e.g., t or F) is computed for each data division, and the proportion of the data divisions with as large a test statistic value as the value for the obtained results determines the significance of the results. For testing hypotheses about experimental treatment effects, random assignment but not random sampling is required. In the absence of random sampling, the statistical inferences are restricted to the subjects actually used in the experiment, and generalization to other subjects must be justified by nonstatistical argument.

Random assignment is the only random element necessary for determining the significance of experimental results by the randomization test procedure; therefore, assumptions regarding random sampling and those regarding normality, homogeneity of variance, and other characteristics of randomly sampled populations are unnecessary. No matter how simple or complex, any statistical test is thus transformed into a distribution-free test when significance is determined by the randomization test procedure. For any experiment with random assignment, the experimenter can guarantee the validity of any test they want to use by determining significance by the randomization test procedure. Chapter 1 summarizes various advantages of the randomization test procedure, including its potential for developing statistical tests to meet the special requirements of a particular experiment and its usefulness in providing for the valid use of statistical tests on experimental data from a single subject.

A great deal of computation is involved in performing a randomization test and for that reason, such a means of determining significance was impractical until recent years when computers became accessible to experimenters. As the use of computers is essential for the practical application of randomization tests, computer programs for randomization tests accompany discussions throughout the book. The programs will be useful for a number of practical applications of randomization tests but their main purpose is to show how programs for randomization tests are written.

Inasmuch as the determination of significance by the randomization test procedure makes any of the hundreds (perhaps thousands) of published statistical tests into randomization tests, the discussion of application of randomization tests in this book cannot be exhaustive. Applications in this book have been selected to illustrate different facets of randomization tests so that the experimenter will have a good basis for generalizing to other applications.

Randomization Tests is dedicated to my wife, Jane Armitage Edgington, because of her encouragement and understanding. It was written during sabbatical leave granted me by the University of Calgary. Many of my colleagues at the university have been helpful but three of them deserve a special thanks: Allan Strain, for writing most of the programs; Otto Haller, for many enjoyable discussions of randomization tests; and Gerard Ezinga, to whom I am grateful for his careful reading of the entire manuscript. Finally, I want to express my indebtedness to Benjamin S. Duran of Texas Tech University, whose critical and sympathetic review was of inestimable value in revising the manuscript.

Authors

Eugene S. Edgington is emeritus professor of psychology at the University of Calgary in Calgary, Alberta, Canada. Dr. Edgington is the author of a book on distribution-free statistical inference and numerous papers on statistics. His special research interests include nonparametric statistics, measurement theory, and classification theory. He is a member of the American Psychological Association. Dr. Edgington received his B.S. (1950) and M.S. (1951) degrees in psychology from Kansas State University, and his Ph.D. (1955) degree in psychology from Michigan State University.

Patrick Onghena is professor of educational statistics at the Katholieke Universiteit Leuven in Leuven, Belgium. Dr. Onghena coauthored several books on methodology and statistics and numerous papers in the scientific literature. His main research interests include nonparametric statistics, single-case experimental designs, meta-analysis, and statistical power analysis. He acts as associate editor of the *Journal of Statistics Education* and of *Behavior Research Methods*. He is a member of the Belgian Statistical Society and the Belgian Federation of Psychologists, and an international affiliate of the American Psychological Association and the American Educational Research Association. Dr. Onghena received his B.S. (1985), M.S. (1988), and Ph.D. (1994) degrees in psychology from the Katholieke Universiteit Leuven.

Table of Contents

1

Statistical Tests that Do Not Require Random Sampling

Randomization tests are a subclass of statistical tests called *permutation tests*. Permutation tests are tests in which the P-value is the proportion of data permutations or configurations providing a test statistic as large as (or as small as) the value for the research results. Randomization tests are permutation tests for randomized experiments, which are experiments where there is a random assignment of research units to treatments for testing null hypotheses about treatment effects. Randomization (random assignment) is the only random element that the experimenter must introduce to permit the use of a randomization test to determine the significance of a treatment effect. Although random sampling can be used in addition to or as a substitute for random assignment, random selection of experimental units is unnecessary; the validity of a randomization test does not depend on how the subjects or other research units are selected.

As statistical tests that do not require random sampling, randomization tests have properties quite different from traditional tests, where statistical inferences are often thought to necessarily be inferences about population parameters based on random sampling. Randomization tests make it possible to draw valid statistical inferences about experimental treatment effects on a group of units that have not been randomly selected. (Nonrandom samples are sometimes called "convenience samples" but this term can be misleading because an experimenter is not likely to simply select any units at hand but will screen potential units to ensure their appropriateness for the experiment.)

In a sense, randomization tests are the ultimate nonparametric tests. To say that they are free from parametric assumptions is a gross understatement of their freedom from questionable assumptions — they are free from the most conspicuously incorrect assumption of all, which is the assumption that the subjects or other research units were randomly drawn from a population.

1.1 Randomization Tests

Randomization tests[1] are statistical procedures based on the random assignment of experimental units to treatments to test hypotheses about treatment
effects. After the results are obtained for the actual assignment, one determines the results that would have occurred for other possible assignments
if the null hypothesis were true. For example, there are $6!/3!3! = 20$ ways to
assign six subjects to two treatments with three subjects per treatment, so
for application of a randomization test the experimental results plus 19
hypothetical outcomes for the alternative possible assignments could be
used. The set of possible assignments is transformed into a *reference set* of
potential results on the basis of the results for the random assignment that
was actually carried out (this reference set is the analog are of the sampling
distribution used with random sampling tests). All possible assignments do
not need to be included in the reference set; however, subsets of the exhaustive reference set can be validly employed to determine significance as well.
As with other permutation tests, the P-value is the proportion of test statistics
in the reference set that are greater than or equal to (or in some cases, less
than or equal to) the obtained test statistic value.

In early discussions of random assignment (e.g., Fisher, 1926), researchers
were primarily concerned with plots of land as the *experimental units* that were
assigned to treatments, but in experiments in psychology, education, biology,
medicine, and other areas employing randomized experiments, the experimental units are commonly persons or animals, parts of a person or animal (e.g.,
left ear or right ear), or groups of people or animals (e.g., families, litters, sets
of twins, teams of horses, or married couples). However, randomization tests
are also applicable to any other experimental units randomly assigned to experimental treatments (e.g., neurons, trees, geographical entities, and so on).

Before proceeding with a discussion of randomization tests in general, it
will be helpful to consider examples of specific applications. In these examples, random assignment is the sole random element that must be introduced
by the experimenter. The experimental units can be selected according to the
requirements of the experimenter because random sampling is not required.
Randomization tests can be free of the assumption of normality or any other
assumption about a sampled population because samples from populations
are not required. The statistical inferences refer to the actual experiment and
the units involved in it.

1.2 Numerical Examples

Example 1.1. An experimenter is interested in determining whether there is
a difference between the effects of treatments A and B on reaction time. If
the two treatments can be shown to have a different effect on the reaction

times of the subjects, it will be theoretically valuable. The experimenter wants to test this null hypothesis:

For every subject in the experiment, the reaction time is the same as it would have been under assignment to the other treatment.

The alternative hypothesis is the complement of the null hypothesis:

For some (one or more) subjects, the reaction time would have been different under assignment to the other treatment.

The requirements of the experimental task being rather complex, the experimenter is very careful in selecting the subjects, finally finding 10 subjects that are suitable. The experimenter expects one of the treatments to provide a shorter reaction time than the other treatment for at least some of the 10 subjects but cannot anticipate which treatment that would be. The experimenter decides to use $|D|$, the absolute difference between means, as the test statistic. Five subjects are randomly assigned to each of the treatments and the experiment is run. (We will designate the subjects as $a, b, c, d, e, f, g, h, i$, and j.)

The experimental results in arbitrary measurement units for the actual assignment are:

A: 10, 8, 12, 11, 6   for subjects e, g, h, b, f

B: 13, 14, 16, 14, 15   for subjects d, c, i, a, j

$|D|$, the absolute difference between means, is 5.0.

To construct the reference set to determine the P-value, the experimenter represents the results that would have occurred under alternative assignments if the null hypothesis were true. There are $10!/5!5! = 252$ distinctive *divisions of subjects* between the two treatments with five subjects per treatment, so under the null hypothesis there are 252 divisions of the 10 reaction time measurements between A and B. (Notice that because there are tied measurements — subjects c and a both have a measurement of 14 — not all *divisions of measurements* are distinctive.) The P-value is the proportion of those 252 divisions of data that have a value of $|D|$ as large as the obtained value of 5.0. With results showing no overlap, the experimenter does not need to actually perform all 252 data divisions and compute all 252 test statistic values. With the five lowest reaction time measurements for one treatment and the five highest reaction time measurements for the other treatment, the value of $|D|$ is a maximum. Of the 251 other data divisions, none could give a larger $|D|$ and only one, namely the one where all A and B measurements are switched, would give the same $|D|$. Therefore, within the reference set of 252 data divisions there are only two with such a $|D|$ as large as 5.0 (the obtained result), so the P-value is 2/252, or about 0.008.

The logical justification for this procedure for determining the significance of a null hypothesis of identity of treatment effects is straightforward. The null hypothesis H_0 is that the reaction time for every subject is the same as it would have been for any other assignment to treatments. The null hypothesis assumes that assignment of a subject to A or B "assigns" the reaction time of that subject to that treatment. The random assignment of subjects to treatments allowed 252 equally probable ways in which the subjects could be assigned. If H_0 were true, a subject's reaction time would have been the same if the subject had been assigned to the alternative treatment. Thus, given the random assignment of subjects in conjunction with H_0, there are 252 equally probable ways in which the 10 subjects and their reaction times could have been divided between the two treatments. If H_0 were true, how likely would it be that the random assignment performed in the experiment would happen to provide one of the two largest values in the distribution of 252 values? The answer, as stated above, is 2/252, or about 0.008.

So the experimenter obtained results significant at the 0.01 level but, like experimenters in general, he is interested in the P-value as a measure of strength of evidence against the null hypothesis rather than just whether it reached some conventional level of significance, such as 0.05, 0.01, or 0.001. So although a |D| as large as 5.0 could occur by random assignment alone through the experimenter performing a random assignment favorable to what was anticipated, the probability of that is only about 0.008, leading to the conclusion that the null hypothesis is false. And if the null hypothesis is false, the alternative hypothesis — which is that some (one or more) of the subjects would have provided a different reaction time measurement under the other treatment — is true, and that is the *statistical* inference that can be drawn. There is no statistical inference that can be drawn about treatment effects in a population because no population was sampled. Although these statistical inferences do not concern any subjects except those 10 subjects, the experimenter believes the careful selection of subjects enables them to make *nonstatistical inferences* about other subjects of that kind, something that would not be feasible for a heterogeneous sample that could have resulted from randomly sampling a population.

If the data had not been such as to allow the experimenter to determine by examination the number of data divisions that would provide such a large test statistic as the one for the obtained data, the experimenter might have had to derive the complete reference set of 252 data divisions, a daunting task prior to the computer age. Today, a computer could determine significance for a randomization test with thousands of test statistic values in a small fraction of a second.

Example 1.2. Let us now consider an example with few enough possible random assignments to make it practical to display hypothetical results for all of them. A comparative psychologist decides to run a pilot experiment with rats prior to a full-size experiment to learn how to conduct the main experiment. As the pilot experiment is simply for the guidance of the experimenter for future experimentation, it is unnecessary to employ a large number of rats.

TABLE 1.1

Hypothetical Results for All Possible Assignments

	A	B	A	B	A	B	A	B	A	B
	6	30	6	18	6	18	6	12	6	12
	12	54	12	54	12	30	18	54	18	30
	18		30		54		30		54	
$\bar{X}$ =	12	42	16	36	24	24	18	33	26	21
$|D|$ =	30		20		0		15		5	

	A	B	A	B	A	B	A	B	A	B
	6	12	12	6	12	6	12	6	18	6
	30	18	18	54	18	30	30	18	30	12
	54		30		54		54		54	
$\bar{X}$ =	30	15	20	30	28	18	32	12	34	9
$|D|$ =	15		10		10		20		25	

For the pilot study, five systematically selected rats are randomly assigned to two treatments, with three rats assigned to treatment A and two to treatment B. These results are obtained from the experiment: A: 6, 12, 18; B: 30, 54. To test the null hypothesis of no differential treatment effect, the value of $|D|$, the absolute difference between means, is used as the test statistic. $|D|$ is computed for all $5!/3!2! = 10$ divisions of the data associated with the possible assignments of rats, providing the results shown in Table 1.1.

The first data division represents the experimentally obtained results. Designate the three rats listed sequentially for Treatment A in the first data division as a, b, and c, and the two rats for Treatment B as d and e. The other nine data divisions represent the results that, under the null hypothesis, would have occurred for alternative assignments of rats a, b, c, d, and e. For example, given the null hypothesis to be true, the last data division represents the outcome that would have resulted from the assignment of rats c, d, and e to Treatment A and rats a and b to Treatment B.

The distribution of 10 $|D|$ values for Table 1.1 is the reference set used for determining significance. Those 10 values in order from high to low are 30, 25, 20, 20, 15, 15, 10, 10, 5, and 0. The value of $|D|$ for the obtained results (which is 30) is the largest of the ten $|D|$'s; consequently, the P-value is $1/10$, or 0.10, the smallest possible for a randomization test with only 10 possible randomizations. The impossibility of getting an extremely small P-value is unimportant because the study was a pilot study for the experimenter's own guidance rather than for the purpose of clearly convincing others in publications that the two treatments produce different results. In the conduct of the experiment, the experimenter gained valuable knowledge on good ways to administer the treatments and to measure the responses. Furthermore, the P-value is small enough to encourage the experimenter to try out further experimental comparisons of treatments A and B.

1.3   Randomization Tests and Nonrandom Samples

A randomization test is valid for any type of sample, regardless of how the sample is selected. This is an extremely important property because the use of nonrandom samples is common in experimentation, and parametric statistical tables (e.g., t and F tables) are not valid for such samples.

Parametric statistical tables are applicable only to random samples, and their invalidity of application to nonrandom samples is widely recognized. For example, Hays (1972, p. 292) in his well-respected introductory statistic book states: "All the techniques and theory that we will discuss apply to random samples, and do not necessarily hold for any data collected in any way." The random sampling assumption underlying the significance tables is that a sampling procedure that gives every possible sample of n individuals within a specified population a certain probability of being drawn. (This probability is equal for all samples in cases of simple random sampling, the most common assumption.) Arguments regarding the "representativeness" of a nonrandomly selected sample are irrelevant to the question of its randomness: a random sample is random because of the sampling procedure used to select it, not because of the composition of the sample. Thus, random selection is necessary to ensure that samples are random.

1.4   The Prevalence of Nonrandom Samples in Experiments

It must be stressed that violation of the random sampling assumption invalidates parametric statistical tables not just for the occasional experiment but for virtually all experiments. A person conducting a poll may be able to enumerate the population to be sampled and select a random sample by a lottery procedure, but an experimenter cannot randomly sample the hypothetical population a scientist is interested in. Few experiments in biology, education, medicine, psychology, or any other field use randomly selected subjects, and those that do usually concern populations so specific as to be of little interest. Whenever human subjects for experiments are selected randomly, they are often drawn from a population of students who attend a certain university, are enrolled in a particular class, and are willing to serve as subjects. Biologists and others performing experiments on animals generally do not even pretend to take random samples, although many use standard hypothesis-testing procedures designed to test null hypotheses about populations. These well-known facts are mentioned here as a reminder of the rareness of random samples in experimentation and of the specificity of the populations on those occasions when random samples are taken. For similar views in early publications on the widespread use of nonrandom samples in experiments, see Cotton (1967), Keppel (1973), Kirk (1968), and Spence et al. (1976).

Despite the acknowledged invalidity of significance tables for *t* tests and analysis of variance (ANOVA) in the absence of random sampling, these remain the most common means of determining significance. The reason is that for routine use of commonly used tests like the *t* test and one-way ANOVA, the validity of the significance tables is seldom challenged. Rarely is the criticism raised that no population whatsoever was randomly sampled. However, the less routine applications of statistical tests call for thoughtful planning, and as soon as thought replaces habit, questions of validity arise. It is for such situations that determination of significance by a randomization test is especially valuable. For example, the application of ANOVA to dichotomous data is likely to arouse concern over validity in situations where application to quantitative data would not do so. The validity issue also comes to the fore if an experimenter wants to apply ANOVA to data from a single-subject experiment or from an N-of-1 clinical trial (see Chapter 11 for an elaboration of this issue). With less common tests, like analysis of covariance and multivariate analysis of variance, any type of application may stimulate discussions of the tenability of the assumptions underlying the statistical tables. Certainly when a new statistical test is introduced, its validity must be demonstrated. As a consequence of the multiplicity of ways in which the validity of P-values can be called into question, it is not unusual for an experimenter to use a less powerful statistical test than the one he has in mind, or even to avoid experiments of a certain type (such as single-subject experiments or N-of-1 trials) to have publishable results. Compromises of this kind can be avoided by using randomization tests for determining significance instead of conventional statistical tables.

1.5 The Irrelevance of Random Samples for the Typical Experiment

Of course, if experimenters were to start taking random samples from the populations about which they want to make statistical inferences, the random sampling assumption underlying significance tables would be met. If the relevant parametric assumptions (such as normality and homogeneity of variance) concerning the populations were met, parametric statistical tables could be employed validly for determining significance. However, the irrelevance of random samples for the typical experiment makes that prospect unlikely.

In most experimentation, the concept of population comes into the statistical analysis because it is conventional to discuss the results of a statistical test in terms of inferences about populations, not because the experimenter has randomly sampled some population to which he wishes to generalize. The population of interest to the experimenter is likely to be one that cannot be sampled randomly. Random sampling by a lottery procedure, a table of random numbers, or any other device requires a finite population, but experiments of a basic nature are not designed to learn something about a particular finite existing population. For example, with animals or human subjects the intention is to draw inferences applicable to individuals already dead and individuals not yet born, as well as those who are alive at the present time. If we were concerned only with an existing population, we would have extremely transitory biological laws because every minute some individuals are born and some die, producing a continuous change in the existing population. Thus, the population of interest in most experimentation is not one about which statistical inferences can be made because it cannot be sampled randomly.

1.6   Generalizing from Nonrandom Samples

Statistical inferences about populations cannot be made without random samples from those populations, and random sampling of the populations of interest to the experimenter is likely to be impossible. Thus, for practically all experiments valid statistical inferences about a population of interest cannot be drawn. In the absence of random sampling, statistical inferences about treatment effects must be restricted to the subjects (or other experimental units) used in an experiment. Inferences about treatment effects for other subjects must be nonstatistical inferences — inferences without a basis in probability. Nonstatistical generalization is a standard scientific procedure. We generalize from our experimental subjects to individuals who are quite similar in those characteristics that we consider relevant. For example, if the effects of a particular experimental treatment depend mainly on physiological functions that are almost unaffected by the social or physical environment, we might generalize from our experiments to persons from cultures other than the culture of our subjects. On the other hand, if the experimental effects were easily modified by social conditions, we would be more cautious in generalizing to other cultures. In any event, the main burden of generalizing from experiments always has been, and must continue to be, carried by nonstatistical rather than statistical logic.

1.7   Intelligibility

Confident use of an unfamiliar statistical test, such as a randomization test, requires that it be understood. The intelligibility of the randomization test procedure is important to the people to whom the experimenter communicates the results of his analysis, many of whom may have little or no formal education in statistics. Those people may include the experimenter's employer, a government agency or other sponsoring organization, scientists interested in the research findings, and the general public. Fortunately, although unfamiliar to many, the freedom of randomization tests from the unrealistic assumption of random sampling makes the justification of the procedure fairly straightforward, as illustrated in Example 1.1 and Example 1.2.

Wilk and Kempthorne (1955) state: "We feel that any mathematical assumptions employed in the analysis of natural phenomena must have an explicit, recognizable, relationship to the physical situation." Randomization tests meet that requirement, whereas random sampling tests do not. The assumption of random sampling of some hypothetical infinite population is not an assumption that refers to a physical situation; furthermore, conceiving of a set of data as if randomly selected from such a population is difficult to understand when no physical process of such sampling from any population occurred.

Keeping questionable assumptions to a minimum by dispensing with the unrealistic assumption of random sampling and specifying the necessity of actual random assignment for the validity of randomization tests clarifies the basic nature of randomization tests to the extent that new randomization tests can readily be developed and validated.

1.8   Respect for the Validity of Randomization Tests

Chapter 14 is devoted to procedures for assessing the validity of randomization tests in general, and the validity of specific tests is dealt with when they are described. The intent of this section is to show the advocacy that randomization tests have attracted because of their validity in the absence of random sampling. As might be expected, statistical tests that are at variance with mainstream statistical tests are commonly open to strong criticism from the statistical "establishment." However, randomization tests have been recognized for more than 50 years as ideal tests of experimental results, and the lack of widespread use was attributed to the extensive computation that frequently would be required to determine significance, but that liability has now been overcome with the advent of high-speed computers. Consequently, as will be illustrated in this section, advocacy of randomization tests is not limited to people with a natural craving for the unorthodox but is widespread among people with a strong reputation for their contributions to conventional statistics.

Randomized experiments in medical research have long been called the "gold standard," the ideal experiment, and when medical experiments are relevant and feasible, financial support frequently depends on the incorporation of appropriate random assignment to treatments. Therefore, it should not be surprising that randomization tests — tests based on random assignment in experiments — have come to be recognized by many in the field of medicine as the "gold standard" of statistical tests for randomized experiments.

Bradley (1968, p. 85) indicated that "eminent statisticians have stated that the randomization test is the truly correct one and that the corresponding parametric test is valid only to the extent that it results in the same statistical decision." Sir Ronald Fisher (1936, p. 59) stated, in regard to determining significance by the randomization test procedure, that "conclusions have no justification beyond the fact that they agree with those which could have been arrived at by this elementary method." In his classical book on experimental design, Kempthorne (1952) gave his own view of the validity of randomization tests: "Tests of significance in the randomized experiment have frequently been presented by way of normal law theory, whereas their validity stems from randomization theory."

In Germany, Jürgen Bredenkamp has expressed strong views in several books and publications (Bredenkamp, 1980; Bredenkamp and Erdfelder, 1993) about the inappropriateness of the assumption of random sampling in psychological experimentation and the consequent need to employ randomization tests. In the same vein, a review by Cotton (1973, p. 168) of Winer's (1971) book on experimental design, while referring to the book as the best of its kind, had a lengthy statement about the pervasiveness of nonrandom samples and the need for reassessing the validity of significance tables for such samples. Part of the statement follows:

> Though Winer (1971, p. 251) mentions randomization tests as a way of avoiding distribution assumptions, he is like most statistical textbook authors in neglecting a more important advantage: randomization tests permit us to drop the most implausible assumption of typical psychological research — random sampling from a specified population. We need to tell our students, again and again, that random sampling occurs infrequently in behavioral research and that, therefore, any statistical tests making that assumption are questionable unless otherwise justified.

Cotton (1973) then pointed out again that randomization tests would permit valid statistical inferences in the absence of random sampling. He followed the lead of Fisher in emphasizing that determining significance by the use of conventional published statistical tables is of questionable validity until it has been shown that the significance so obtained would agree closely with that given by randomization tests. In the following passage, Cotton (1973, p. 169) used the expression "randomization theory" to refer to the theory of randomization tests:

> A rigorous approach by Winer would have been to indicate for which of his many designs the F tests are indeed justified by randomization theory and those for which an investigator should either make exact probability calculations, perform distribution-free tests which give approximately accurate significance levels, or perform F tests without pretending to meet their assumptions.

A Statistical Task Force of prominent statisticians (Brillinger, Jones, and Tukey, 1978) was established to advise the United States Government on the statistical analysis of weather modification data. In its endorsement of the use of randomization tests, the Task Force made the following statement:

> The device of judging the strength of evidence offered by an apparent result against the background of the distribution of such results obtained by replacing the actual randomization by random-izations that might have happened seems to us definitely more secure than its presumed competitors, that depend upon specific assumptions about distribution shapes or about independence of the weather at one time from that at another. (Page D-l)

In a later passage, the Task Force again stressed the unrealistic nature of parametric assumptions in regard to weather modification and again pointed to the necessity of doing "re-randomizations" (i.e., performing randomization tests):

> Essentially every analysis not based on re-randomization assumes that the days (or storms) behave like random samples from some distribution. Were this true in practice, weather could not show either the phenomenon of persistence (autocorrelation) or the sys-tematic changes from month to month within a season which we know do occur. In an era when re-randomization is easily accessible, and gives trustworthy assessments of significance, it is hard indeed to justify using any analysis making such assumptions. (Page F-5)

Spino and Pagano (1991) referred to a talk by Tukey (1988), called "Ran-domization and Re-randomization: The Wave of the Past in the Future," in which he revealed his continued belief that the application of randomization tests (which he refers to as "re-randomization") provides the best results for randomized experiments in general, not just randomized weather mod-ification experiments. If the use of significance tables always provided close approximations to the significance values given by randomization tests, the fact that randomization tests are the only completely valid procedures to use for finding significance with nonrandom samples would be of little more than academic interest. However, the significance tables often provide values that differ considerably from those given by randomization tests, and so the consideration of validity is a practical issue. Of course, the P-values given by the two methods are sometimes very similar. In cases of large differences, the difference is sometimes in one direction and sometimes in the other. Too many factors affect the closeness of the significance pro-vided by the two methods to allow the formulation of a general rule for the experimenter to tell him how much he would be in error by using significance tables instead of randomization tests. Some of the factors are the absolute sample sizes, the relative sample sizes for the different treatments, the sample distribution shape, the number of tied measurements, and the type of statistical test.

1.9 Versatility

Randomization tests are extremely versatile as a consequence of their potential for ensuring the validity of existing statistical tests and for developing new special-purpose tests. Randomization tests provide the opportunity to develop new customized test statistics and to analyze data from designed experiments involving unconventional random assignment procedures. Furthermore, intelligibility to nonstatisticians is a property of randomization tests that enables experimenters to use their potential and thereby acquire greater flexibility in planning their experiments and analyzing the data.

The idea of a researcher possessing both the expertise and the time to develop a new statistical test to meet his special research requirements may sound farfetched, as indeed it could be if a special significance table had to be constructed for the new test. The original derivation of the sampling distribution of a new test statistic preparatory to formulation of a significance table may be time consuming and demands a high level of mathematical ability. However, if the researcher determines significance by means of reference sets based on random assignment, not much time or mathematical ability is necessary. In the following chapter, there are numerous examples of new statistical tests with significance based on the randomization test procedure.

The versatility of randomization tests is related to increased power or sensitivity of statistical testing in a number of ways. That is, there are several ways in which the versatility can increase the chances of an experimenter getting significant results when there are actual differences in treatment effects. For example, ANOVA and other procedures that are more powerful than rank order tests are sometimes passed over in favor of a rank order test to ensure validity, whereas validity can be guaranteed by using randomization tests without the loss of power that comes from transforming measurements into ranks. In cases where there is no available rank order counterpart to a parametric test whose validity is in question, a randomization test can be applied to relieve the test of parametric assumptions, including the assumption of random sampling. The possibility of ensuring the validity of any existing statistical test then provides the experimenter with the opportunity to select the test that is most likely to be sensitive to the type of treatment effect that is expected. Furthermore, the experimenter is not limited to existing statistical tests; the use of randomization tests allows the development of statistical tests that may be more sensitive to treatment effects than conventional ones. The experimental design possibilities are also greatly expanded because new statistical tests can be developed to accommodate radically different random assignment procedures.

In light of the multitude of uses for randomization tests, little more can be done in this book than to present a few of the possible applications to illustrate the versatility of this special means of determining significance. These applications are to statistical tests of various types, and experimenters and others should encounter no difficulty in generalizing to other statistical tests.

1.10 Practicality

In Example 1.2, there were only 10 permutations of the data; consequently, the amount of time required to determine significance by using a randomization test was not excessive. However, the number of data permutations mounts rapidly with an increase in sample size, so that even for fairly small samples, the time required with the use of a desk calculator may be prohibitive. However, computers make randomization tests practical. Given the widespread accessibility of personal computers and the availability of randomization test programs (as discussed in the following chapters), there is no longer a computational reason not to use randomization tests. Randomization tests are now within reach of any researcher who has access to a computer.

To provide experimenters with the background for making efficient use of computers for the performance of randomization tests, Chapter 3 gives suggestions for minimizing computer time and the time to write programs. For many of the statistical tests used to illustrate the randomization test procedure of determining significance, suggested computer programs accompany the discussion to help the experimenter understand how to write his own programs or how to tell a programmer his requirements.

1.11 Precursors of Randomization Tests

Unlike conventional parametric tests, randomization tests determine P-values directly from experimental data without reference to tables based on infinite continuous probability distributions, so the procedure is easily understood by persons without a mathematical background. The mathematical procedures that were precursors of randomization tests were those associated with combinatorial mathematics, which deals with combinations and permutations as well as other configurations of elements in finite sets.

Biggs (1979) pointed out that combinatorial mathematics was applied in the first century, and that Boethius knew how to compute the number of combinations of n things taken two at a time in the 5th or 6th century. Furthermore, Biggs learned that "Pascal's triangle" and its properties (commonly reported as developed in about 1650 by Blaise Pascal) were anticipated by a Chinese mathematician, Chi Shih-Chieh, in the 14th century and possibly by the Persian poet Omar Khayyam in about 1050 A.D.

The sign test is a nonparametric test, which Pascal could have produced from his grasp of the binomial distribution underlying Pascal's triangle but presumably did not. Arbuthnott (1710) devised a sign test but not for application in experiments. A reference set for the sign test can be generated by switching (permuting) data within pairs with the direction of difference being used in the test statistic. Thus, it was an early rank test where the data within each pair could be ranked 1 and 2 to indicate the direction of difference.

An example will show how Pascal's triangle and the binomial coefficients it represents permit direct computation of the P-value for a randomization test for paired comparisons with dichotomous data, sometimes called the sign test.

Example 1.3. An experimenter has 10 sets of twins as experimental units in which it is randomly determined within each set which twin will receive the A treatment and which will receive the B treatment. The experimenter wants to test the directional null hypothesis that none of the subjects would give a larger measurement under A than under B, and the alternative hypothesis to be "accepted" if the null hypothesis is rejected is that some (one or more) of the subjects would have yielded larger measurements under A than under B. The experimenter runs the experiment and the experimental outcome is that in 8 of the 10 sets of twins, the A measurement is larger than the B measurement. There were $2^{10} = 1024$ possible assignments, so if the experimenter generated a reference set of the outcomes for all $2^{10} = 1024$ assignments to determine the significance, he would find the proportion of possible outcomes that showed eight or more pairs where the A measurement is larger than the B measurement. Generating the reference set is unnecessary because tables derived from the binomial distribution necessarily provide the same significance value for this type of experiment, and in this instance give a probability of 0.055 of getting as many as eight pairs where the A measurement is larger than the B measurement.

Gambling in the 17th century is commonly said to have inspired much of the development of finite probability computations, and their applications in dice and card games involved decisions based on the knowledge of combinations and permutations. If randomized experiments and the concept of statistical tests had existed at that time, modern randomization tests could readily have been developed.

It is instructive to note that prominent mathematicians in the 17th century sometimes advocated erroneous probability calculation procedures for what we today would regard as simple problems. Swartz (1993) discussed procedures advocated by Leibniz and D'Alembert (respected 17th century mathematicians) for determining probabilities in throwing dice and flipping coins. Leibniz and D'Alembert both regarded the probability of throwing a 12 with two dice to be the same as the probability of throwing an 11 because each of these outcomes was conceived as the only way of achieving that outcome — with two 6s for 12 and a 5 and a 6 for 11. Maybe if one die were red and the other green, they would have thought of two ways that exist to get an 11. D'Alembert made a similar error in his "Croix ou Pile" in stating that the probability of at least one head in two tosses of a coin is 2/3. He was implicitly regarding only these three possible outcomes as equally probable: 2 heads, 2 tails, and 1 head and 1 tail. Randomization tests are as new to statisticians and experimenters as probabilistic procedures for gambling were for mathematicians and gamblers in the days of Leibniz and D'Alembert, so it is essential that randomization tests be studied intensively to minimize the chances of comparable misunderstanding of them.

Another landmark article was published by André in 1883 on the development of a recursion formula for a runs test making use of combinatorial logic. The test statistic was the number of changes from increasing to decreasing in magnitude and from decreasing to increasing in magnitude in an ordered series of observations. The formula generated the number of changes in a series of N numerical values for all N! permutations of the series, and the proportion of the N! permutations providing as few (or as many) changes as the number in the obtained results would be the P-value. The formula was modified much later (Edgington, 1961) to apply to the number of rises and falls in a sequence rather than the number of changes, and a table for N less than or equal to 25 was derived and reprinted in Bradley (1968) and Gibbons (1971), who call the modified runs test a "runs up and down test." Like the sign test and rank tests, it is a permutation test that permits deriving the P-value by directly permuting the research results, and significance tables for various sample sizes have been produced in that manner.

However, it remained for Fisher (1935) to demonstrate a permutation test whose every use required determination of significance by permuting the data because a significance table was impossible for the application he had in mind. He discussed a statistical test of data from a hypothetical experiment. It was a test of paired data that could have been tested by use of a sign test, but instead of using the number of differences in a given direction as the test statistic, Fisher used a test statistic dependent on the actual measurements. The test statistic was the difference between the sums of A and B measurements. The sign test and André's run test could be regarded as primitive permutation tests that permitted significance tables to be derived but Fisher's test seems to be the first permutation test employing a reference set that had to be derived separately for each different set of data. To demonstrate a statistical test whose reference set must be generated from the obtained data is a considerable step in the evolution of randomization tests and other permutation tests. In fact, some people regard Fisher's test as the first permutation test.

Details of Fisher's permutation test will be discussed in the following chapters because it has repeatedly been misrepresented. Fisher's test frequently has been discussed as if it — like modern randomization tests — was based on random assignment alone. To the contrary, at several points in his discussion Fisher referred to testing the null hypothesis of identity of populations and, in fact, random sampling of populations was required for his test. It is Pitman who should be regarded as the first expositor of randomization tests.

Pitman gave several examples of permutation tests as well as a rationale applicable both to randomization tests and nonexperimental permutation tests. After providing a rationale for permutation tests of differences between populations, he showed that permutation tests could be based on random assignment alone, stressing their freedom from the assumption of random sampling and giving an example in each of the three papers (Pitman, 1937a; 1937b; 1938). This demonstration is of great importance to the vast number of experimenters who use random assignment of subjects or other units that have been systematically, not randomly, selected.

A large number of rank tests were developed shortly after the articles by Pitman. Rank tests tended to be represented as testing hypotheses about populations on the basis of random samples, even in regard to experimental research. Those rank permutation tests have been used widely because of their readily available significance tables. The applicability of rank tests to experimental data on the basis of random assignment alone tended to be overlooked, as was the possibility of permuting raw data in the same way as the ranks are permuted to have a more sensitive test.

Over many years, Kempthorne made important contributions to the field of randomization tests. He and a few colleagues carefully examined the link between randomization procedures and statistical tests applied to the experimental results. Kempthorne (1952; 1955) and his colleagues completed considerable work on "randomization analysis," which concerned investigation of the distributions of analysis of variance F and its components under null and non-null hypotheses for various experimental designs, i.e., for various forms of random assignment. The distributions generated by the various randomization procedures were those that would constitute reference sets for randomization tests using F or related test statistics in the analysis of actual experimental data. For a randomization test, instead of producing a reference set of alternative outcomes by simply rearranging (permuting) the data, Kempthorne proposed a two-stage process for producing that reference set. The first stage involves dividing the experimental units (e.g., subjects) in all possible ways for the random assignment procedure, which provides a "reference set" of assignments regardless of the null hypothesis to be tested. The second stage is to change each possible assignment into an outcome for that assignment under the null hypothesis to provide a reference set of possible outcomes under the null hypothesis for alternative random assignments to treatments. This two-stage approach, called the *randomization-referral* approach in Chapter 14, is extremely useful for understanding the rationale of randomization tests.

When the number of permutations of data is too large to generate the complete reference set for the possible assignments, "random sampling" of the complete reference set can be employed by using a computer program that randomly rather than systematically produces a reference set. If the data are permuted randomly n times, a reference set consisting of the n data permutations plus the obtained results can be used for valid determination of the P-value. The proportion of those $(n + 1)$ elements of the reference set with a test statistic as large as the value for the obtained results is the P-value. Dwass (1957) was the first to provide a rationale for that random procedure, which yielded what he termed "modified randomization tests" based on random samples of all possible data permutations. For example, the randomly generated reference set might consist of only 1000 of the data permutations, consisting of the experimental results plus 999 random permutations of those results. The randomization test procedure proposed by Dwass is not a Monte Carlo approximation to a "true randomization test" but is a completely valid test in its own right, even when the number of data permutations in the random subset is a very small portion of the number associated with all possible assignments.

Another procedure for determining significance on the basis of only a subset of the complete reference set was proposed by Chung and Fraser (1958). Unlike Dwass's subset, theirs was a systematic (nonrandom) subset. It is doubtful that the Chung and Fraser approach, based on a permutation group rationale, will be as useful as the random procedure for reducing the computational time for a randomization test, but their permutation group approach is of considerable theoretical importance and is fundamental to the general randomization test theory presented in Chapter 14. Their approach provides a sound theoretical justification for many applications of randomization tests to complex experimental designs, thus providing a basis for a substantial expansion of the scope of applicability of randomization tests. Bailey (1991) provided an extensive discussion of the role of permutation groups in randomization theory.

1.12 Other Applications of Permutation Tests

This is a book on randomization tests, not permutation tests employing the same permutations or data divisions in situations where there is no random assignment to experimental treatments. Since publication of the first edition of this book (Edgington, 1980) — which contained many computer programs — interest in randomization tests has been considerable, but there also have been many publications (including several books) that have appeared in which the programs are applied to data from nonexperimental research. When random sampling is involved in nonexperimental research, theoretical justification by Pitman (1937a) would provide support for statistical inferences about the sampled populations, but in the vast majority of applications in nonexperimental research neither random sampling nor random assignment is present.

That is not to say that the vast number of such nonexperimental applications of permutation tests is inappropriately applied but that the computation of P-values for those applications does not provide statistical inferences about causal relationships, whereas randomization tests do. Nevertheless, because such applications use some of the same computational procedures as randomization tests and are popular and useful, they deserve additional discussion, which appears in Chapter 3.

1.13 Questions and Exercises

1. In pointing out the importance of using randomization tests to analyze data from weather modification, Tukey used the term "re-randomization" to designate the randomization test procedure. Why would "re-randomization" be a misnomer for the way the randomization test for the data in Table 1.1 was performed?

2. If D'Alembert made the same sort of mistake he made in determining the probability of an 11 in a toss of two dice, what would he have determined as the probability of exactly two heads in three tosses of a coin? Explain.

3. Many consider Fisher to have conducted the first permutation test when he determined significance by permuting data within pairs, even though many years earlier significance based on permuting within pairs had already been employed. Explain.

4. Prove that for a total of N subjects assigned to two treatments, the maximum number of randomizations is when the number of subjects is equal.

5. Given four subjects assigned to A and three subjects to B, with a single measurement from each subject, what is the maximum possible P-value? What is the minimum possible P-value?

6. Who in the 1930s provided the first justification for determining statistical significance in the absence of random sampling?

7. How many different lineal sequences of five people, consisting of three men and two women, are possible, given the restriction that each woman is next to at least one man in every sequence?

8. An experimenter has three sets of twins as experimental units in which it is randomly determined within each set which twin will receive the A and which the B treatment. The experimenter uses a dichotomous outcome variable, with 0 representing a failure and 1 representing a success. He wants to test the null hypothesis that for every subject in the experiment, the score is the same as it would have been under assignment to the other treatment, and the alternative hypothesis is that some (one or more) of the subjects would have yielded a different score under A than under B. The experimenter runs the experiment and the experimental outcome is that all three subjects that received treatment A scored a 1 and all three subjects that received treatment B scored a 0. Perform a randomization test by enumerating all possible outcomes and using the difference between the sums as the test statistic.

9. Take the same experiment as in Question 8 but now use the difference between the means as your test statistic. Do you arrive at a different P-value? Why would that be?

10. Which classical nonparametric test could be used for the design and data in Question 8 and Question 9? Use a published table to get the P-value of this test. Does it coincide with your computations for Question 8 and Question 9?

NOTES

[1]The first occurrence of the term *randomization* is in Fisher's 1926 paper on "The Arrangement of Field Experiments" (David, 1995; 1998). According to David (2001), the term *randomization test* makes its first appearance in Box and Andersen's 1955 paper on "Permutation Theory in the Derivation of Robust Criteria and the Study of Departures from Assumption," in which it is used as an alternative for *permutation test*. However, the term *randomization test* is already prominently present in Moses's 1952 paper on "Non-parametric Statistics for Psychological Research" (the term even appears as a section title on page 133) and in Kempthorne's 1952 classical textbook, "The Design and Analysis of Experiments" (also as a section title, page 128), suggesting an even older (possibly common) source.

REFERENCES

André, D. Sur le nombre de permutations de *n* éléments qui présentent *s* séquences [On the number of permutations of *n* elements that provide *s* runs], *Comptes Rendus (Paris)*, 97, 1356–1358, 1883.

Arbuthnott, J. An argument for Divine Providence, taken from the constant Regularity observ'd in the Births of both Sexes, *Philosophical Transactions*, 27, 186–190, 1710.

Bailey, R.A. Strata for randomized experiments, *J. R. Statist. Soc. B.*, 53, 27–78, 1991.

Biggs, N.L. The roots of combinatorics, *Historia Math.* 6(2), 109–136, 1979.

Box, G.E.P. and Andersen, S.L. Permutation theory in the derivation of robust criteria and the study of departures from assumption, *J. R. Statist. Soc. B*, 17, 1–34, 1955.

Bradley, J.V. *Distribution-Free Statistical Tests*, Prentice-Hall, Englewood Cliffs, NJ, 1968.

Bredenkamp, J. *Theorie und Planung Psychologischer Experimente* [Theory and Planning of Psychological Experiments], Steinkopff, Darnstadt, 1980.

Bredenkamp, J. and Erdfelder, E. Methoden der Gedächtnispsychologie [Methods in the Psychology of Memory], in Enzyklopädie der Psychologie, Themenbereich C, Serie II, Band 4, *Gedächtnis* [Memory], Albert, D. and Stapf, K.H. (Hrsg.), Hogrefe, Göttingen, 1993.

Brillinger, D.R., Jones, L.V., and Tukey, J.W. *The Management of Weather Resources. Volume II. The Role of Statistics in Weather Resources Management.* Report of the Statistical Task Force to the Weather Modification Board, Department of Commerce, Washington, DC, 1978.

Chung, J.H. and Fraser, D.A.S. Randomization tests for a multivariate two-sample problem, *J. Am. Statist. Assn.*, 53, 729–735, 1958.

Cotton, J.W. *Elementary Statistical Theory for Behavior Scientists*, Addison-Wesley, Reading, MA, 1967.

Cotton, J.W. Even better than before, *Contemp. Psychol.*, 18, 168–169, 1973.

David, H.A. First (?) occurrences of common terms in mathematical statistics, *The American Statistician*, 49(2), 121–133, 1995.

David, H.A. First (?) occurrences of common terms in probability and statistics: A second list, with corrections, *The American Statistician*, 52(1), 36–40, 1998.

David, H.A. First (?) occurrence of common terms in statistics and probability, in *Annotated Readings in the History of Statistics*, David, H.A. and Edwards, A.W.F. (eds), Springer, New York, 2001, Appendix B and 219–228.

Dwass, M. Modified randomization tests for nonparametric hypotheses, *Ann. Math. Statist.*, 28, 181–187, 1957.

Edgington, E.S. Probability table for number of runs of signs of first differences in ordered series, *J. Am. Statist. Assn.*, 56, 156–159, 1961.

Edgington, E.S. *Randomization Tests*, Marcel Dekker, New York, 1980.

Fisher, R.A. The arrangement of field experiments, *J. Min. Agri. G.B.*, 33, 503–513, 1926.

Fisher, R.A. *The Design of Experiments*, Oliver and Boyd, Edinburgh, 1935.

Fisher, R.A. The coefficient of racial likeness and the future of craniometry, *J. R. Anthropol. Inst.*, 66, 57–63, 1936.

Gibbons, J.D. *Nonparametric Statistical Inference*, McGraw-Hill, New York, 1971.

Hays, W.L. *Statistics for the Social Sciences* (2nd ed.), Holt, Rinehart & Winston, New York, 1972.

Kempthorne, O. *The Design and Analysis of Experiment*, Wiley, New York, 1952.

Kempthorne, O. The randomization theory of experimental inference, *J. Am. Statist. Assn.*, 50, 946–967, 1955.

Keppel, G. *Design and Analysis: A Researcher's Handbook*, Prentice-Hall, Englewood Cliffs, NJ, 1973.

Kirk, R.E. *Introductory Statistics*, Brooks/Cole, Belmont, CA, 1968.

Moses, L.E. Non-parametric statistics for psychological research, *Psychol. Bull.*, 49, 122–143, 1952.

Pitman, E.J.G. Significance tests which may be applied to samples from any populations, *J. R. Statist. Soc. B.*, 4, 119–130, 1937a.

Pitman, E.J.G. Significance tests which may be applied to samples from any populations. II. The correlation coefficient, *J. R. Statist. Soc. B.*, 4, 225–232, 1937b.

Pitman, E.J.G. Significance tests which may be applied to samples from any populations. III. The analysis of variance test, *Biometrika*, 29, 322–335, 1938.

Spence, J.T., Cotton, J.W., Underwood, B.J., and Duncan, C.P. *Elementary Statistics* (3rd ed.), Appleton-Century-Crofts, New York, 1976.

Spino, C. and Pagano, M. Efficient calculation of the permutation distribution of trimmed means, *J. Am. Statist. Assn.*, 86, 729–737, 1991.

Swartz, N. The obdurate persistence of rationalism, in *Vicinae Deviae: Essays in Honour of Raymond Earl Jennings*, Hahn, M. (ed.), Simon Fraser University, Burnaby, British Columbia, Canada, 1993.

Tukey, J.W. Randomization and rerandomization: the wave of the past in the future, Seminar given in honor of Joseph Ciminera, Philadelphia Chapter of the American Statistical Association, Philadelphia, June 1988.

Wilk, M.B., and Kempthorne, O. Fixed, mixed, and random models, *J. Am. Statist. Assn.*, 50, 1144–1167, 1955.

Winer, B.J. *Statistical Principles in Experimental Design* (2nd ed.), McGraw-Hill, New York, 1971.

2

Randomized Experiments

Simple applications of randomization tests can be made on the basis of a general understanding of the role of random assignment. However, to take full advantage of the versatility of randomization tests the experimenter requires a more extensive discussion of some aspects of random assignment (randomization) than is provided in standard statistics books. The employment of randomization tests instead of conventional parametric tests permits the use of a variety of methods of random assignment that otherwise would be inappropriate. The utilization of randomization tests thus helps "put the horse before the cart": instead of the experimenter's efforts being expended on designing an experiment to fit the requirements of a statistical test, the experimenter incorporates randomization in the manner desired and can rest assured that a randomization test can be devised to fit the design employed.

The multiplicity of randomization procedures that randomization tests can accommodate enables experiments to be performed that otherwise would be neglected because of the pressure in many fields for experiments to be randomized experiments. For example, in the field of medicine *randomized clinical trials* are frequently called the "gold standard" for experimentation on the effects of therapeutic treatments on patients, which discourages experimenters whose needs (or desires) cannot be met by conventional random assignment procedures. Randomization tests thus can further the use of randomized experiments, and randomized experiments can yield benefits that nonrandomized nonexperimental research cannot.

2.1 Unique Benefits of Experiments

Randomization tests are tests of null hypotheses concerning experimental treatment effects in randomized experiments. A fundamental difference between nonexperimental research and experimental research lies in experiments being useful for inferring *causal* relations, whereas with other research one can only determine association among variables. Statistical tests on nonexperimental data do not permit statistical inferences about causation. Inferring from nonexperimental research that variation in some variable X leads to specified changes in variable Y might be risky, as has been illustrated frequently in medicine when the nonexperimental research is followed up with randomized experiments.

Showing that manipulation of X produces certain changes in Y demonstrates more than a causal relationship between variables. An experiment also shows a practical way to produce those changes in Y and this may lead to technological advances. For example, even when we already know or have convincing theoretical support for the notion that particular chemical changes in cells will cure a disease, it may be of considerable value to demonstrate experimentally how those changes can be produced by administering certain medication. The adage, "everybody talks about the weather but nobody does anything about it," is definitely a reflection of the desire to control — not just understand — the weather, and that attitude underlies advances in technology.

2.2   Experimentation without Manipulation of Treatments

In laboratory experiments, a treatment may be manipulated to change temperature, humidity, light intensity, or other characteristics that can be altered by adjusting a laboratory device. However, it should be noted that experiments can be conducted without altering a treatment. Those experiments can employ assignment of experimental units to preexisting conditions, such as assigning subjects in a memorization experiment to one list of words or to a different list using preexisting lists instead of changing the words on one list to form another list. Random assignment of subjects to the lists constitutes a randomized experiment. Another instance of experimentation without manipulation of treatments is when random assignment of subjects to alternative hospitals is done to determine whether something other than susceptibility of subjects to a certain disease accounts for an excessive number of outbreaks of the disease in a particular hospital.

Ecology is a science concerned with the interaction between organisms and their natural environment. Randomized experiments can be carried out to determine the effect of environmental differences without altering the environment. For instance, certain species of rodents on a plateau may be known to live substantially longer than members of the same species in the plain below the plateau. Perhaps the greater longevity of rodents on the plateau is due to natural selection over many generations of hardier rodents. An experimenter without the inclination or means to conduct a number of laboratory experiments to simulate differences between the two locations decides to conduct a simple randomized experiment. The experimenter randomly assigns to the two locations tagged newborn rodents of the same species from a locality other than the plateau or the plain below it. Transmitters attached to the rodents transmit heartbeats, enabling the experimenter to determine when one dies. Admittedly, the two locations under comparison could differ in temperature, light, soil, and many other aspects, as well as altitude, so significant results from a randomization test would not suggest a specific cause of the difference in longevity. Nevertheless, it would provide a statistical inference that environmental differences had some effect on longevity; it would indicate that more than "survival of the fittest" was involved in the longevity differences.

2.3   Matching: A Precursor of Randomization

Attempts to match units assigned to different treatments has a long history. Long before random assignment became standard experimental practice, it was recognized that apparent differences in treatment effects could simply be the result of differences in the subjects or other experimental units assigned to the treatments. To control for the influence of differences among experimental units, attempts were made to systematically "equate" the experimental units assigned to different treatments. Characteristics of experimental units that were likely to be associated with responses to the treatments would be singled out, and those characteristics that could readily be observed or measured would be used for matching. Two such characteristics in both human and animal studies are sex, a categorical variable, and age, a quantitative variable. Procedures to systematically match subjects for sex could be to have all subjects be of the same sex or to have the sexes be in the same proportion for the various treatments. For age and other quantitative variables, matching in some cases might consist of making the average age of subjects assigned to one treatment identical to the average for any other treatment. Of course, precise matching is necessarily limited to a small number of characteristics because matching on one characteristic often will conflict with matching on others.

2.4   Randomization of Experimental Units

Random assignment of experimental units to treatments is an alternative to matching. Randomization is a means of simultaneously controlling all unit variables, even the unknown attributes of the units. Randomization can equate the expected values of unit attributes for the treatments. For example, in research involving human participants randomization ensures that the age, sex, intelligence, weight, or any other subject variable averaged over all

TABLE 2.1

Alternative Assignments of Subjects and Their Ages

Treatment A		Treatment B	
Subjects	**Ages**	**Subjects**	**Ages**
a, b	20, 22	c, d	28, 35
a, c	20, 28	b, d	22, 35
a, d	20, 35	b, c	22, 28
b, c	22, 28	a, d	20, 35
b, d	22, 35	a, c	20, 28
c, d	28, 35	a, b	20, 22
Mean:	26.25		26.25

possible assignments of subjects among treatments would be the same for all treatments. The following example illustrates the equating of "expected" (average) values that between-subject random assignment would produce. The example deals with expected values for age but, as will be explained, the expected values for any other variable would also be equated by random assignment.

Example 2.1. Four subjects designated as *a*, *b*, *c*, and *d* are 20, 22, 28, and 35 years old, respectively. Two are randomly assigned to treatment A, leaving the remaining two to take treatment B. The six possible assignments of subjects and their ages are shown in the rows of Table 2.1.

Thus, on the average for all six equally probable assignments, the mean age for A subjects and B subjects is the same, namely 26.25, which is the average age of the four subjects. It can be seen that each possible pair of subjects occurs once for each treatment, which ensures that the sums (and means) will be the same for the two treatments and that the means (expected values) would be the same, no matter what subject variable was listed. If there were three treatments with equal sample sizes, every possible combination of subjects would occur equally often for each of the three treatments and that equality would hold for any number of treatments with an equal number of subjects for the treatments.

An important caveat is that although conventional random assignment procedures ensure that the expected values of the measurements will be the same for all unit attributes, this is not the case for certain random assignment procedures that randomization tests can accommodate, and which will be discussed in subsequent chapters.

2.5   Experimental Units

The term *sampling unit* is used in statistics to refer to subjects or other elements that are selected from a population with random sampling. For example, if there is random selection of cities from a population of cities in a country, the sampling unit is a city, or if houses are selected randomly from a population of houses in a city, the sampling unit is a house. A sampling unit may be an individual element, like a person or a tree, or it may be a group of elements, like a family or a forest.

The term *experimental unit* refers to analogous units associated with random assignment. An experimental unit is a unit that is assigned randomly to a treatment. For example, in many experiments in psychology, education, biology, and medicine the unit that is assigned is a person or an animal or a group of persons or animals (see Chapter 1). Whether it consists of one or many individual elements, the experimental unit is randomly assigned in its entirety to the treatment.

2.6 Groups as Experimental Units

Sometimes the experimental unit is a collection or group of entities that is assigned to a treatment. A number of distinct entities, not just one, can constitute a single experimental unit. Families may be randomly assigned to alternative family therapy techniques or entire hives of bees may be assigned to alternative treatments.

For practical reasons, an educational psychologist may randomly assign entire classes to instructional procedures instead of individual students. For a randomization test, the nondirectional null hypothesis would be that for each class, the "class measurement" (such as the mean score of students in a class) is invariant over all possible assignments. If there were five classes, two of which were to be assigned to one procedure and three to another, there would be five experimental units consisting of the five classes along with predetermined treatment times, perhaps extending over several weeks. No matter how many students were in each class, the total number of possible assignments would be only $5!/2!3! = 10$, the number of ways of dividing five things into two groups with two in a specified group and three in the other. Thus, although a large number of students could be involved, the smallest possible significance value that could be obtained by a randomization test is $1/10$.

2.7 Control over Confounding Variables

Before the use of randomization to control for unit differences, it was recognized that the apparent differences in treatment effect could be due to external factors known as *confounding variables*, which are systematically associated with the experimental treatments, not with the experimental units. For example, one treatment might be in a noisy environment that reduced its effectiveness. Systematic attempts might be made to lower the noise to the level of the other treatment conditions. If the noise could not be reduced in the noisy environment, noise might be increased in the other conditions. In any event, the objective in effect would be to equate the noise level among treatments.

Some confounding variables may necessarily be associated with some treatments and not others. For example, a fixed association would arise when equipment that is permanently fixed in different locations and available at different times must be used for producing the alternative treatment conditions. However, the confounding effects of some variables can be controlled by systematically or randomly associating levels of those variables with the subjects or other experimental units. In some experiments (e.g., in psychology), there may be good reason to control for confounding effects of the experimenter whenever more than one experimenter is used in a single experiment. That control can be gained by systematic or random association of the subjects with the experimenters prior to random assignment of the subjects and their experimenters to treatments. Example 2.2 is an illustration of the systematic assignment of subjects to experimenters to control for experimenter effects.

Example 2.2. For practical reasons, it is necessary to use two experimenters, E_1 and E_2, in an experiment comparing the effectiveness of treatments A and B. Six subjects are used: S_1, S_2, S_3, S_4, S_5, and S_6. The first three subjects chose E_1 to be their experimenter and the other three chose E_2. There are thus six experimental units formed, each consisting of a subject and the experimenter who will give the experimental treatment. Three of the six units are randomly selected for treatment A and the remaining three are assigned to treatment B. The following results are obtained:

A		B	
$(S_2 + E_1)$	13	$(S_1 + E_1)$	5
$(S_3 + E_1)$	17	$(S_4 + E_2)$	8
$(S_6 + E_2)$	9	$(S_5 + E_2)$	8

Because the experimenter is randomly assigned to the treatment along with the subject for that experimenter, the confounding effects of the experimenter variable are controlled.

Thus, for some confounding variables we can exercise statistical control over their effects by associating a particular level of the variable systematically with a participant (or any other component of the experimental unit) to form a *composite experimental unit* in which levels of those confounding variables are assigned randomly to a treatment along with the participant. The following example illustrates similarities of, and differences between, systematic and random control over confounding variables.

Example 2.3. To determine whether a new drug is more effective than a placebo, an experimenter requires two nurses to inject 20 patients, with each nurse injecting 10 of them. Because the injection technique, appearance, attitude, and other characteristics of the nurses may have some effect on the response to the injection, it is decided to control for the nurse variable. One way is to systematically associate 10 of the patients with each nurse before assignment to treatment. Another way is to randomly divide the patients between the two nurses, prior to randomly assigning every *nurse + patient* unit to drug or placebo. Both procedures permit valid randomization tests of treatment effects but they may not be equally useful to the experimenter.

If compatibility of patient and nurse is of great importance, the systematic matching of nurses and patients to ensure a high degree of compatibility might be better prior to random assignment of *nurse + patient* units to treatments. Systematic assignment of nurses to patients provides 20 *nurse + patient* units to be randomly assigned to drug and placebo with 10 units for each treatment. For a randomization test, the results and test statistics (such as differences between means) would yield a reference set of $20!/10!10! = 184,756$ equally probable outcomes under the null hypothesis.

Suppose there was interest in the nurse effect as well as the drug effect because of interest in learning whether the nurse variable would otherwise have been a confounding variable. For that purpose, random assignment of patients to nurses can be employed with 10 patients assigned to each nurse; then the *nurse + patient* units would be randomly assigned to treatments. To perform a randomization test of nurse effects, there would be a random assignment of the 10 patients for each nurse into 5 for the drug treatment and 5 for placebo. Thus, instead of a reference set of $20!/10!10!$ elements to test for treatment effects, there would be $(10!/5!5!)^2 = 63,504$ elements. As will be shown in the chapter on factorial designs, the test of treatment effect with the random association of nurses with patients (prior to random assignment of *nurse + patient* units to treatment) is conducted as a 2×2 factorial experiment with two levels of the nurse factor and two levels of the treatment factor. The random association of additional confounding variables with patients to determine their effects would require adding dimensions to the factorial design.

2.8   Between-Subject and Within-Subject Randomization

The most common types of random assignment that are carried out are the random assignment of subjects or other entities to treatments and, for repeated-measures designs, the random assignment of the sequential order of treatments.

The common practice in repeated-measures experiments of randomizing the sequence in which different treatments are given provides some control over the treatment time as a confounding variable. However, better control over the confounding effects of treatment time would be gained by specifying in advance, for each subject or other experimental unit, k treatment times for the k treatments and then randomly assign those treatment times to the treatments. That procedure controls other potentially confounding effects of treatment time than just the sequence in which treatments are taken. Sometimes the time a treatment is given can be a confounding variable even in a between-subject experiment. If all subjects are given their treatments at essentially the same time, time will not be a confounding variable. Alternatively, if all subjects randomly assigned to the A treatment had to take it early in the morning and all assigned to the B treatment take it in the afternoon, the time of day could often be a confounding variable. An effective method of controlling for morning versus afternoon confounding would be to construct composite experimental units for assignment, as done in Example 2.3 to control for the possible confounding effect of nurses. In controlling for the differential effect of morning and afternoon, the composite units to be randomly assigned to treatments would be *subject + time-of-day*, where *time-of-day* is either morning or afternoon.

2.9   Conventional Randomization Procedures

Although there are many possible types of random assignment that can be employed in experiments, only a few types are commonly used. In this section, we will discuss the two most common random assignment procedures: those for the typical independent-groups (between-subject) and repeated-measures (within-subject) experiments. Those are the random assignment procedures that are assumed by t and F tables in determining significance of experimental results. The rationale can be extended to more complex experimental designs.

Independent-groups experiments are those in which each participant takes only one treatment. One-way analysis of variance (ANOVA) and the independent t test are frequently used to analyze data from such experiments. Random assignment is conducted with sample-size constraints, usually with equal numbers of experimental units for each treatment. However, whether an equal or unequal numbers of N subjects are assigned to k treatments, the number of different possible assignments will be $N!/n_1!n_2!\ldots n_k!$. For example, there are $10!/3!3!4! = 4200$ possible assignments of 10 subjects to treatments A, B, and C, with three subjects for A, three for B, and four for C. That is the number of assignments when the number of subjects to be assigned is specified for each treatment. However, when there is a division of 10 subjects among three treatments with four subjects in any of the treatments and three subjects for each of the other two, there are three times as many possible assignments.

In the standard repeated-measures experiment in behavioral science and medicine, each participant takes all treatments. Statistical procedures that are frequently used for such experiments are the correlated t test and repeated-measures ANOVA. Suppose there are five patients, each of whom will take treatments A, B, and C. The order of treatments is randomized for each patient separately. Altogether, there are $(3!)^5 = 7776$ possible assignments because there are 3! possible assignments for each patient, each of which could be associated with 3! possible assignments for each of the other patients. In general, for N units each taking all k treatments, there are $(k!)^N$ possible assignments.

Random assignment conventions almost universally followed by researchers are:

1. More than one unit is assigned to a treatment.
2. The number of units assigned to each treatment is fixed by the experimenter, not determined randomly.
3. The treatment-assignment possibilities are the same for all units:
    a. for independent-groups experiments, any unit can be assigned to any one of the treatments; and
    b. for repeated-measures experiments, each unit is assigned to all treatments.

In the next section, it will be demonstrated that random assignment for randomization tests can follow those conventional procedures if desired but need not. For each of the conventions, randomization tests do not follow any of them.

2.10 Randomization Procedures for Randomization Tests

Randomization tests can be used to determine significance for experiments involving the common independent-groups or repeated-measures random assignment procedures described in Section 2.9, but they are not restricted to such applications. They can be used with any random assignment procedure whatsoever, not just conventional procedures. Randomization tests based on various unconventional random assignment procedures are distributed throughout this book; they are indeed new tests, not just new ways of determining significance for old tests. By allowing the experimenter alternatives to conventional random assignment, such randomization tests provide more freedom in the conduct of an experiment.

In the previous section, the first random assignment convention given is to assign more than one unit to a treatment. This is commonly required for conventional parametric tests because they need an estimate of the within-treatment variability. However, as will be demonstrated in later chapters, randomization tests can be applied validly even if there is only one unit in one or more treatments.

In a way, a single-subject experiment is a repeated-measures experiment with only one subject. However, in a single-subject experiment, unlike the standard repeated-measures experiment, the subject takes the same treatment many more times (repeatedly or alternating) and that makes a test of a treatment effect with a single subject possible. After deciding in advance for each of the treatments how many times the subject will take that treatment, the experimenter can specify the treatment times to be randomly assigned to treatments. Significance is determined by a randomization test in the same way as if the N measurements represented one measurement from each of N randomly assigned subjects in a between-subject experiment. This is one of the random assignment procedures to be discussed in Chapter 11.

Another type of unconventional random assignment (which will be discussed later) is one in which, contrary to the second random assignment convention given in Section 1.9, the sample size for each treatment is determined randomly rather than being fixed in advance. For instance, in a single-subject experiment with 120 successive blocks of time, one of the middle 100 blocks may be selected at random for the beginning of the block to serve as the point where a special treatment will be introduced, that treatment remaining in effect over the subsequent blocks. All blocks of time before the special treatment intervention are control blocks, and all blocks after the intervention are experimental blocks. The random assignment procedure randomly divides the 120 blocks of time into two sets: those for the pre-intervention (control) condition and those for the post-intervention (experimental) condition. The blocks are by no means independently assigned to the treatments: if a particular block is assigned to the post-intervention condition, then every block after it must also be assigned to that treatment. With this special random assignment procedure, the number of treatment blocks assigned to a treatment is determined randomly and this completely determines which blocks are assigned to that treatment. This is a procedure appropriate for certain single-subject experiments, and significance can validly be determined in such cases by a randomization test.

The third random assignment convention listed in Section 1.9 is that treatment possibilities will be the same for all units but that requirement can be ignored when significance is determined by the randomization test procedure. For instance, a randomization test can accommodate data from a special between-subject experiment involving treatments A, B, C, and D in which a restricted-alternatives assignment procedure is used, whereby the possible alternative treatment assignments vary from one subject to the next. Some of the subjects are assigned randomly to any one of the four treatments, while others are assigned randomly within certain subsets of the four treatments. Some subjects may be willing to be assigned to A, B, or C but not to D, so that D is not a possible assignment for them. Some subjects might be willing to participate in the experiment if they are assigned to A or B but not otherwise; so for those subjects, their random assignment is restricted accordingly. Restricted-alternatives random assignment is feasible also for repeated-measures experiments. Some subjects may take all four treatments while others take only two or three, and random assignment of treatment times to treatments for each subject is made according to the alternatives available to that subject. The use of randomization tests for restricted-alternatives random assignment is discussed in Chapter 6.

2.11 Further Reading

It would be difficult to attribute the different ideas presented in this chapter to specific books or articles. However, major sources of inspiration are listed below. These books and articles can also be considered as additional material for the reader interested in the history, applications, and advanced possibilities of random assignment procedures.

Boik, R.J. Randomization, in *Encyclopedia of Statistics in Behavioral Science*, Everitt, B.S. and Howell, D.C. (eds.), John Wiley & Sons, Chichester, UK, 2005, 4, 1669.

Donner, A. and Klar, N. *Design and Analysis of Cluster Randomization Trials in Health Research*, Arnold, London, 2000.

Edgington, E.S. Randomized single-subject experimental designs, *Behav. Res. Ther.*, 34, 567, 1996.

Finch, P. Randomization — I, in *Encyclopedia of Statistical Sciences*, Kotz, S. and Johnson, N.L. (eds.), John Wiley & Sons, New York, 1986, 7, 516.

Fisher, R.A. *The Design of Experiments*, Oliver & Boyd, Edinburgh, 1935.

Kempthorne, O. *The Design and Analysis of Experiments*, John Wiley & Sons, New York, 1952.

Kempthorne, O. The randomization theory of experimental inference, *J. Am. Stat. Assoc.*, 50, 946, 1955.

Kempthorne, O. Why randomize?, *J. Stat. Plan. Inference*, 1, 1, 1977.

Rosenberger, W.F. and Lachin, J.M. *Randomization in Clinical Trials*, John Wiley & Sons, New York, 2002.

2.12 Questions and Exercises

1. Define "experimental unit."

2. Random assignment of a group of participants to two treatments equates the two treatments with respect to age, height, IQ, and all conceivable quantitative participant variables, whether known or unknown. What is the mathematical sense in which the treatments become "equated"?

3. The famous statistician R.A. Fisher (a heavy smoker) contended that the vast amount of "evidence" that was used to support the notion that smoking increased the likelihood of lung cancer and other diseases associated with smoking was inconclusive. What type of evidence do you think he said was necessary? How could he account for the strong association between various diseases and smoking that scientific research revealed?

4. Define "confounding variable."

5. An apparent difference in the effect of two treatments could be due to differences between the experimenters administering the treatments. What random assignment procedure could you apply to control for such experimenter effects?

6. What is restricted-alternatives random assignment?

7. When two treatments are given to each participant in a repeated-measures experiment, control over the effect of the sequential order of the treatments can be achieved by randomizing order within each participant. How can an experimenter control for differences in the time of the treatments, not just the sequence?

8. Describe a dice-tossing procedure for randomly assigning patients to two treatments with four patients assigned to Treatment A and two to Treatment B.

9. Six families are to be randomly assigned to two treatments with three families for each treatment. Two of the six families live in the same neighborhood and must be assigned to the same treatment. How many different assignments are possible?

10. Describe a procedure of assignment to two treatments that makes use of both matching and random assignment.

3

Calculating P-Values

3.1 Introduction

There are two basic methods of permuting experimental results to calculate P-values by means of a randomization test: the systematic, often called exact, method; and the random, often called Monte Carlo, method. Even with fairly small sample sizes, the number of data permutations required for both basic methods makes randomization tests impractical unless a computer is used. This chapter is concerned with calculating P-values when randomization tests are performed by a computer, but first it will be necessary to consider calculation for randomization tests without the use of a computer to determine what it is that the computer must do for the systematic method and two forms of the Monte Carlo method.

3.2 Systematic Reference Sets

Example 3.1. An experimenter designs an experiment involving two treatments, A and B. It is expected that the A measurements will tend to be larger than the B measurements, and so a one-tailed test incorporating that prediction is chosen. There are only four subjects readily available who are appropriate for the experiment. As the administration of a treatment for an individual subject takes considerable time, the experimenter decides to systematically associate a treatment time with each subject so that the experimental unit will be a subject plus the predetermined treatment time. The letters *a*, *b*, *c*, and *d* will be used here to indicate the four subjects and their treatment times. Two subjects are to be assigned to treatment A and two to treatment B, and so the experimenter writes the four designations on slips of paper, mixes them up in a small container, and draws out two slips, without replacement, to determine which two subjects will take treatment A. (The remaining two subjects, of course, take treatment B.) Use of that random drawing technique leads to the selection of experimental units a and c for treatment A, with the two remaining units, b and d, to take treatment B. The following results are obtained and $D = \bar{A} - \bar{B}$, because $\bar{A}$ was predicted to be larger than $\bar{B}$:

	A			B	
Unit	Measurement		Unit	Measurement	
a	6		b	3	
c	7		d	4	
	$\bar{A} = 6.5$			$\bar{B} = 3.5$	
		$D = 3$			

The random assignment procedure determines the equally probable assignments. There are $4!/2!2! = 6$ of them. H_0 of identical treatment effects implies that wherever a unit is assigned, its measurement goes with it. Under H_0, there is thus a data permutation associated with each possible subject assignment, providing the following six data permutations:

	A	B	A	B	A	B	A	B	A	B	A	B
	3	6	3	4	3	4	4	3	4	3	6	3
	4	7	6	7	7	6	6	7	7	6	7	4
$\bar{X} =$	3.5	6.5	4.5	5.5	5	5	5	5	5.5	4.5	6.5	3.5
$D =$		−3		−1		0		0		1		3

For a test where it had been predicted that $\bar{A}$ would be greater than $\bar{B}$, the proportion of the data permutations having as large a value of D as 3 would be determined. The data permutation representing the obtained results is the only one meeting the requirement, so the proportion is 1/6. The one-tailed P-value associated with the obtained results is 1/6, or about 0.167.

If there had been no prediction of the treatment that would provide the larger mean, a two-tailed test would have been appropriate. The proportion of the data permutations with as large a value of $|D|$ as 3, the obtained value, is determined. This proportion is 2/6, or about 0.333. The P-value associated with a two-tailed test is thus about 0.333.

Use of a one-tailed "two-sample t-test" and corresponding entry in the t table to determine significance would have suggested that H_0 could be rejected at the 0.05 level of significance. The use of t tables with such small samples is inappropriate and would be unacceptable to most researchers.

Example 3.2. There are two subjects, a and b, in an experiment in which each subject takes both experimental treatments. Each subject has two treatment times, making four experimental units. The designations a_1, a_2, b_1, and b_2 will be used to specify the two subjects and their treatment times, where subscript "2" refers to the later treatment time for a subject. Treatment A is expected to provide the larger mean. It is randomly determined (as by tossing a coin) which treatment time for subject *a* is for treatment A and which is for treatment B. Then it is determined randomly for subject *b* which treatment time goes with each treatment. The random assignment leads to subject *a* taking treatment B first and subject *b* taking treatment A first, with the following results:

	A		B		
Unit	Measurement	Unit	Measurement		A − B
a_2	10	a_1	8		2
b_1	7	b_2	6		1
	$\bar{A} = 8.5$		$\bar{B} = 7$		D = 1.5

The observed mean difference test statistic gives a D value of 1.5 for the above results. There were two possible assignments for subject *a*, each of which could be associated with either of two possible assignments for subject *b*, making four possible assignments. Under H_0, the measurement for any experimental unit is the same as it would have been if that unit had been assigned to the other treatment; in other words, at each treatment time a subject's response was the same as it would have been at that treatment time under the other treatment. There are thus four data permutations that can be specified for the four possible assignments. The four data permutations are:

	A	B	A	B	A	B	A	B
	8	10	8	10	10	8	10	8
	6	7	7	6	6	7	7	6
$\bar{X} =$	7	8.5	7.5	8	8	7.5	8.5	7
D =		−1.5		−0.5		0.5		1.5

Although there is the same number of experimental units as in Example 3.1, there are only four possible assignments because of the random assignment procedure, and consequently only four data permutations instead of six. For example, the following permutation is not used:

A	B
8	6
10	7

It is not used because it would be associated with an impossible assignment, namely the assignment where subject *a* took treatment A at both of *a*'s treatment times and subject *b* took treatment B at both of *b*'s treatment times. Because the random assignment procedure required each subject to take both treatments, the only data permutations to be used in testing H_0 are those where, for each subject, one of the two obtained measurements is allocated to one treatment and the other measurement to the other treatment.

Inasmuch as the obtained results (represented by the last of the four permutations) show the largest D value, a one-tailed test where the direction of difference was correctly predicted would give a P-value of 1/4, or 0.25.

3.3   Criteria of Validity for Randomization Tests

The validity of randomization tests for various types of applications has been pointed out previously. However, it must be stressed that the use of a randomization test procedure does not guarantee validity. A randomization test is valid only if it is properly conducted. In light of the numerous test statistics and random assignment procedures that can be used with randomization tests, it is essential for the experimenter to know the basic rules for the valid execution of a randomization test. Before dealing with the validity of randomization test procedures, it will be useful to specify criteria of validity for statistical testing procedures in general. Within the decision-theory model of hypothesis testing, which requires a researcher to set a level of significance in advance of the research, the following criterion is appropriate:

> *Decision-Theory Validity Criterion*: A statistical testing procedure is valid if the probability of a type I error (rejecting H_0 when true) is no greater than α, the level of significance, for any α.

For instance, the practice of determining significance of a one-tailed *t* test in accord with the obtained, rather than the predicted, direction of difference between the means is an invalid procedure because the probability of rejecting H_0 when it is true is greater than α. The above criterion, which is implicit in most discussions of the validity of statistical testing procedures, unfortunately is expressed in terms associated with the decision-theory model: "rejection," "type I error," and "α." This can create the impression that only within the Neyman-Pearson decision-theory framework of hypothesis testing can one have a valid test. Restriction of validity to situations with a fixed level of significance may be satisfactory for quality control in industry but it is unsatisfactory for scientific experimentation.

Interest in levels of significance set in advance is by no means universal. For the numerous experimenters who are interested in using the smallness of a P-value as an indication of the strength of evidence against H_0 (an interpretation of P-values inconsistent with the decision-theory approach), a more general validity criterion is required. An operationally equivalent validity criterion that does not use decision-theory terminology (Edgington, 1970) follows:

> *General Validity Criterion*: A statistical testing procedure is valid if, under the null hypothesis, the probability of a P-value as small as P is no greater than P, for any P.

For instance, under the null hypothesis the probability of obtaining a P-value as small as 0.05 must be no greater than 0.05, obtaining one as small as 0.03 must be no greater than 0.03, and so on. Obviously, the two criteria of validity are equivalent: for any procedure, they lead to the same conclusion regarding validity. Therefore, the general validity criterion can be used by experimenters interested in the decision-theory approach. It is useful also to experimenters who do not set levels of significance in advance but instead use the smallness of the P-value as an indication of the strength of the evidence against H_0. (However, experimenters often report their results as significant at the smallest conventional α level permitted by their results.)

It is useful to know the conditions that must be satisfied for certain testing procedures to meet the general validity criterion. Depending on the testing procedure, the conditions or assumptions may include random sampling, random assignment, normality, homogeneity of variance, homoscedasticity, or various other conditions. In the following discussion, we will consider components of a valid randomization test procedure.

Randomization tests permute data in a manner consistent with the random assignment (randomization) procedure, so random assignment is a requirement. Some people might call random assignment an assumption but it does not have to be assumed; it is under the direct control of the experimenter. However, the other condition presupposed by randomization tests, experimental independence of the experimental units, is an assumption. Two units are experimentally independent if one unit does not influence the measurement of the other unit. The following example will show how P-values can be spuriously small when there is lack of experimental independence.

Example 3.3. Ten rats are assigned randomly to treatments with five rats being assigned to a treatment group that is vaccinated against a certain disease and the other five being assigned to a control group that is not vaccinated. Unknown to the experimenter, one of the rats already has the disease when it was assigned and the disease spread throughout the group to which it was assigned. That rat was assigned to the control group, and even though the rats in that group were kept in separate cages, the disease was transmitted by a laboratory assistant who sometimes handled the rats. Thus, all of the control rats caught the disease. The disease was not transmitted to any of the vaccinated rats because they were housed in a separate room and served by a different laboratory assistant. A randomization test based on the prediction of a higher rate of disease for the control group would give a P-value of $1/252$ because there are $10!/5!5! = 252$ ways to divide five diseased and five healthy rats between the control and vaccinated groups, and in only one of those divisions are all five diseased rats in the control group and all five healthy rats in the experimental group. To see how misleading the P-value can be, consider that the chance of assigning the rat that already had the disease to the control group and having the disease spread to the other control animals was $1/2$, and that assignment alone could have resulted in the disease in all of those animals, providing a P-value of $1/252$ even if the vaccination had no effect whatsoever. With absolutely identical treatments, carryover effects of this type can result in a high probability of getting a spuriously small P-value.

Aside from contagious diseases, there are many other ways in which experimental independence can be violated by allowing the responses of an experimental unit to be dependent on the other units assigned to that treatment condition. For instance, in an experiment on the effects of semistarvation, if the food-deprived laboratory animals are not kept isolated from each other, the behavior of one or two animals in the group may greatly influence the behavior of the others, violating the assumption of experimental independence and invalidating statistical tests.

Sometimes it is difficult to isolate units from each other sufficiently to ensure experimental independence. For example, in a comparison of different teaching methods, it is likely that students within classrooms influence the learning of other students in the same classrooms. Hundreds of students may be assigned at random to the classrooms taught by the different methods but the results as a whole may be strongly dependent on just a few students. If one or two students in a classroom — by their enthusiasm, boredom, penetrating questions, or other means — strongly influence the learning of the rest of the students in their class, the lack of experimental independence could make the results almost worthless.

Experimenters should isolate experimental units from each other, or by some other means reduce the chances of communication or other interaction that may result in one unit's measurements being influenced by other experimental units. Although it may be impossible to completely rule out interaction among units, minimizing the opportunity for interaction should be a consideration in designing an experiment.

Therefore, for drawing statistical inferences about treatment effects, randomization tests depend on randomization and experimental independence. In experiments, frequently the units can be isolated well enough to make experimental independence plausible, but that can be difficult to accomplish in repeated-measures experiments where each unit receives more than one treatment. Then, experimental independence over time can be an issue. In that case, careful attention should be paid to the timing and the events between treatment times. For example, if remembering the results from a previous task causes problems, then the use of distracting tasks between treatment administrations might be a solution. Or to take another example, short-term physiological effects that carry over from an earlier treatment to a later one might be controlled by allowing more time between treatments. An experimenter can do various things that are likely to minimize opportunity for the units being affected by other units but experimental independence remains an assumption.

Randomization and experimental independence are necessary but not sufficient for the validity of randomization tests. What is acceptable or unacceptable with respect to the permuting of data is dependent upon the randomization employed and the null hypothesis tested. (As will be shown, other null hypotheses than just the null hypothesis of no treatment effect can be tested.) Chapter 14 deals with the validity of randomization tests in general, and individual chapters concern the validity of the particular randomization tests involved.

3.4   Randomization Test Null Hypotheses

Traditionally, randomization tests have been represented as testing the null hypothesis of no differential treatment effect on any of the experimental units, and the discussion in early editions of this book considered only that null hypothesis. However, other hypotheses can be tested by the use of the randomization test procedure.

Tests of null hypotheses regarding a specified magnitude of effect (such as the hypothesis that for no subject would the treatment A effect be greater than the B effect by as much as 2 milliseconds) are discussed in Chapter 13. The alternative, complementary hypothesis would be that for some subjects, a time measurement would be 2 milliseconds or greater under A than under B. For situations where the null hypothesis of no differential effect of treatments might be of little interest, the null hypothesis of a specified magnitude of effect (other than zero) might be.

Tests of quantitative relationships are described in Chapter 12. An example of a null hypothesis for a quantitative relationship in an independent-groups experiment would be the hypothesis that for treatment levels 5, 10, and 15, the response of every subject would be proportional to the magnitude of the treatment level. The alternative, complementary hypothesis would be that the quantitative relationship did not hold for at least some of the subjects; for some subjects, the response would not be proportional to the treatment level.

A valid randomization test can be devised for testing any null hypothesis for which the experimental randomization, in conjunction with experimental results, permits construction of the experimental outcomes that would have resulted under any of the alternative assignments under the null hypothesis. For the simplest null hypothesis, that of no treatment effect, all data permutations would contain the same measurements — although rearranged — but for null hypotheses relating to magnitude or type of effect, some data permutations would contain transformed measurements.

3.5   Permuting Data for Experiments with Equal Sample Sizes

There are special techniques for permuting data that can be used with equal sample sizes, techniques that use only a subset of the reference set of data permutations associated with all possible assignments. In earlier numerical examples involving a design with equal sample sizes, there is an even number of data permutations and they can be paired so that one member of a pair is the mirror image of the other member in the sense that the measurements are divided into two groups in the same way with the treatment designations reversed. In Example 3.1, the first permutation was A: 3, 4; B: 6, 7, whereas the last permutation was A: 6, 7; B: 3, 4. Obviously, when there are equal sample sizes for two treatments, all data permutations can be paired in this fashion; if there was one for which there was no such mirror image, it would indicate that a possible assignment had been overlooked. Because $|D|$ only depends on which measurements are in each of the two groups and not at all on the treatment designations associated with the two groups of measurements, both members of each mirror-image pair must give the same value of $|D|$. If we divide all data permutations into two subsets, where each subset contains one member of a mirror-image pair, then the proportion of D statistics that have an absolute value as large as some specified value is the same for each of the subsets. Consequently, the proportion of D statistics — in either of the subsets — that have an absolute value as large as some value is the same as the proportion for the entire set. If significance was determined by using one of the two subsets, the same two-tailed P-value for D would be obtained as if the entire set of data permutations was used. However, the amount of computation would be cut in half.

Mirroring of the same kind occurs with designs in which there are more than two treatments. Any nondirectional test statistic for more than two treatments could be employed. For example, we could use the sum of squares between groups, SS_B, as our test statistic because SS_B, like $|D|$, is a nondirectional test statistic. Given equal sample sizes, the P-value for SS_B based on a subset of all data permutations would be the same value as the P-value computed from the entire set. For k treatments with equal sample sizes, a subset of $1/k!$ of all permutations with each data division represented only once in the subset (not several times with different treatment designations) is all that is necessary to determine the P-value that would be obtained by using the entire set of data permutations. Of course, this can result in a considerable savings of computational time.

3.6   Monte Carlo Randomization Tests

Random data permutation is a method that uses a random sample of all possible data permutations to determine the P-value. (Alternatively, random data permutation can be regarded as a method that uses data permutations associated with a random sample of possible assignments.) It serves the same function as systematic data permutation with a substantial reduction in the number of permutations that must be considered. Instead of requiring millions or billions of data permutations, as might be necessary for the systematic data permutation method for many applications of randomization tests, the random data permutation method may be effective with as few as 1000 data permutations. The following discussion is based on Edgington (1969a; 1969b), and the procedure for determining the P-value is the same as that of Dwass (1957) and Hope (1968).

Suppose the experimenter decides that the time and money to deal with all data permutations in the relevant set or subset is too great to be practical. He can then proceed in the following way, using the random permutation method: He performs 999 random data permutations, which is equivalent to selecting 999 data permutations at random from the reference set that would be used with the systematic method. Under H_0, the data permutation representing the obtained results is also selected randomly from the same set. Thus, given the truth of H_0, we have, say, 1000 data permutations that have been selected randomly from the same set, one of which represents the obtained results. We determine the P-value as the proportion of the 1000 test statistic values that are as large as the obtained value. That this is a valid procedure for determining significance can be readily demonstrated. Under H_0, the obtained test statistic value (like the 999 associated with the 999 data permutations selected at random) can be regarded as randomly selected from the set of all possible test statistic values, so that we have 1000 randomly selected values, one of which is the obtained value. If all 1000 test statistic values were different values, so that they could be ranked from low to high with no ties, under H_0 the probability would be 1/1000, or 0.001, that the obtained test statistic value would have any specified rank from 1 to 1000. So the probability that the obtained test statistic value would be the largest of the 1000 would be 0.001. If some values were identical, the effect of tied values would be that the probability could be less than 0.001 but no greater. Admitting the possibility of ties, the probability is no greater than 0.001 that the obtained test statistic value would be larger than all of the other 999 values. Similarly, the probability is no greater than 0.002 that it would be one of the two largest of the 1000 values, and so on. In general, when H_0 is true, the probability of getting a P-value as small as P is no greater than P, and so the method is valid.

A randomization test using the random data permutation method, although employing only a sample of the possible data permutations, is therefore valid: When H_0 is true, the probability of rejecting it at any α level is no greater than α. However, if H_0 is false and there is an actual treatment effect, random data permutation could be less powerful than systematic data permutation because of an increased discreteness in the reference set and a larger minimum P-value. Increasing the number of data permutations used with the random data permutation method makes the difference in power between a randomization test employing random data permutation and a randomization test employing random data permutation negligible.

Another random data permutation method is sometimes advocated and implemented in statistical software (e.g., Senchaudhuri, Mehta, and Patel, 1995). This method will be called *Monte Carlo P-value estimation*. In Monte Carlo P-value estimation, the obtained data are permuted or rearranged in the same manner as the random data permutation method described previously, but instead of including the obtained test statistic as an element of the reference set this estimation procedure excludes the obtained results from the reference set. The P-value that is reported then is the proportion of the elements in the reference set (which does not include the obtained results) that have as large a test statistic value as the value for the obtained results. This P-value represents an unbiased estimate of the value the systematic randomization test procedure would give, but the resulting testing procedure is not valid according to the criterion presented in Section 3.3. For example, as explained previously, random selection of 999 elements from the systematic reference set would give a valid P-value of 1/1000 if the obtained results had the largest test statistic value. By contrast, the Monte Carlo P-value estimation procedure that selected the same 999 elements would give a "P-value" of 0/999 because the obtained results would not be counted in either the numerator or denominator. In fact, the Monte Carlo procedure for estimating the systematic P-value will always calculate a P-value with both the numerator and denominator smaller by 1 than the valid procedure for determining statistical significance.

For reference sets that are large, which is commonly the case when Monte Carlo estimation procedures are employed, the difference in P-values between a randomization test employing random data permutation and the Monte Carlo P-value estimation procedure will be small. For example, if the Monte Carlo estimated P-value is 5/999, the random data permutation procedure P-value would be 6/1000. What makes the Monte Carlo P-value estimation procedure invalid for the determination of statistical significance is not that it gives P-values smaller than would be possible for the systematic method; after all, random data permutation procedures could also sometimes do that. The invalidity of the Monte Carlo estimation procedure is that although it provides unbiased estimation of the systematic P-value, there is no exact control over the Type 1 error rate — the probability of getting a P-value as small as p when the null hypothesis is true, is slightly greater than p (see also Onghena and May, 1995).

The computer programs in this book for random data permutation permit the experimenter to specify the number of data permutations to be performed. If N permutations are specified, the computer will treat the obtained results as the first data permutation and randomly permute the data an additional $(N-1)$ times. Using 1000 data permutations does not provide the sensitivity and accuracy given by several thousand but the sensitivity and accuracy are already substantial. For example, the probability is 0.99 that an obtained test statistic value that would be judged significant at the 0.01 level by using systematic data permutation will be given a P-value no greater than 0.018 by random data permutation with 1000 random permutations. Also, the probability is 0.99 that an obtained test statistic value that would be found significant at the 0.05 level (according to systematic data permutation) will be given a P-value no greater than 0.066 by random data permutation using 1000 random permutations.

Because the systematic and random data permutation methods permute the data differently, different computer programs are required. However, the distinction between systematic and random data permutation should be recognized as a technical one that has practical importance for performing a randomization test but does not affect the interpretation of a P-value given by a randomization test. Nevertheless, some experimenters believe that the distinction between computing P-values on the basis of systematic and random data permutation is sufficiently important to merit specifying which procedure was used to determine significance in a research report. As the sensitivity and accuracy of a random data permutation procedure relative to a systematic procedure generating all relevant data permutations are a function of the number of random data permutations employed, that number should be provided in a report of results of a randomization test whenever a random permutation procedure is used.

3.7   Equivalent Test Statistics

Two test statistics that must give the same P-value for a randomization test are called *equivalent test statistics*. The test statistic D is an equivalent test statistic to $\bar{A}$ for a one-tailed test where treatment A is expected to provide the larger mean because as A increases D must increase, providing the same ranking of the two test statistics over the data permutations and therefore the same P-value. It takes less time to compute a simpler, equivalent test statistic to t, F, r, or some other conventional test statistic for every data permutation to determine the significance value for the more complex one. It will be proven in later chapters that D is an equivalent test statistic to t and $|D|$ is an equivalent test statistic to $|t|$, the two-tailed t test statistic.

Two test statistics are equivalent if and only if they are perfectly monotonically correlated over all data permutations in the set. Expressed in terms of correlation, there will be a perfect positive or negative rank correlation between the values of the two test statistics over the set of data permutations used for determining the P-value. In the following chapters, equivalent test statistics for a number of common test statistics will be given.

Most of the examples of derivations of simpler equivalent test statistics that will be given involve several steps. The first step is to simplify a conventional parametric test statistic somewhat, and each successive step simplifies the test statistic resulting from the preceding step. Various techniques, resulting both in linearly and nonlinearly related equivalent test statistics, will be employed in those chapters. The test statistic for a randomization test can be simpler than an equivalent parametric test statistic because for a randomization test, the test statistic must reflect only the type of treatment effect that is anticipated — unlike parametric test statistics, which must also involve an estimate of the variability of the component reflecting the anticipated effect. For example, the numerator of t, which is the difference between means, is what is sensitive to a treatment effect; the denominator is necessary for parametric tests as an estimate of the variability of the numerator. For other parametric test statistics as well, the denominator very frequently serves the function of estimating the variability of the numerator under the null hypothesis, whereas the randomization test procedure generates its own null distribution of the measure of effect, such as a difference between means, making the denominator irrelevant.

3.8   Randomization Test Computer Programs

The amount of computation required for randomization tests made them impractical in the days when computation had to be done by hand or by mechanical desk calculators. The number of data permutations required is large even for relatively small samples if systematic data permutation is employed. For example, an independent t test with only seven subjects in each of two groups requires consideration of $14!/7!7! = 3432$ possible assignments, of which half (i.e., 1716) must be used for the determination of significance. Because a t value or a value of an equivalent test statistic must be computed for each of the 1716 permutations, the computational requirements are considerable. Thus, even for small samples the amount of computation for a randomization test is great. However, although several thousand computations of a test statistic is an excessive number when the computations are done by primitive methods, in the modern age of computers performing a few thousand computations is not at all expensive or time consuming. Furthermore, the possibility of using random data permutation to base a randomization test on a random sample of all possible data permutations makes randomization tests practical even for large samples.

Modern personal computers can run systematic or random data permutation programs and many programs of both kinds are available for personal computers. A package of programs and illustrative FORTRAN codes are available on the accompanying CD. This package and other packages that can be used to compute randomization test P-values are presented in Chapter 15.

To avoid excessive computing time, it is important to know in advance how many data permutations would be generated by a systematic program. It is useful to have the computer perform computation of that number as soon as the sample sizes are entered and to have the computer programmed to switch to a random permutation procedure automatically if the number of data permutations exceeds some user-specified value.

Random data permutation tests are valid in their own right, not just approximations to systematic tests, and their sensitivity for a reasonable number of data permutations is comparable to that of the corresponding systematic test. But if an experimenter wants the "latest model," there are a number of variations of random data permutation programs that incorporate procedures for speeding up the determination of statistical significance. Chapter 14 can serve as a guide in evaluating the validity of the new procedures.

Mehta and Patel (1983), Pagano and Tritchler (1984), Dallal (1988), Baker and Tilbury (1993), and others have speeded up systematic tests substantially, and those developments will be of interest to people who want "exact" P-values and not just ones that are validly produced, as by random data permutation. Their fast programs are for *t*-test types of comparisons in which the total of the measurements for one of the treatments can be the test statistic. For many applications, the practical sample size might be many times as large as that for which a regular systematic program would be feasible. Not only does that permit many users to use an "exact" test instead of a random test for their data but it also has value for theoreticians because it allows them to examine exact rather than random test statistic distributions for large samples.

3.9   Writing Programs for Randomization Tests

For many applications of randomization tests, the experimenter can use the programs and code available on the accompanying CD. The language used for the programs is FORTRAN but the programs can be easily translated into another computer language. For either understanding the programs on the CD or writing new programs, it is necessary to know what the programs require the computer to do.

All computer programs for randomization tests must specify the performance of the following operations:

1. Compute an obtained test statistic value, which is the value for the experimental results.
2. Systematically or randomly permute the data.
3. Compute the test statistic value for each data permutation.
4. Compute the P-value. The P-value is the proportion of the test statistic values, including the obtained value, that are as large as (or, where appropriate, as small as) the obtained test statistic value.

Figure 3.1 shows the basic structure of the computer programs on the CD and represents the steps a computer goes through in determining the P-value of a randomization test if the data permutations are to be generated one at a time and compared one at a time to the obtained data permutation (experimental results) to see whether to increment a counter. However, it should be noted that some fast and "clever" algorithms make such individual comparisons and incrementing of a counter unnecessary for several types of randomization tests. Those algorithms have a structure that is much more complex than the structure in Figure 3.1. For the technical details, the interested reader is referred to Mehta and Patel (1983), Pagano and Tritchler (1984), Mehta, Patel, and Senchaudhuri (1988), Mehta (1992), and Baker and

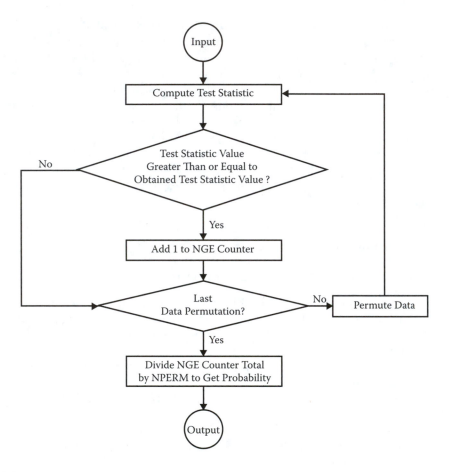

FIGURE 3.1
Flowchart for randomization tests.

Tilbury (1993). For a basic understanding of the randomization test rationale
and for simple applications, the "brute force" approach as exemplified in
Figure 3.1 suffices.

For systematic data permutation, it is essential that each data permutation
in the set (or subset) of data permutations associated with the relevant
possible assignments be determined. The problem in making certain that all
data permutations have been produced once and only once by a computer
is that which confronts the experimenter if data permutations are done by
hand. Imagine the difficulty in checking several thousand data permutations
to see if all possible permutations are represented. Obviously, some sort of
systematic listing procedure is required. In a program using systematic data
permutation, the counterpart of systematic listing is specification of a systematic sequence for generating data permutations. Different types of random assignment require different sets of data permutations, and so there are

correspondingly different procedures for systematically producing the data permutations. For each type of random assignment, it must be determined how one can systematically produce the possible permutations.

Example 3.4. We have five subjects, two of which are randomly assigned to A and three to B. There will be $5!/2!3! = 10$ data permutations. We obtain the following results: A: 12, 10; B: 9, 12, 8. How can we list the 10 data permutations systematically to make sure each one is listed once and only once?

We want a procedure that can be generalized to cases where there are so many data permutations that a definite sequential procedure of listing the data permutations is the only way to be certain that the list is correct. When there are two treatments, all we need is a systematic listing procedure for the measurements allocated to one of the treatments because the remaining measurements are allocated automatically to the other treatment. Thus, what is needed for this randomization test where there are only two subjects assigned to treatment A is a listing procedure for the 10 possible pairs of measurements for treatment A. Such a procedure results from associating an index number with each of the five measurements, systematically listing the pairs of index numbers, then translating the index numbers back into the measurements with which they are associated. We will use the integers 1 to 5 as index numbers representing the measurements in the order in which they were listed:

Measurements	12	10	9	12	8
Index numbers	1	2	3	4	5

The list of all possible pairs of index numbers that can be associated with the measurements allocated to treatment A for the 10 data permutations is (1, 2), (1, 3), (1, 4), (1, 5), (2, 3), (2, 4), (2, 5), (3, 4), (3, 5), and (4, 5). These index numbers are in ascending order within the two values for each permutation; then those pairs of index numbers are listed in ascending order. This procedure pairs each index number with every other one and eliminates the possibility of listing the same combination of index numbers with the index numbers in two different orders. Translating the index number list into a list of measurements for treatment A produces the following: (12, 10), (12, 9), (12, 12), (12, 8), (10, 9), (10, 12), (10, 8), (9, 12), (9, 8), and (12, 8). For each of these 10 pairs of measurements for A, the remaining three measurements are associated with the B treatment, providing the 10 data permutations.

In Section 3.5, it was pointed out that when there are equal sample sizes for *t* tests or ANOVA, it is necessary to compute only $1/k!$ of the possible data permutations (where k is the number of treatments) because of the redundancy due to mirror-image data permutations that have the same partitions of the data but transposed treatment designations. Listing procedures have been incorporated into the equal-N programs that list only $1/k!$ nonredundant data permutations. When random data permutation is employed, it is not necessary to work up a systematic listing procedure because only a sample rather than all of the data permutations are generated. Random sampling either with or without replacement is acceptable.

The programs given in this book will accommodate a large number of applications, but effective utilization of the potential of randomization tests can only be achieved if experimenters make additional applications that require new programs. Any randomization test program for testing the general null hypothesis of no differential effect of treatments can follow the basic format given in Figure 3.1, so that with a slight modification a program can be changed into a new one. If the calculation of a P-value by a randomization test is the only acceptable procedure, or if use is made of a new test statistic or customized design for which there is no alternative method of calculating the P-value, the expense of modifying a program or writing an entirely new one is justifiable even if the new program is used only once. However, many new programs can be employed repeatedly, either by the same experimenter or by others.

A computer program created for one test can be changed to be used for a quite different test simply by modifying the test statistic and portions of the program for computing the test statistic. For example, we have been illustrating the use of a computer program with a difference between means as the test statistic, and a change in the test statistic to a difference between median results in a randomization test sensitive to a somewhat different type of treatment effect. To determine the P-value for this new test statistic, its value would be computed for the experimentally obtained results and for each of the data permutations that otherwise would be used for determining the significance of a difference between means. The data permutation procedure is based on the random assignment procedure used by the experimenter, but the test statistic in a program for a certain kind of random assignment can be any test statistic whatsoever, enabling an experimenter to devise a test statistic that is sensitive to the type of treatment effect that is expected.

Computer programs for calculating P-values by data permutation for conventional experimental designs fall into two major classes: programs for experiments with independent groups, where each experimental unit could be assigned to any treatment, and programs for experiments of a repeated-measures type, where each experimental unit is subjected to all treatments. The two types of random assignment account for the bulk of random assignment that is performed in experiments so frequently that there is need for only two basic programs, each of which can be employed with a number of different test statistics.

Other types of random assignment are sometimes used and they require special types of programs and special procedures for permuting the data; but for each random assignment procedure, a single basic program is sufficient for many applications, modifications in the program being made through changes in the test statistic.

3.10 How to Test Systematic Data Permutation Programs

After a systematic data permutation program has been written, it should be checked to see if it gives correct P-values. Data sets for which the P-value can be determined without the use of a computer program can be used to test programs. It is useful to test programs with small sets of data because then the data permutations can be easily listed and the P-value determined readily without a computer program. A computer program for systematic data permutation will give the same value if the program is written correctly.

However, large sets of data can sometimes be useful to test programs for systematic data permutation. The data may be allocated to treatments in such a way as to provide the maximum value of the test statistic, and the probability value then would be the number of permutations giving the maximum value divided by the total number of permutations. For example, with completely distinctive, nonoverlapping sets of measurements with five measurements in each group, the two-tailed P-value for D should be 2/252 (about 0.008) because two permutations give the maximum value and there are 10!/5!5! = 252 data permutations. Data configurations that provide the next-to-maximum value of a test statistic also are convenient to use to test programs. Accompanying the systematic data permutation programs in the following chapters are sample data that can be used to test the programs. All or some of the data permutations and test statistic values should be printed out as a double check on the program.

3.11 How to Test Random Data Permutation Programs

The correctness of a random data permutation program is more difficult to assess because of its random element. Rarely will a random sample of the relevant data permutations yield exactly the same P-value as systematic data permutation but the correspondence should be close. It has been shown elsewhere that the closeness of correspondence can be expressed probabilistically (Edgington, 1969a). Given a systematic randomization test P-value for a data set and the number of permutations used for random data permutation, the probability that the value given by random data permutation will be within a certain distance of the systematic randomization test P-value can be determined. This fact can be used to test random data permutation programs. The larger the number of data permutations used for random data permutation, the greater the probability that the obtained P-value will be within a certain distance of the value given by systematic data permutation, so it is useful in testing random data permutation programs to use a large number of data permutations. Table 3.1 is based on 10,000

TABLE 3.1

99% Probability Intervals for Random Data Permutation P-Values
Based on 10,000 Data Permutations

Systematic Data Permutation P-Value	99% Probability Interval for Random Data Permutation
0.01	0.0075–0.0127
0.02	0.0165–0.0237
0.03	0.0257–0.0345
0.04	0.0350–0.0452
0.05	0.0445–0.0557
0.06	0.0540–0.0662
0.07	0.0635–0.0767
0.08	0.0731–0.0871
0.09	0.0827–0.0975
0.10	0.0923–0.1078
0.25	0.2389–0.2612
0.50	0.4871–0.5130

permutations. A new random permutation program can be tested in the following way:

1. Determine the systematic randomization test P-value associated with a set of test data. (Some test data and the corresponding systematic randomization test P-values are given within each of the chapters in this book.)

2. Determine the P-value for the test data with the random data permutation program to be tested, using 10,000 data permutations.

3. Use Table 3.1 to see whether the P-value given by random permutation falls within the 99% probability interval associated with the systematic randomization test P-value.

4. Repeat the above procedure with new data if the first set of data gives a P-value outside the 99% probability interval. If the P-value again falls outside the 99% probability interval, the program should be examined very carefully.

Table 3.1 is based on 10,000 data permutations to give probability intervals small enough to be effective in testing new programs. Of course, once a program has been checked out, it may be used with fewer or more than 10,000 data permutations if the experimenter desires.

As with the testing of systematic permutation programs, it is desirable to double-check a random permutation program by having all or some of the data permutations and their test statistic values printed out. For this check, the number of random permutations for conducting the test could be 20 or 30 instead of 10,000.

3.12 Nonexperimental Applications of the Programs

Although developed for experimental applications, randomization test programs have been and still are extensively used for nonexperimental applications as well. The programs on the CD and other commercially available or free "permutation test" or "exact test" computer programs obviously cannot distinguish between experimental and nonexperimental applications. They cannot know how the data were collected, which design was used, and whether there was random assignment or not.

However, the rationale for permuting data in nonexperimental research differs from that for permuting data when there has been experimental randomization, and this is not nearly as straightforward. It should not be thought that nonexperimental use of a program that can be used to determine the P-value for treatment effects for a randomization test applied to experimental data automatically endows that test with the properties of a randomization test. In research reports and in nonparametric books, the term "randomization test" is sometimes used to refer to any permutation test, and the application of their "randomization test" to data from nonexperimental research on nonrandomly selected subjects is presented as if random sampling was unnecessary, as indeed would be the case for a true randomization test. Such a practice overlooks an important property of permutation tests: The computational procedure must be supplemented with random sampling or random assignment to provide valid tests of hypotheses.

In nonexperimental research, random sampling is required when the purpose is to test hypotheses, but other uses of P-values as probabilistic measures of strengths of relationship between naturalistic variables do not require random sampling, and those uses seem to hold the same appeal for users of permutation tests that they have long held for users of parametric tests. Winch and Campbell (1969) proposed an interesting rationale for statistical "tests" in certain situations where there is neither random sampling nor random assignment, illustrating their approach by a permutation test example. Their proposal was to consider a P-value in those situations as not being for the purpose of disproving a hypothesis but nonetheless for helping the researcher make research decisions. For example, suppose we compare several hundred rural and urban people by the permutation test procedure of repeatedly dividing the data between the two groups and computing a difference between means with respect to some measure of physical fitness. If the one-tailed P-value is, say, 0.40, indicating that random division of the data gives such a large difference 40% of the time, one would be less likely to pursue the investigation further than if the P-value was 0.02. As it is not an experiment, no statistical inferences can be drawn about the cause of the difference, no matter how small a P-value is obtained, but a small P-value is more likely than a large value to result in the researcher continuing to research that area, developing hypotheses to account for the differences, or following up with experiments to test hypotheses. When researchers refer to a "chance" difference or relationship, often they seem to be referring to how often arbitrary ordering, dividing, pairing, or some other random rearrangement would lead to such a strong relationship. For those experimenters, a natural statistical procedure to use may be a permutation test because its rearrangement of the data may closely match their concept of randomness of relationship.

There is opportunity for fruitful interchange of ideas between experimental and nonexperimental users of permutation tests. Because the same computational procedures can be used by both types of users, programming advances are mutually beneficial. Computer scientists have recognized the potential for the development of new permutation tests and more efficient programs and have contributed useful programming ideas to randomization test users and other users of permutation tests. Any test statistic whatsoever can be incorporated into a permutation test, and that has stimulated exploration by both types of users into the development of test statistics sensitive to certain types of relationships. And a research design may well incorporate both experimental and nonexperimental components, as in the case of randomized blocks, where the effect of the blocking variable — such as sex or age — cannot be determined experimentally but where a permutation test could be useful in determining whether the blocking variable is related strongly enough to merit its use in blocking.

3.13 Questions and Exercises

1. Five subjects are randomly assigned to treatments A and B, with three assigned to A and the remaining two to B. The mean of A minus the mean of B is the test statistic. Give five numerical measurements for which the smallest possible P-value is 0.20 for any permutation of the data.

2. In conventional decision-theory statistical testing, what is a Type 1 error?

3. For a repeated-measures experiment in which each of five subjects takes both the A treatment and the B treatment, what is the smallest possible P-value for a randomization test?

4. In this chapter, a distinction is made between two conditions necessary for the validity of a randomization test: "random assignment" and "experimental independence." The distinction that is stressed is that random assignment is not an "assumption" in the sense that normality and homogeneity of variance are for parametric tests, whereas experimental independence definitely is an assumption in the same sense as those parametric assumptions are. Explain.

5. There are several reasons why an experimenter might want to carry out a randomization test even when the null hypothesis of no differential effect of two treatments is considered by the experimenter as not worth testing. Give one reason.

6. Give an example of a quantitative law or relationship that could be tested by a randomization test.

7. Why are equal-N programs less important today than they were many years ago?

8. What makes one test statistic an equivalent test statistic to another for a randomization test? In other words, what is meant by saying that two test statistics are equivalent?

9. Are the P-value determination procedures in the random data permutation programs recommended in this chapter valid or do they simply give P-values close to those that a valid program would give?

10. Give one or more fundamental differences between randomization tests and other permutation tests.

REFERENCES

Baker, R.D. and Tilbury, J.B., Rapid computation of the permutation paired and grouped t tests, *Appl. Stat.*, 42, 432, 1993.

Dallal, G.E., PITMAN: A FORTRAN program for exact randomization tests, *Comp. Biomed. Res.*, 21, 9, 1988.

Dwass, M., Modified randomization tests for nonparametric hypotheses, *Ann. Math. Stat.*, 28, 181, 1957.

Edgington, E.S., *Statistical Inference: The Distribution-Free Approach*, McGraw-Hill, New York, 1969a.

Edgington, E.S., Approximate randomization tests, *J. Psychol.*, 72, 143, 1969b.

Edgington, E.S., Hypothesis testing without fixed levels of significance, *J. Psychol.*, 76, 109, 1970.

Hays, W.L., *Statistics for the Social Sciences* (2nd ed.), Holt, Rinehart & Winston, New York, 1972.

Hope, A.C.A., A simplified Monte Carlo significance test procedure, *J. R. Statist. Soc.*, 30, 582, 1968.

Mehta, C.R., Comment: An interdisciplinary approach to exact inference for contingency tables, *Stat. Sci.*, 7, 167, 1992.

Mehta, C.R. and Patel, N.R., A network algorithm for performing Fisher's exact test in rxc contingency tables, *J. Am. Stat. Assoc.*, 78, 427, 1983.

Mehta, C.R., Patel, N.R., and Senchaudhuri, P., Importance sampling for estimating exact probabilities in permutational inference, *J. Am. Stat. Assoc.*, 83, 999, 1988.

Onghena, P. and May, R., Pitfalls in computing and interpreting randomization test p-values: A commentary on Chen & Dunlap, *Behav. Res. Methods Instrum. Comput.*, 27, 408, 1995.

Pagano, M. and Tritchler, D., On obtaining permutation distributions in polynomial time, *J. Am. Stat. Assoc.*, 78, 435, 1984.

Senchaudhuri, P., Mehta, C.R., and Patel, N.R., Estimating exact p-values by the method of control variates or Monte-Carlo rescue, *J. Am. Stat. Assoc.*, 90, 640, 1995.

Winch, R.F. and Campbell, D.T., Proof? No. Evidence? Yes. The significance of tests of significance, *Am. Sociol.*, 4, 140, 1969.

4

Between-Subjects Designs

4.1 Introduction

This is the first of several chapters dealing with the applications of randomization tests. These chapters serve a dual function: The primary function is to illustrate randomization test principles by specific instances. A second function is to provide applications useful to experimenters. Early randomization test programs and packages produced in the 1960s and 1970s were BASIC or FORTRAN programs. Programs are readily available from commercial and private sources in modern computer languages and are more efficient, but a discussion of the early programs will help in understanding modern programs and the randomization test rationale underlying them. The programs on the CD accompanying this book therefore are in FORTRAN. The randomization test procedure is best understood if it is first considered in relation to familiar statistical tests. Therefore, the early chapters concern applications to common analysis of variance (ANOVA) and t tests, providing the foundation for later applications to more complex and less familiar tests.

When the question of violations of parametric assumptions arises, determination of P-values by means of a randomization test is a ready answer. This answer applies to the simple tests considered in this chapter as well as to the more complex tests in some of the later chapters. It is true that in the vast majority of applications of ANOVA and independent t tests, the use of F or t tables for determining the P-value goes unchallenged; other researchers and editors are not likely to question the tenability of the parametric assumptions. (However, as the practicality of randomization tests is becoming better known, this attitude seems to be changing.) Although the proportion of objections to the use of the parametric significance tables is small, those tables are used so frequently that even a small proportion of objections can be a considerable number. For example, the experimenter who applies those tests to dichotomous data is likely to encounter the objection that the data must be taken from a continuum. Also, if the sample sizes are very small the experimenter may encounter objections. In such cases, the determination of the P-value by data permutation (using programs in this chapter) is a solution to the difficulty. Even inequality of sample sizes can cause problems. Studies of "robustness" have examined the validity of F and t tables when the parametric assumptions under-lying the tables are not met, and they have shown repeatedly that parametric assumptions are important considerations when sample sizes are unequal. After discussing ANOVA with unequal sample sizes, Keppel (1973, p. 352) stated:

> This analysis should entail an increased concern for violations of the assumptions underlying the analysis of variance. As we have seen, the F test is relatively insensitive to violations of these assumptions, but only when the sample sizes are equal. Serious distortions may appear when these violations occur in experiments with unequal sample sizes.

This view expressed by Keppel is consistent with that of many researchers (and editors), so the experimenter would be well-advised to use a randomization test to determine the P-value when it is necessary to use unequal sample sizes.

For one-way ANOVA and the independent t test, the random assignment to treatments is performed in the following way. Typically, the experimental unit is a subject. The number of experimental units to be assigned to each treatment is fixed by the experimenter. From the subjects to be used in the experiment, the experimenter randomly selects n_1 for the first treatment, n_2 of the remaining subjects for the second treatment, and so on. Any subject could be assigned to any of the treatments, and it is randomly determined (with sample-size constraints) which treatment a subject takes. Of course, the experimental unit could be a cluster or group of subjects along with treatment times, and could include the experimenter or technician who administers the treatment to which the unit is assigned.

The systematic procedure to be described in this chapter for determining the P-value of t for an independent t test by means of the randomization test procedure is one described by Fisher (1936). Fisher explained how to derive a reference set of the test statistic by repeatedly mixing and dividing two hypothetical sets of sample data. The application was nonexperimental and required random sampling of two populations to test their identity. Pitman (1937) described the same computational procedure and considered its application to both experimental and nonexperimental data. He showed that the procedure of determining the P-value could be applied in experi-mental situations where "the two samples together form the whole popula-tion" (p. 129) and indicated the practical value of this area of application where random sampling is not required. It is Pitman's rationale for the application of randomization tests to nonrandom samples that is of major importance in the typical experimental setting.

ANOVA and t tests are used for detecting differences between experimen-tal treatments where it is expected that some treatments will be more effective than others. The independent t test can be regarded as a special case of one-way ANOVA — the case where there are only two treatments. However, because t tests were developed before ANOVA, they tend (for historical reasons) to be used for statistical analysis, although ANOVA can accommodate data from experiments where there are only two treatments as well as from those with more treatments.

The *t* test and ANOVA are basic components of more complex tests, such as factorial and multivariate ANOVA. This chapter will discuss the use of randomization tests to determine the P-value of *t* or *F*. Random selection of experimental subjects from a population is not required, making randomization tests especially useful for the numerous experimental situations in which the experimenter has not selected subjects at random from a population. Given random assignment of subjects to treatments, a valid test of treatment effect can be carried out by computing *t* or F in the standard way and then determining the P-value by the randomization test procedure. However, it must be stressed that in the absence of random sampling there can be no statistical inferences about populations. The statistical inferences that these randomization tests can provide are, like those for other randomization tests, inferences about treatment effects on the subjects in the experiment.

The experimental randomization associated with these tests is that for conventional one-way analysis of variance and independent *t* tests. After deciding on N subjects for an experiment, the experimenter randomly selects n_A of them for the A treatment, then n_B of the N − n_A subjects for treatment B, and so on.

For ANOVA, the randomization test null hypothesis is no differential effect of the treatments on any of the subjects, which we will call the null hypothesis of no treatment effect. According to the null hypothesis, the measurement for each of the subjects is invariant (unchanging) over alternative assignments. The alternative hypothesis — the hypothesis for which supporting evidence is sought in the experiment — is the complement of the null hypothesis. For example, for ANOVA with three treatments a small P-value supports the hypothesis that the null hypothesis is not true, that for at least one subject the measurement would have been different under at least one of the two other treatments to which the subject could have been assigned. In parametric analysis of variance discussions, it is stressed that rejection of the general null hypothesis of no differences between treatments does not imply that all treatments are different. And with randomization tests, "rejection" does not imply that there are any two treatments that would have different effects on all of the subjects. To draw conclusions about the subjects affected or the treatments involved would require reliance on nonstatistical inferences, inferences based on other knowledge or beliefs, rather than simply relying on the P-value. Of course, the same is true for generalizing beyond the experimental setting and subjects in the experiment.

4.2   One-Way ANOVA with Systematic Reference Sets

We will first consider a randomization test based on systematic permutation that can be used to determine the P-value of results for a comparison of two or more treatments, whether the number of subjects per treatment is equal or unequal. In the case of two treatments, the P-value provided by this procedure is the same as the two-tailed value for the *t* test, which will be discussed later.

A systematic way of listing alternative outcomes under the null hypothesis is necessary for the derivation of the systematic reference set to ensure that all data permutations (divisions of data among treatments) are considered once and only once. We will now consider how this can be done for one-way ANOVA, making use of a numerical example.

Example 4.1. Only seven subjects are readily available for our experiment comparing the effects of three different drugs (A, B, and C) on time to solve a puzzle. Our assignment procedure is the random selection of two of the seven subjects for drug A, then two of the remaining five subjects for drug B, and leaving the last three subjects to take drug C. The number of possible assignments the assignment procedure provides is $(7!/2!5!) \times (5!/2!3!) = 7!/2!2!3! = 210$. (The general formula for the number of assignments of N subjects to k experimental treatments is $N!/n_1!n_2!...n_k!$.)

Under the null hypothesis of no differential effect of the drugs on time to solve the puzzle, random assignment of a subject to take a particular drug "assigns" the subject's measurement to that treatment. Thus, when all subject assignments are listed those various partitions of subjects among drugs can readily be transformed into partitions of the experimental results to provide 210 *F* values in the systematic reference set. In the following list, the numbers from 1 to 7 are used to designate the individual subjects. Starting with the experimental results, the successive partitions of the 7 subjects are represented in Table 4.1.

Within each treatment, the subject numbers are always in ascending order because we are not concerned with the order of subject assignments within a treatment. For example, for one-way analysis of variance, "1 2 |3 6| 4 5 7" represents the same assignment to treatments as "2 1 |3 6| 4 7 5," so to avoid listing the same assignments ("randomizations") more than once we restrict assignment designations by having the subject numbers in ascending

TABLE 4.1

Successive Partitions of Subjects

A	B	C
1 2	3 4	5 6 7
1 2	3 5	4 6 7
1 2	3 6	4 5 7
1 2	3 7	4 5 6
1 2	4 5	3 6 7
1 2	4 6	3 5 7
	.	
	.	
	.	
6 7	3 4	1 2 5
6 7	3 5	1 2 4
6 7	4 5	1 2 3

order within treatments. Notice that if the vertical bars between treatments were removed, we would have seven-digit sequences in ascending order of magnitude from "1 2 3 4 5 6 7" to "6 7 4 5 1 2 3." Because of the constraining order of subject numbers within treatments, the entire list does not contain $7! = 5040$ different seven-digit sequences but only $7!/2!2!3! = 210$ seven-digit sequences. The P-value is the proportion of those 210 data permutations in the reference set with a value of F as large as the value for the experimental results (the first data permutation).

We will now transform some of the nine partitions of subjects in Table 4.1 into data permutations representing outcomes under the null hypothesis, given alternative random assignments. The two subjects actually assigned to treatment A are given the first two numbers, namely 1 and 2, then the two subjects that were assigned to B are designated with the next two numbers, and the remaining subjects assigned to C are designated by the last three numbers. Suppose the experimental results associated with the subjects were as follows:

Treatments	A		B		C		
Subject #	1	2	3	4	5	6	7
Data	16	18	18	23	15	14	10
$F = 4.84$							

This is the first data permutation and associated F value in the list of 210 possible outcomes that comprises the systematic reference set. The second and third data permutations in the reference set can be derived from the second and third subject partitions in Table 4.1, which are "1 2 |3 5| 4 6 7" and "1 2 |3 6| 4 5 7":

Second Data Permutation							
Treatments	A		B		C		
Subject #	1	2	3	5	4	6	7
Data	16	18	18	15	23	14	10
$F = 0.05$							

Third Data Permutation							
Treatments	A		B		C		
Subject #	1	2	3	6	4	5	7
Data	16	18	18	14	23	15	10
$F = 0.97$							

And the last, the 210th data permutation, based on the 210th subject number partition into "6 7 |4 5| 1 2 3," is the following:

210th Data Permutation							
Treatments	A		B		C		
Subject #	6	7	4	5	1	2	3
Data	14	10	23	15	16	18	18
$F = 2.57$							

The P-value for the obtained F is the proportion of the 210 F values that are as large as the F for the observed results shown in the first data permutation.

All 210 subject number partitions are now distinctive but the F values associated with those 210 subject partitions are not all different. Subject 2 and subject 3 both had a measurement of 18, so for some data permutations in the list of 210 different subject partitions necessarily give the same F values (or values of any other test statistic) because of those tied values. For example, the first subject partition, which is "1 2 | 3 4 | 5 6 7," gives the same data permutation, namely "16 18 | 18 23 | 15 14 10," as the data permutation associated with subject partition "1 3 | 2 4 | 5 6 7," where the assignment of subject 2 and subject 3 are switched. And for the same reason, the second subject partition "1 2 | 3 5 | 4 6 7" gives the same data permutation and value of F as subject partition "1 3 | 2 5 | 4 6 7."

The utilization of the CD Program 4.1 gives a P-value of 10/210, or about 0.0476, which is statistically significant at the 0.05 level. Therefore, we have found support for the hypothesis that the measurements of the subjects did show some variation over the different drugs. Given statistical significance, we use what we know of various known side-effects of the drugs to see if that will give a clue whether only one of the drugs differed from the other two in effectiveness or whether all three were different. But no theoretical or empirical findings favor one possibility over the other. However, our method of carefully screening subjects to select the seven experimental subjects causes us to believe it is plausible that most or all of the subjects would have responded differentially. Also, nonstatistical considerations lead us to postulate that other physically fit subjects in the same age group and of comparable intelligence would respond in a similar manner.

In justifying our permuting of data as if random assignment of subjects to drugs was, under the null hypothesis, comparable to randomly assigning the measurements of the subjects to drug treatments, we were acting as though independent assignment of subjects rendered the measurement of a subject to be independent of the assignment of other subjects. The independence of the measurements of different subjects is sometimes called an "assumption" of a randomization test, but unlike many parametric "assumptions" it is to a considerable degree under the control of the experimenter. In our experiment, as in many other experiments, the situation was physically arranged to limit the likelihood of interaction between the subjects.

What constitutes a treatment — in this case, a drug treatment — is an important factor to control. When we conclude that the different drugs did not have the same effect, we are thinking of the only variation in the treatment being the drug administered and not the time or the person who administered it. Any systematic association of such factors with treatments means that they are part of the drug treatment and might very well have led to significant findings even when the drugs per se would have identical effects. Therefore, it is desirable to hold such influences constant over treatment levels or randomly assign to the subjects the times and the persons giving the drug, so that random assignment of a subject to a treatment randomly assigns the experimental unit consisting of a subject plus that subject's treatment plus a particular person administering the drug.

4.3   A Simpler Test Statistic Equivalent to *F*

Table 4.2 is a summary table for one-way ANOVA, where k is the number of treatments and N is the total number of subjects. The *sum of squares between groups* (SS_B) and the *sum of squares within groups* (SS_W) are defined as:

$$SS_B = \sum \left(\frac{T_i^2}{n_i} \right) - \frac{T^2}{N} \qquad (4.1)$$

$$SS_W = \sum X^2 - \sum \left(\frac{T_i^2}{n_i} \right) \qquad (4.2)$$

where T_i and n_i refer to the total and the number of subjects, respectively, for a particular treatment, and where T and N refer to the grand total and the total number of subjects for all treatments, respectively.

It will now be shown that $\Sigma(T_i^2/n_i)$ is an equivalent test statistic to F for one-way ANOVA for determining the P-value by systematic listing of all data permutations. That is, it will be shown that using it as the test statistic will give the same randomization test P-value as using F. The formula for F in Table 4.2 is equivalent to $(SS_B/SS_W) \times [(N - k)/(k - 1)]$. Inasmuch as $[(N - k)/(k -1)]$ is a constant multiplier over all data permutations, its elimination has no effect on the rank of the data permutations with respect to the test statistic value and therefore no effect on the P-value. Thus, SS_B/SS_W is an equivalent test statistic to F. The *total sum of squares* (SS_T) is a constant over all data permutations and $SS_T = SS_B + SS_W$; consequently, SS_B and SS_W must vary inversely over the list of elements of the reference set, which in turn implies that SS_B and SS_B/SS_W show the same direction of change over all data permutations. Therefore, SS_B is an equivalent test statistic to SS_B/SS_W and, of course, also to F. The final step is to show that $\Sigma(T_i^2/n_i)$ is an

TABLE 4.2

Summary Table for One-Way ANOVA

Source of Variation	Sum of Squares	Degrees of Freedom	Mean Square	F
Treatment	SS_B	$k - 1$	$SS_B/(k - 1)$	$\dfrac{SS_B/(k-1)}{SS_W/N-k}$
Error	SS_W	$N - k$	$SS_W/(N - k)$	
Total	SS_T	$N - 1$		

equivalent test statistic to SS_B. The grand total T and the total number of subjects N are constants over all data permutations, and so T^2/N in Equation 4.1 is a constant whose removal would change each test statistic value by the same amount, leaving the rank of the test statistic unchanged. Thus, the term $\Sigma(T_i^2/n_i)$ in Equation 4.1 is an equivalent test statistic to SS_B and, therefore, also to F.

For determining the P-value by the use of a randomization test, the computation of $\Sigma(T_i^2/n_i)$ as a test statistic will give the same P-value as the computation of F. It is easier than F to compute and so it will be used to determine the P-value of F for the general-purpose (equal-or-unequal sample size) computer programs for one-way ANOVA. For equal sample size ANOVA, ΣT_i^2 is an equivalent test statistic to $\Sigma(T_i^2/n_i)$ and F, and will be the test statistic used for determining the P-value.

Instead of using the notation $\Sigma(T_i^2/n_i)$, hereafter we will omit the subscripts and use the simplified notation $\Sigma(T^2/n)$. Similarly, instead of using ΣT_i^2, we will use ΣT^2 to refer to the sum of the squared totals for the individual treatments. We will now consider the principal operations to be performed by a computer in finding the P-value of F for one-way ANOVA by systematic data permutation:

Step 1. Arrange the experimental results in tabular form (grouping the subjects and their measurements according to treatment) and designate the subjects with numbers 1 to N, beginning with subjects assigned to treatment A.

Step 2. Generate a systematic sequence of partitions of the subject numbers so that there are n_A numbers for A, n_B numbers for B, and n_C numbers for C.

For each of these partitions of subject index numbers, perform Step 3 and Step 4 on the measurements associated with the index numbers. The first permutation represents the obtained results, and so Step 3 and Step 4 for the measurements associated with the first permutation provide the test statistic value for the obtained results. There will be $N!/n_A!n_B!...n_k!$ permutations of the N measurements for the k groups.

Step 3. Compute $T_A, ...T_k$, the totals for all of the treatments.

Step 4. Add $T_A^2/n_A, ...T_k^2/n_k$ to get $\Sigma(T^2/n)$, the test statistic.

Step 5. Compute the probability for F as the proportion of the $N!/n_A!n_B!...n_k!$ permutations (including the first permutation) that provide as large a value of $\Sigma(T^2/n)$ as the obtained test statistic value, the value computed for the first permutation.

Program 4.1 on the CD goes through analogous steps to determine the P-value of F for one-way ANOVA by means of a randomization test based on systematic data permutation. Program 4.1 can be tested by using the following data, for which the P-value is 2/1260, or about 0.0016:

A	1	2		
B	3	4	5	
C	7	8	9	10

TABLE 4.3

Experimental Results

Treatment A		Treatment B		Treatment C	
Subject Number	Measurement	Subject Number	Measurement	Subject Number	Measurement
1	6	3	9	6	17
2	8	4	11	7	15
		5	9	8	16
				9	16

Example 4.2. An experimenter randomly assigns nine subjects to three treatments, with two subjects to take treatment A, three to take treatment B, and four to take treatment C. For the obtained results, given in Table 4.3, $T_A = 14$, $T_B = 29$, and $T_C = 64$. The test statistic value for the obtained results therefore is $(14)^2/2 + (29)^2/3 + (64)^2/4 = 1402.333$. The second element in the systematic reference set (not shown) is the one where treatment A has the same measurements but treatment B contains the measurements for index numbers 3, 4, and 6, and treatment C contains the measurements for index numbers 5, 7, 8, and 9. The test statistic value for that rearrangement of data is $(14)^2/2 + (37)^2/3 + (56)^2/4 = 1338.333$.

There are $9!/2!3!4! = 1260$ data permutations for which test statistic values are computed, and the obtained test statistic value is found to be larger than any of the other 1259 test statistic values. Thus, the probability for the observed results is $1/1260$, or about 0.0008, which is the P-value that the randomization test would have provided for F if it had been computed for each of the 1260 data patterns.

4.4 One-Way ANOVA with Equal Sample Sizes

In the preceding examples of determining the P-value of F for one-way ANOVA by systematic data permutation, there were unequal sample sizes. In the last example, there were nine subjects assigned to the three treatments, so it would have been possible to have equal sample sizes, but in the earlier example with only a total of seven subjects, it would not have been possible. Why would the experimenter not increase the sample size by adding two more subjects? After all, in random-sampling discussions of t or F it is pointed out that the more subjects, the greater the chance of detecting a treatment effect. If indeed an experimenter selected subjects at random from a large population, that might make sense but for the typical experiment where subjects are not selected at random but are carefully screened to ensure appropriateness for an experiment, only a few subjects may be available. Thus, for experiments where subjects should meet certain selection criteria, relaxing those criteria to have a larger sample can actually reduce the power of the experiment to detect treatment effects. Alternatively, small samples may be necessary because of the time and expense of running individual subjects. Small samples for experiments, unlike sample sizes for estimation (as in taking polls), can often be useful in experiments and those samples do not always fit the equal sample size recommendation of advocates of parametric statistical tests.

There are two advantages of equal sample size randomization tests for ANOVA that were especially important when desk calculators or even main-frame computers were required for the computation. One advantage is that for equal sample sizes, there is an equivalent test statistic to F that involves less computation for an equal number of subjects per treatment. That equiv-alent test statistic is ΣT_2, whereas the test statistic for the general Program 4.1 for obtaining the P-value of F is $\Sigma(T^2/n)$. For situations with equal sample sizes, n is a constant divisor over all data permutations, making ΣT^2 an equivalent test statistic to $(\Sigma T^2)/n$ and thus equivalent to F. Use of the simpler test statistic simplifies the program and reduces the computational time somewhat for the thousands or tens of thousands of computations of the test statistic for even relatively small sample sizes.

Another shortcut that saves time when the sample sizes are equal is to use a subset of the reference set of data partitions. In Section 3.5, it was explained that to determine the P-value by systematic data permutation for a nondi-rectional test statistic like F or $|t|$, it is necessary to compute only $1/k!$ of the total number of the data partitions for k treatments with equal sample sizes. The reason is because for each division of the data into k groups, there are $k!$ ways of assigning the k treatment designations to those groups, all of which must give the same test statistic value. We will now consider a data listing procedure for generating a subset of data partitions that will give the same P-value as the set of all partitions in the reference set when the sample sizes are equal.

Example 4.3. The value of ΣT^2 for a difference between three groups of equal size is the same for the data permutation associated with the division of subject numbers into "3 7 8 | 2 5 9 | 1 4 6" as for any case where the measurements are separated into three groups in the same way but with transposition of the treatment labels. For example, the data permutation associated with the partition of subject numbers into "2 5 9 | 1 4 6 | 3 7 8" would give the same value of ΣT^2. Inasmuch as ΣT^2 is not affected by the direction of the difference between means, clearly the test statistic value is the same for one partition of subjects as for the other. In fact, for the three sets of measurements, there are $3! = 6$ different ways of parti-tioning the subjects among the three treatments, each of which gives the same value of ΣT^2 as the others. Described in Section 4.2, the systematic listing procedure for listing all possible data permutations would provide a list that includes all six of these. Table 4.4 shows part of the list of index number partitions. The asterisks indicate the partitions that involve the separations of subjects into "1 4 6 | 2 5 9 | 3 7 8." A different listing

TABLE 4.4

List of Subject Number Partitions

A			B			C		
1	2	3	4	5	6	7	8	9
1	2	3	4	5	7	6	8	9
.	.	.	.	.	.	.	.	.
*1	4	6	2	5	9	3	7	8
.	.	.	.	.	.	.	.	.
*1	4	6	3	7	8	2	5	9
.	.	.	.	.	.	.	.	.
2	5	9	1	4	3	6	7	8
*2	5	9	1	4	6	3	7	8
.	.	.	.	.	.	.	.	.
*2	5	9	3	7	8	1	4	6
2	5	9	4	6	7	1	3	8
.	.	.	.	.	.	.	.	.
*3	7	8	1	4	6	2	5	9
.	.	.	.	.	.	.	.	.
*3	7	8	2	5	9	1	4	6
.	.	.	.	.	.	.	.	.
7	8	9	4	5	6	1	2	3

procedure will now be described that will require listing only 1/6 of all partitions when the three groups have equal sample sizes to get a distribution of ΣT^2 for computing the P-value that is equivalent to the distribution based on all data permutations in the larger reference set of 9!/3!/3!/3!, which is 1680.

Like the procedure described in Section 4.2, the listing procedure to eliminate redundancy arranges index numbers in ascending order within groups for each partition of index numbers, and the partitions of index numbers are listed sequentially in ascending order of magnitude of the subject numbers where the N subject numbers for a particular partition are regarded as an N-digit number. To provide the restriction required to eliminate the redundancy, only those partitions are listed where the smallest index numbers for treatments are in ascending order from A to C. Thus, of the six data partitions with asterisks, only one would be listed, namely "1 4 6 | 2 5 9 | 3 7 8." That is the only one of the six partitions where the three-digit numbers formed by the index numbers are in ascending order from A to C. Such a restriction permits every data permutation to be listed once and only once; relabeling the group designations for a particular data permutation is not permitted. This then provides the one-sixth of the partitions of the measurements when group designations are ignored; therefore, duplication of ΣT^2 values resulting from the same data permutation being simply relabeled is eliminated. (The obtained results, with subject partition "1 2 3 | 4 5 6 | 7 8 9," is first in the nonredundant list.)

To generalize, the above listing procedure permits the P-value of F to be determined by the use of a randomization test employing only $1/k!$ of all possible data permutations when all k treatments have equal sample sizes. Program 4.2 on the included CD determines the P-value of F for a randomization test with equal sample sizes. This program can be tested by applying it to the following data, for which the P-value is $18/1680$, or about 0.0107:

A	1	2	4
B	3	5	6
C	7	8	9

4.5   One-Way ANOVA with Random Reference Sets

Reducing the number of data permutations and test statistic computations to be performed through the use of a listing procedure that omits redundant data permutations (ones that are "mirror images" of patterns already listed) makes the use of a randomization test to determine the P-value of F for one-way ANOVA more practical than otherwise — but there are two disadvantages. First, such a procedure can be employed only when the sample sizes are equal. Second, although performing only $1/k!$ of the data patterns is a considerable reduction in the amount of work required, there is still a lot of work if the total number of elements is large. Thus, to make the determination of the P-value for one-way ANOVA by a randomization test procedure practical for relatively large samples, an alternative procedure is required. An alternative method of permuting that is very practical, even for large sample sizes, is the random permutation method, a method wherein a fixed number of data permutations are randomly selected and the P-value is based upon those data permutations. Although it is most helpful to use random permuting of data when the number of possible data patterns is large, it can also be useful when the total number of possible data permutations is only a few thousand. The random permutation procedure incorporated into the programs for one-way ANOVA and the independent t test in this chapter is an analog of the procedure of random selection described by Green (1963). There is no need for a separate random permutation procedure for the special case where the sample sizes are equal. Thus, the test statistic to be employed for the random ("Monte Carlo") method is that of the systematic data permutation method that is appropriate for both equal and unequal samples: $\Sigma(T^2/n)$.

Use a random number generation algorithm that will select n_A index numbers without replacement from the N index numbers to assign the corresponding measurements to treatment A. From the remaining index numbers, randomly select without replacement n_B numbers to assign the corresponding measurements to treatment B. Continue assigning until all of the index numbers have been assigned to treatments. Each assignment of index numbers to every treatment provides a data permutation. Where NPERM is the requested number of permutations, perform (NPERM − 1) permutations and for each result compute $\Sigma(T^2/n)$, the test statistic. Compute the probability for F as the proportion of the NPERM data permutations (the NPERM − 1 performed by the computer plus the obtained data configuration) that provide a test statistic value as large as the obtained test statistic value.

Program 4.3 on the included CD goes through those six steps for a randomization test to determine the P-value of F by random permutation. This program can be tested with the following data:

A	1	2		
B	3	4	5	
C	6	7	8	9

The P-value for systematic data permutation is 6/1260, or about 0.0048.

4.6 Analysis of Covariance

Analysis of covariance is a form of analysis of variance with statistical control over the effect of extraneous variables. It consists of an ANOVA performed on transformed dependent variable measures that express the magnitude of the dependent variable relative to the value that would be predicted from a regression of the dependent variable on the extraneous variable. For instance, consider a comparison of the effects of two drugs on body weight. An ANOVA performed to test the difference between weights of the subjects in the two treatment groups might fail to give significant results simply because the large variability of the pre-experimental body weight made the statistical test relatively insensitive to the difference in the drug effects. One way to control for the effect of pre-experimental body weight is by applying ANOVA to difference scores, that is, to the gain in weight. Generally speaking, applying ANOVA to difference scores for a randomization test — although valid — is regarded as less suitable than employing a randomization test based on analysis of covariance for such an application. Unlike the application of analysis of variance to difference scores, analysis of covariance takes into consideration the degree of correlation between the control or concomitant variable (such as pre-experimental body weight) and the dependent variable (such as final body weight), thereby tending to provide a more powerful test than would result from analysis of the difference scores.

Analysis of covariance also can control for the effects of an extraneous variable in situations where there are no relevant difference scores, as when the extraneous variable is different in type from the dependent variable. For example, one may want to control for differences in intelligence in a comparison of various methods of learning. Subjects might differ considerably in intelligence within and between groups, and insofar as variation in intelligence results in variation in learning performance that is not a function of the experimental treatments, it is desirable to remove this effect. Analysis of covariance can be used to increase the power of the test by controlling for differences in intelligence. Analysis of covariance would compare learning performance under the alternative methods, where the learning performance of a person is expressed relative to the performance expected of a person of that level of intelligence, where the "expected" performance is that determined by the regression of performance on intelligence.

For drawing statistical inferences about randomly sampled populations, analysis of covariance involves assumptions in addition to those of ANOVA, such as linearity of regression and a common slope of regression lines over various treatment groups, so that even when users are not concerned about the absence of random sampling, there might be a need for a nonparametric procedure. Let us then consider how analysis of covariance can be performed with the P-value determined by the randomization test procedure. First, before conducting the experiment, a measure of the concomitant variable (e.g., intelligence) is taken. (By taking the measurement at that time, it is ensured that the concomitant variable is not affected by the treatment.) Then subjects are assigned randomly to the alternative treatments and measures of the dependent variable are taken. An analysis of covariance then is carried out on the experimental results in the same manner as the procedure used with the random-sampling model to provide the obtained analysis of covariance F. That test statistic reflects the difference between treatments when there is control over the effect of the concomitant variable. The data are then permuted and for each data permutation the test statistic F is again computed, thereby providing the reference set of data permutations to which the obtained F is referred for determination of the P-value.

4.7   One-Tailed *t* Tests and Predicted Direction of Difference

For one-tailed *t* tests, the ANOVA programs are not particularly useful. The one-tailed probability values are not given directly by the programs and they cannot always be derived from the two-tailed probabilities that are given. For probability combining (a technique discussed in Chapter 7 that can be quite useful), it is essential that there be a P-value associated with a one-tailed *t* test, even when the direction of difference is incorrectly predicted. Let us now consider what such a P-value means.

In classical parametric statistics, Student's *t* test statistic is defined as

$$t = \frac{\bar{A} - \bar{B}}{s\sqrt{\frac{1}{n_A} + \frac{1}{n_B}}} \tag{4.3}$$

with $\bar{A}$ and $\bar{B}$ as the means of the two groups, n_A and n_B as the number of observations in each group, and s as the square root of the pooled variance estimator computed from the variances of the two groups, $s_A{}^2$ and $s_B{}^2$:

$$s = \sqrt{\frac{(n_A - 1)s_A^2 + (n_B - 1)s_B^2}{n_A + n_B - 2}} \tag{4.4}$$

It is conventional in computing a value of t, where $(\bar{A} - \bar{B})$ is the numerator, to designate as treatment A the treatment predicted to provide the larger mean. In that way, a positive value of t will be obtained if the direction of difference between means is correctly predicted and a negative value of t will be obtained if the obtained direction is opposite to the predicted one. Thus, a measure of the extent to which a value of t is consistent with the predicted direction of effect is its magnitude, taking into consideration the sign associated with the value of t. Therefore, we define the one-tailed probability for t for a randomization test as the probability, under H_0, of getting such a large value of t as the obtained value. If the direction of difference between means is correctly predicted, the obtained t will have a positive value and the proportion of data permutations with such a large t may be fairly small, resulting in a small P-value, but if the direction of difference is incorrectly predicted, the t value will be negative and the P-value will tend to be large.

4.8 Simpler Equivalent Test Statistics to t

In testing the difference between treatment effects, the t test must use t as the test statistic because the t tables used to determine statistical significance are based on the distribution of t under the null hypothesis. However, randomization tests use the obtained data to generate their own theoretical distributions of the test statistic under the null hypothesis, and this facility to develop their own reference sets instead of relying on fixed tables of test statistics allows users of randomization tests to choose any test statistic they desire. Test statistics sensitive to detecting differences in the magnitude of effect for two treatments are employed when random assignment is of the type used with the independent t test. From Equation 4.3, it can easily be shown that the A mean minus the B mean, where A is the treatment predicted to provide the larger mean, is an equivalent test statistic to t. The following example shows several test statistics that are equivalent for the two-tailed null hypothesis of identical treatment effects, providing the same randomization test P-value for this type of design.

Example 4.4.

Treatment A	Treatment B
14	9
10	7
10	6
8	5

Larger mean = 10.5
| Difference between means | = 3.75
$t = 2.47$

The larger mean and the absolute difference between means are equivalent two-tailed test statistics because the same eight measurements are divided between the treatments for all 8!/4!4! = 70 elements of the reference set. As the mean of measurements for one of the treatments increases, the mean of the other treatment must decrease, thereby increasing the difference between means.

4.9 Tests of One-Tailed Null Hypotheses for *t* Tests

The importance of one-tailed tests is widely recognized. In experimental research, findings indicative of a direction of difference counter to that which theory or conventional wisdom would suggest can be quite important. Thus, a directional test is sometimes worthwhile when a test of the hypothesis of no differential effect is unimportant. For example, an experimenter might believe that exposure to a substance widely regarded as carcinogenic would, in special circumstances or for certain types of subjects, inhibit the development of cancer. Being able to perform a randomization test to support the complementary hypothesis that the substance inhibited the development of cancer for even one subject could be important in such a case.

In Chapter 3, it was already indicated that a randomization test null hypothesis need not be simply one of no differential treatment effect on the experimental units but can be that of no differential treatment effect on the "measurements" associated with the experimental units, where the term "measurements" includes ratios of treatment and response magnitudes or other quantitative expressions of relationship between treatments and responses. The reference sets for testing such null hypotheses will be called *precise reference sets* because they represent the precise outcomes for alternative random assignments under the null hypothesis.

Precise reference sets for one-tailed null hypotheses are not possible because the null hypothesis A ≤ B is not precise; it contains an inequality. A randomization *t* test has a precise reference set providing exact P-values for tests of the two-tailed null hypothesis. For tests of one-tailed null hypotheses, there can be no precise reference set of test statistic values. It is so common-place to regard the use of a one-tailed test statistic as constituting a one-tailed test in the sense of its testing a one-tailed null hypothesis that it is essential that the distinction between one-tailed test statistics and tests of one-tailed null hypotheses be clear in the following discussion. The expression "test of a one-tailed null hypothesis" will be used instead of the more ambiguous expression "one-tailed test," which frequently refers to a test of a two-tailed null hypothesis using a one-tailed test statistic.

Example 4.5. Suppose we predict $\bar{A} > \bar{B}$ for an experimental design in which six subjects are randomly assigned to treatment A and treatment B, with three subjects for each treatment, and obtain the following results:

	A	B
	7	6
	9	8
	12	5
Totals	28	19
$\bar{A} - \bar{B}$	3	

Because we want the smallness of our P-value to indicate support for the alternative hypothesis A > B for some subjects, we are interested in a test of the one-tailed H_0: A ≤ B for all subjects, not in a test of the two-tailed H_0: A = B for all subjects. We derive data permutations from the above results by use of a procedure described by Edgington (1969) in which we generate all 20 data permutations as for a test of the precise null hypothesis of no treatment effect and then attach "or less" to any of the obtained B measurements that have been transferred to A and "or greater" to any of the obtained A measurements that have been transferred to B. The modified data permutations then truthfully, although imprecisely, represent results associated with our one-tailed null hypothesis. Only the obtained data permutation shows exact values. For example, under the one-tailed null hypothesis, one of the modified data permutations derived from the preceding obtained results would be:

	A	B
	6 or less	5
	12	7 or more
	8 or less	9 or more
Totals	26 or less	21 or more
$\bar{A} - \bar{B}$	1.662	or less

Every data permutation except the obtained will have at least one A value with "or less" appended and at least one B value with "or more" appended. Thus, the test statistic value for any data permutation will be exactly the same as it would have been for the precise reference set except for the appended "or less." All of the randomizations that gave a T_A or $(\bar{A} - \bar{B})$ value less than the obtained value in the precise reference set consequently would give a value less than the obtained for the modified reference set, and so the P-value for the modified reference set could be no larger than the value for the precise reference set.

Previously, it was pointed out that for a two-tailed test the difference between means is an equivalent test statistic to t, providing the same randomization test P-value. However, they are not equivalent test statistics for the one-tailed test. The "or more" and "or less" appendages can increase or decrease the variability within treatments so much that a change in the means in one direction may be associated with a change in t in the opposite direction.

4.10   Unequal-N One-Tailed Null Hypotheses

When the sample sizes are equal, the distribution of t under systematic data permutation is symmetrical about 0, there being associated with every positive t value a negative t with the same absolute value. If an absolute value of t of 2.35 has a P-value of 0.10 by data permutation, then 5% of the t's are greater than or equal to +2.35, and 5% are less than or equal to −2.35. If we have correctly predicted the direction of difference between means (thereby obtaining a positive value of t), the one-tailed probability will be 0.05, half of the two-tailed value. But suppose we predicted the wrong direction. What is the one-tailed probability in that case? In other words, what proportion of the data permutations gives a t value greater than or equal to −2.35? We know that 95% are greater than −2.35 but we do not know what percentage are greater than or equal to −2.35 because we do not know how many t's equal to −2.35 are in the distribution given by data permutation. Consequently, although we can halve the two-tailed P-value given by Program 4.2 with equal sample sizes to determine the one-tailed P-value for a correct prediction of the direction of difference between means, we cannot determine the one-tailed P-value for an incorrectly predicted direction of difference from the two-tailed P-value.

However, when sample sizes are unequal not even the one-tailed P-value for a correct prediction of the direction of difference between means can be determined by halving the two-tailed P-value. The proportion of test statistics with a value of $|t|$ as large as the obtained $|t|$ and with the same direction of difference between means as the obtained direction is not necessarily half of the proportion of $|t|$'s that are as large as the obtained $|t|$.

Example 4.6. Consider the three permutations of the measurements 2, 3, and 5 for treatment A and treatment B, where A has one measurement and B has two:

	A	B	A	B	A	B
	2	3	3	2	5	2
		5		5		3
$|D| =$		2		0.5		2.5

If the third permutation was the obtained permutation, the computed P-value would be 1/3 for the two-tailed t test. If we were to halve that value because we correctly predicted that the A mean would be larger than the B mean, we would obtain a P-value of 1/6, when in fact the smallest possible P-value for any test statistic is 1/3 because only three assignments are possible. For the above situation, the two-tailed P-value would be the same as the one-tailed P-value for a correctly predicted direction of difference between means. With unequal sample sizes, we cannot obtain a one-tailed P-value from the two-tailed P-value but must compute it separately.

The following computer programs for the independent t test will accommodate either equal or unequal sample sizes. For a two-tailed test, the test statistic used is $\Sigma(T^2/n)$, which is equivalent to $|t|$. For a one-tailed test, the test statistic used is T_L, the total of the measurements for the treatment predicted to give the larger mean, which is an equivalent test statistic to t with the sign considered.

We will now list the steps to be performed by a computer in determining both two-tailed and one-tailed P-values for unequal or equal sample sizes:

Step 1. Partition the experimental data, grouping the measurements according to treatment, where the first treatment is the treatment predicted to have the larger mean and the second treatment is the treatment predicted to have the smaller mean. Assign index numbers 1 to N to the measurements, beginning with the measurements for the first treatment.

Step 2. Perform Step 5 for the obtained data to compute the obtained one-tailed test statistic value.

Step 3. Perform Step 5 to Step 7 for the obtained data to compute the obtained two-tailed test statistic value.

Step 4. Generate a systematic sequence of permutations of the index numbers associated with the measurements, with n_1 index numbers for treatment 1 and n_2 for treatment 2. For each of those permutations perform Step 5 to Step 7 on the measurements associated with the index numbers.

Step 5. Compute T_1, the total of the measurements for treatment 1, to get the one-tailed test statistic value.

Step 6. Compute T_2, the total of the measurements for treatment 2.

Step 7. Add T_1^2/n_1 and T_2^2/n_2 to get $\Sigma(T^2/n)$, the two-tailed test statistic value.

Step 8. Compute the two-tailed P-value by determining the proportion of the permutations that provide as large a two-tailed test statistic value as the obtained two-tailed value.

Step 9. Compute the one-tailed P-value by determining the proportion of the permutations that provide a one-tailed test statistic value as large as the obtained one-tailed value.

Program 4.4 can be used to obtain both two-tailed and one-tailed P-values for t for equal or unequal sample sizes by systematic permutation. The program can be tested with the following data, for which the one-tailed P-value is 1/35, or about 0.0286, and the two-tailed P-value is 2/35, or about 0.0571:

Treatment 1	4	5	6	7
Treatment 2	1	2	3	

4.11   Fast Alternatives to Systematic Data Permutation for Independent t Tests

The number of permutations for systematic data permutation does not increase as rapidly with an increase in number of subjects per group for the t test as for ANOVA where there are several groups, but the increase is rapid enough to restrict the utility of the systematic permutation method to relatively small sample sizes. With only 10 subjects for each of the two treatments, the number of permutations is 184,756 and with 12 subjects per treatment, the number of permutations is 2,704,156. Thus, to determine the P-value by means of a randomization test an alternative method is necessary for the t test, even when the sample sizes are not very large.

Fast alternatives to systematic tests for two-group designs using a t test statistic have been developed by Pagano and Tritchler (1984) and Baker and Tilbury (1993), as mentioned in Chapter 3. Instead of generating data permutations and computing test statistic values one at a time, the fast procedures in effect derive frequency distributions of sample totals (the test statistics) in a matter of seconds. The P-values are exact values, the values that would be given by systematic data permutation if they could be employed with data where the number of data permutations was quite large. However, for certain applications random data permutation is still the preferred method for t tests with large sample sizes, so it will be described here.

4.12   Independent t Test with Random Reference Sets

For determining the P-value for a two-tailed t test by random permutation, Program 4.3 can be used because F and $|t|$ are equivalent test statistics. When the groups are equal in size, the P-value given by Program 4.3 can be divided by 2 to give the one-tailed P-value where the direction of difference has been correctly predicted. The justification for halving the two-tailed P-value to get the one-tailed P-value under random permutation with equal sample sizes is not the same as that of halving the two-tailed P-value when there is systematic permutation. With systematic permutation, halving the two-tailed P-value when sample sizes are equal is a simple way of getting exactly the same P-value as would have been obtained by directly determining the proportion of the permutations providing as large a value of ΣT^2 as the obtained value, with the difference between means in the same direction as for the obtained data. But with the random permutation method — in the sampling distribution of, say, 1000 data permutations — there is no assurance that half of the data permutations giving a test statistic as large as the obtained value will have a difference between means in one direction and half in the other. Consequently, the halving of the two-tailed P-value does not necessarily provide the proportion of the 1000 data permutations that would give a test statistic as large as the obtained value with the obtained direction of difference between means.

The validity of halving the two-tailed P-value for random permutation tests with equal sample sizes can be demonstrated through consideration of the interpretation that is to be placed on the one-tailed P-value. We want the procedure for determining the one-tailed P-values to be such that if the null hypothesis of identical treatment effects is true, the probability of getting a P-value as small as P is no greater than P. For example, in the long run no more than 5% of the one-tailed P-values should be as small as 0.05. The validity of the two-tailed random permutation P-value is shown in Section 3.6, which demonstrates that the P-value associated with any test statistic value can be determined in the manner in which it was determined for ΣT^2 for the random permutation method. For example, we can conclude that when H_0 is true, the probability of getting a two-tailed P-value as small as 0.10 is no greater than 0.10. With equal sample sizes, $(\bar{A} - \bar{B})$ is symmetrically distributed around 0. Because the random sampling distribution of data permutations is a random sample from the population of all data permutations, for any difference between means the probability that it is in the predicted direction is 1/2. Thus, for cases where a two-tailed P-value is as small as 0.10, the probability is 1/2 that the difference is in the predicted direction when H_0 is true. Consequently, under H_0 the probability of getting a two-tailed P-value by random permutation as small as 0.10 with the predicted direction of difference is half of 0.10, or 0.05. This then justifies the determination of a one-tailed P-value by random permutation by means of dividing the two-tailed probability by 2 in the cases where the sample sizes are equal.

Program 4.3, which is for one-way ANOVA with the P-value determined by random data permutation, can be used to provide P-values for the independent t test whether the sample sizes are equal or unequal. If the sample sizes are equal, halving the P-value given by the program gives the one-tailed P-value for the t test if the direction of difference has been correctly predicted. But for unequal sample sizes, with random permutation as with systematic permutation, it is necessary to determine the one-tailed P-value for a correctly predicted direction of difference directly rather than by halving the two-tailed P-value. The test statistic for the two-tailed P-value is $\Sigma(T^2/n)$ and for the one-tailed P-value is T_1, the total of the measurements for the treatment predicted to have the larger mean.

For the reasons given in Section 4.10 for systematic data permutation, one-tailed P-values where the direction of difference was incorrectly predicted (useful P-values for probability combining) cannot be derived from the two-tailed P-values given by Program 4.3 for either equal or unequal sample sizes but must be directly computed. The following steps are required to determine one-tailed P-values for t by random permutation for equal or unequal sample sizes:

Step 1. Arrange the research data in the form of a table with the measurements grouped by treatment, where the first treatment (called treatment 1) is the treatment predicted to have the larger mean and the second treatment (treatment 2) is the treatment predicted to have the smaller mean. Assign index numbers 1 to N to the measurements, beginning with the measurements for treatment 1.

Step 2. Perform Step 5 for the obtained data to compute the obtained one-tailed test statistic value.

Step 3. Perform Step 5 to Step 7 for the obtained data to compute the obtained two-tailed test statistic value.

Step 4. Use a random number generation algorithm that will select n_1 index numbers without replacement from the N index numbers to assign the associated n_1 measurements to treatment 1; the remaining measurements are for treatment 2. Each assignment of n_1 measurements to treatment 1 constitutes a single permutation of the data. Perform (NPERM − 1) permutations and for each permutation go through Step 5 to Step 7.

Step 5. Compute T_1, the total of the measurements for treatment 1, to get the one-tailed test statistic value.

Step 6. Compute T_2, the total of the measurements for treatment 2.

Step 7. Add T_1^2/n_1 and T_2^2/n_2 to get $\Sigma(T^2/n)$, the two-tailed test statistic value.

Step 8. Compute the two-tailed P-value by determining the proportion of the data permutations that provide as large a two-tailed test statistic value as the obtained two-tailed value given by Step 3.

Step 9. Compute the one-tailed P-value by determining the proportion of the NPERM data permutations that provide as large a one-tailed test statistic value as the obtained one-tailed value computed in Step 2.

Program 4.5 on the included CD performs these operations. It can be tested with the following data, for which the one-tailed P-value is 1/35, or about 0.0286, and the two-tailed P-value is 2/35, or about 0.0571:

Treatment 1	4	5	6	7
Treatment 2	1	2	3	

4.13 Independent *t* Test and Planned Comparisons

Where k is the number of treatments for one-way analysis of variance, there are $(k)(k-1)/2$ pairs of treatments that could be subjected to an independent t test. For example, for five treatments there are $(5)(4)/2 = 10$ pairs of treatments. If an experiment is conducted for the purpose of applying t tests to a small number of specified pairs of treatments rather than to all possible pairs, the comparisons that are made are called *planned comparisons*. When an experiment is performed to provide the basis for planned comparisons, the computation of F and the determination of its P-value are unnecessary. Let us see how a t test for a planned comparison is performed when the P-value is determined by a randomization test.

Suppose we have randomly assigned subjects to each of five treatments, A, B, C, D, and E. One of the planned comparisons is of treatment A and treatment C to see if they have the same effect. The test is performed as if A and C were the only treatments in the experiment. The A and C measurements are repeatedly divided between A and C to determine the P-value of t by data permutation. Thus, Program 4.4 or Program 4.5 could be employed to give the one-tailed and two-tailed P-values for the test. Other planned comparisons within the five treatments would be conducted in the same way.

4.14 Independent *t* Test and Multiple Comparisons

When comparisons are not planned before an experiment but t tests are instead applied to all (or virtually all) possible pairs of treatments, such comparisons are called *post hoc comparisons*. In the belief that these comparisons should be treated differently from planned comparisons, special "multiple comparisons" procedures were developed for post hoc comparisons.

Before special multiple comparisons procedures were developed, it was common to determine the P-value for unplanned comparisons in the same way as for planned comparisons, but it was generally assumed that t tests would be conducted only if the overall F was significant at the 0.05 level.

That method of determining the P-value for unplanned multiple comparisons is still used by some experimenters.

A number of alternative procedures have been proposed for determining the P-value for multiple comparisons. Miller (1966) presents a large number of parametric and nonparametric techniques. Experimental design books (e.g., Keppel, 1973; Myers, 1966; Winer, 1971) frequently present several parametric procedures for the reader to consider. Many of the commonly used procedures require reference to a table of the studentized range statistic — which, like the t table, is based on assumptions of random sampling, normality, and homogeneity of variance. The validity of those procedures thus depends on the tenability of the assumptions underlying the table of the studentized range statistic. Therefore, it is recommended that a distribution-free multiple comparisons procedure be employed. *Fisher's modified Least Significant Difference* (LSD) is such a procedure because it does not require reference to the studentized range statistic table or any other table.

Winer (1971) discusses Fisher's modified LSD procedure to control for a *per experiment error rate.* (It is also known as a *Bonferroni procedure,* and that is the way it will be designated in the following discussion.) We will now consider how to apply that procedure when we determine P-values by data permutation. To use the modified LSD procedure, we set a per experiment error rate, α. The per experiment error rate is the probability of falsely rejecting at least one individual-comparison H_0 in an experiment.

Example 4.7. An experimenter sets a per experiment error rate of 0.05 for multiple comparisons of four treatments. For four treatments, there are six possible treatment pairs. The P-value is determined for each of the six comparisons by use of the randomization test procedure, where for each comparison only the data for the compared treatments are permuted. Each P-value is then multiplied by 6 (the number of comparisons) to give the adjusted P-value. If the adjusted value is as small as 0.05, the t value for that comparison is significant at the 0.05 level. The randomization test P-values and the adjusted P-values for the six comparisons of treatments A, B, C, and D follow:

Comparison	A-B	A-C	A-D	B-C	B-D	C-D
P-value	0.021	0.007	0.017	0.047	0.006	0.134
Adjusted P-value	0.126	0.042	0.102	0.282	0.036	0.804

Only the A-C and the B-D treatment comparisons provide an adjusted P-value as small as 0.05, and so those are the only cases where the H_0 of identical effects for a pair of treatments would be rejected at the 0.05 level.

To use the Bonferroni procedure with randomization tests, we can set a per experiment error rate or level of significance, such as 0.05 or 0.01, compute the P-value of t by data permutation for each of the $(k)(k-1)/2$ pairs of treatments, and then adjust those values by multiplying each value by $(k)(k-1)/2$ to get the final adjusted P-values. Unlike some multiple comparisons procedures, this procedure can be used either with equal or unequal sample sizes.

Any multiple comparisons procedures should be given careful consideration to determine whether they are necessary because the P-values are drastically affected by such procedures. Notice that in this example, if the adjustment in the P-values had not been made there would have been five comparisons instead of just two with significance at the 0.05 level. It should also be noted that no unadjusted P-value could lead to an adjusted value of 0.05 or less unless it was smaller than 0.01. When there are more than four treatments, the adjustment has an even greater effect on the P-values. For example, with 10 treatments there are 45 possible treatment pairs, and so an unadjusted P-value would have to be about as small as 0.001 to provide significance at the 0.05 level for the final, adjusted value. Krauth (1988) attributed to Holm (1979) a multiple comparisons procedure similar to Bonferroni's but of a sequential nature: the P-values for n treatment comparisons are ordered from smallest to largest, and the adjusted P-values are determined by multiplying the smallest P-value by n (as for the Bonferroni procedure), the next P-value by $(n - 1)$, the third by $(n - 2)$, and so on, stopping as soon as the adjusted P-value is greater than some preset α. Some multiple comparisons procedures alter the P-value more than others, but the effects of Scheffé's procedure, Tukey's "honestly significant difference" procedure, and other relatively uncontroversial procedures are comparable to those of the Bonferroni procedure. Alt (1982) referred to a number of studies comparing Bonferroni's and Scheffé's multiple comparisons procedures to determine the circumstances in which one was superior to the other in the sense of providing smaller adjusted P-values.

If logic or editorial policy requires the use of multiple comparisons procedures, they must be used but there are cases where use of a few planned comparisons makes more sense. If multiple comparisons procedures are to be employed, the researcher should realize that with a large number of treatments, more subjects will be needed per treatment than with a small number of treatments to achieve the same sensitivity or power for individual comparisons.

The preceding discussion has been in terms of preset levels of significance, which is the way multiple comparisons are almost always presented. However, it should be noted that the Bonferroni approach described controls for Type 1 error rate in such a way that the smallness of the adjusted P-values can be interpreted for any individual comparison as if that P-value was the ordinary randomization test P-value for a t test involving only those two treatments. For more information on the vices and virtues of correcting for multiple testing within a randomization test framework, the reader may be interested to consult Petrondas and Gabriel (1983) and Westfall and Young (1993).

4.15 Loss of Experimental Subjects

It has been noted frequently in experimental design books that a statistical test can become invalid if the treatments are such that there is differential dropping out, in the sense that subjects might drop out if assigned to one treatment but not if assigned to another. The term "dropping out" refers to the case where a subject assigned to a treatment does not participate enough to provide a measurement for that treatment. For instance, a rat may be unable to finish an experiment because it becomes sick or dies or does not reach a certain criterion during the training trials. Whenever some subjects originally assigned to treatments do not provide measurements under those treatments, it is necessary to consider whether such dropping out biased the statistical test carried out on the remaining subjects. For example, this would be the case if a surgical treatment caused the weaker rats to die, leaving only the stronger rats in the group to be compared in respect to physical activity with rats taking another treatment.

However, there are instances where we are quite sure that the subjects that are lost from the experiment would also have been lost if they had been assigned to a different treatment. In such cases, it is clear that the disease or other reason for dropping out of the experiment would have existed even if the subject had been assigned to one of the alternative treatments. We then feel confident that there is no bias resulting from some assigned subjects not finishing the experiment. We must go further and consider how the data permutation is to be performed when some assigned subjects drop out without providing data. It will be assumed in the following example that the subjects that were lost after assignment to treatments would have been lost if they had been assigned to any of the other treatments.

Example 4.8. We have three treatments and have randomly assigned five subjects to each treatment. One subject assigned to treatment A and one assigned to treatment B dropped out, and so we are left with data from only 13 of the subjects, as shown in Table 4.5.

Although we have only 13 subjects, we must consider the possible assignments for all 15 subjects to perform data permutation. There are $15!/5!5!5! = 756,756$ possible assignments. For each of these possible assignments, there is a data permutation based on the null hypothesis and the assumption that the dropouts (indicated in Table 4.5 by dashes) would have dropped out wherever they were assigned. For instance, the data permutation in Table 4.6 is associated with one of the assignments where the two subjects that dropped out are represented as having been assigned to treatment C. The data permutations would then not have the same sample size allocations of data from one data permutation to the next. Although it would be valid to compute F

TABLE 4.5

Experimental Results

A	B	C
5	8	4
8	8	7
6	12	9
9	10	8
—	—	11

TABLE 4.6

Data Permutation

A	B	C
5	8	4
8	8	7
6	12	9
9	10	—
8	11	—

on the basis of such data permutations, the variation in sample size allocations of data from one data permutation to the next would prevent determination of P-values by use of the computer programs given in this chapter.

An alternative procedure that permits use of the standard programs is easier. The data can be fed into the computer as though the random assignment had been of four subjects to treatment A, four to treatment B, and five to treatment C, and the data would be permuted accordingly with no variation in sample size allocation from one data permutation to the next. Instead of $15!/5!5!5! = 756{,}756$ data permutations being involved, there would be $13!/4!4!5! = 90{,}090$, a subset of the total number consisting of those data permutations corresponding to possible assignments when one of the two subjects that dropped out was assigned to treatment A and one to treatment B.

The P-value is the proportion of the data permutations in the subset — of which the obtained data permutation is a member— that provides as large a value of F as the obtained data. When H_0 is true, the probability that the obtained F will be within the upper p percent of the Fs in the subset is no greater than p percent, so this procedure of determining the P-value is valid. Therefore, the computer programs in this chapter can be validly employed for determining the P-value when there has been a loss of subjects after assignment (if the subjects would have been lost under any alternative assignment) by treating the data in the same way as if the data came from an assignment where all subjects provided data.

4.16 Ranked Data

When more precise measurements are available, it is unwise to degrade the precision by transforming the measurements into ranks for conducting a statistical test. That transformation has sometimes been made to permit the use of a nonparametric test because of the doubtful validity of the parametric test, but it is unnecessary for that purpose because a randomization test provides a P-value whose validity is independent of parametric assumptions without reducing the data to ranks. For a researcher determining the P-value by a randomization test, there is no purpose in transforming the data into ranks for the statistical analysis. However, there are occasions when the only available data are rankings, and for those occasions the t test and ANOVA programs used for ordinary data can be used on the ranks to determine the P-value.

An alternative to using data permutation to determine the P-value for ranked data for the independent t test is to use the Mann-Whitney U test, which has significance tables. The significance tables are based on the permutation of ranks. The test statistic U and alternative test statistics, like the sum of ranks for one of the treatments, are equivalent test statistics to t for ranks given data permutation. Thus, for ranked data with no tied ranks reference to the U tables will give the same P-value as would be obtained by determining the P-value of t (for the ranks) by means of a randomization test using systematic data permutation. The Mann-Whitney U tables have been constructed for a wide range of sample sizes, and the only conceivable advantage in carrying out a randomization test on ranks rather than to use the U tables is in analyzing ranks where there are tied ranks. Because the significance tables for U are based on permutation of ranks — with no tied ranks — they are only approximately valid when there are tied ranks.

The Kruskal-Wallis analysis of variance procedure is a nonparametric procedure for testing the significance of difference between two or more independent groups, and the published significance tables for the test statistic H are based on permutation of ranks for small sample sizes. H is an equivalent test statistic to F for one-way ANOVA under data permutation of ranks; so for ranked data with no tied ranks, reference to the H tables will give the same P-value as would be obtained by using a randomization test to determine the P-value of F by systematic data permutation. However, for small sample sizes where the P-values shown in the table have been determined by permutation of the ranks, some tables (e.g., Siegel, 1956) show P-values for no more than three groups. For larger samples, the significance tables for H are not based on data permutation but on the chi-square distribution and thus provide only an approximation to the P-value that would be given by data permutation. Consequently, to find the P-value of F for ranked data, data permutation can usefully complement the H tables for both small and large samples.

4.17 Dichotomous Data

It is not necessary to have a quantitative dependent variable to use the computer programs in this chapter. Provided the random assignment has been conducted in the standard way for independent-groups experiments, the programs can be used to determine the P-value by data permutation even for a qualitative, dichotomous dependent variable. For instance, a response can simply be designated as falling into one of two categories (such as "lived" or "died," or "correct" or "incorrect") but still permit the use of the t test or ANOVA. This is made possible by assigning a value of 1 to every response in one of the categories and a value of 0 to every response in the other category.

The 0s and 1s are the data for the experiment that are permuted to determine the P-value of t or F, as the case may be. Suppose we assigned a 1 to each correct response and a 0 to each incorrect response. Then the difference between treatment means to which the t test and ANOVA are sensitive is a difference between the proportions of correct responses for the different treatments because the mean of the 0s and 1s for a treatment is simply the proportion of correct responses for that treatment. Thus, by assigning 0s and 1s to the two types of responses we are able to carry out a meaningful test on the relative frequency of correct responses over treatments.

The programs for t and F in this chapter can be employed with dichotomous data by assigning 0s to responses in one category and 1s to responses in the other category and determining the P-value for F or t. It would be simpler to compute contingency χ^2 and find the P-value in the chi-square table, but chi-square tables only approximate the P-value given by systematic data permutation. When there are small expected frequencies and just one degree of freedom, one should "correct for continuity" by using the corrected chi-square formula: $\chi^2 = \Sigma[(|o - e| - 0.5)^2/e]$. The absolute difference between the observed and expected frequencies is reduced by 0.5 before being squared, reducing the chi-square value and making it more conservative than the value given by the standard chi-square formula. Even with the correction for continuity, the P-value of chi-square as given in chi-square tables corresponds only approximately to the value given by data permutation. Without correction for continuity, the discrepancy may be considerable even when the expected values are relatively large.

A second reason for using data permutation with 0s and 1s to determine the P-value is that with only two treatment groups, one-tailed tests can be carried out whether the sample sizes are equal or unequal. On the other hand, chi-square tables provide only two-tailed P-values and therefore are not as sensitive when a directional prediction is relevant. There is a procedure called *Fisher's exact test* that gives the same one-tailed P-value as data permutation although it determines the value by direct computation, but the computation can sometimes be considerable without computational aids. Furthermore, Fisher's exact test does not give two-tailed P-values and is restricted to two treatments.

Determining the P-value by permuting dichotomous data is not a very efficient procedure. It would be better to program computational procedures for determining the P-value in the systematic data permutation programs or separately. Such computational procedures are incorporated into many programs.

4.18 Outliers

Outliers are extremely high or extremely low measurements and are so called because they lie far outside the range of the rest of the measurements. When the outlier is a low measurement, it tends to produce a negative skewness in the distribution of data; when it is a high measurement, a positive skew-ness is produced.

It might seem that the addition of a high outlier to the set of measurements with the higher mean, or a low outlier to the set with the lower mean, would tend to increase the value of *t* because it increases the difference between means but that is not necessarily the case. The addition of such an outlier can increase the within-group variability so much that it more than offsets the increase in the between-group difference, actually causing the *t* value to decrease.

Consequently, the presence of an outlier — even when it increases the difference between means — can sometimes reduce the value of *t* and thereby reduce the P-value based on *t* tables. Because the P-value determined by a randomization test is dependent upon the size of the obtained *t* value relative to its size under other data permutations and not on the absolute size, the presence of outliers does not have the same amount of depressing effect on the randomization test P-value. Therefore, randomization tests can be more likely than *t* tables to detect differences when there is a very extreme measurement. We will now illustrate this point with actual data from an experiment conducted by Ezinga (1976) that has been discussed similarly elsewhere (Edgington and Ezinga, 1978).

Example 4.9. There were 10 subjects for each of two treatments and an independent *t* test was used as a two-tailed test. Table 4.7 shows the data

TABLE 4.7

P-Values of *t* for a Skewed Distribution

	Treatment A	Treatment B
	0.33	0.28
	0.27	0.80
	0.44	3.72
	0.28	1.16
	0.45	1.00
	0.55	0.63
	0.44	1.14
	0.76	0.33
	0.59	0.26
	0.01	0.63
$\bar{X}$ =	0.412	0.995

Independent *t* test: $t = 1.78$, 18 degrees of freedom

P-value by *t* table: 0.092 (two-tailed)

P-value by randomization test: 0.026 (two-tailed)

Hypothetical case of maximum difference (no overlap)

Independent *t* test: $t = 2.61$, 18 degrees of freedom

P-value by *t* table: 0.018 (two-tailed)

P-value by randomization test: 0.00001 (two-tailed)

Source: Data from Ezinga's (1976) study.

and the P-value associated with $t = 1.78$ for 18 degrees of freedom given by the t table along with the P-value given by random data permutation using 9000 data permutations. Although the P-value given by a randomization test was 0.026, the results according to the t table would not be significant at the 0.05 level. (The exact P-value based on the parametric t distribution is 0.092.) The effect of the outlier, 3.72, is so great that even if all of the small measurements were in treatment A and all of the large ones in treatment B, with no overlap, the t value would be only 2.61, which would still not be significant at the 0.01 level. (The P-value based on the t table is 0.018.) On the other hand, the P-value given by a randomization test would be about 0.00001.

The set of data in Table 4.7 is just one of 11 sets with extreme outliers. Eleven sets of data out of 71 sets on which an independent t test was conducted were so skewed that the presence of the highest measurement doubled the range of the measurements. They all showed randomization tests to be more powerful for determining the P-value than the t distribution underlying the t table. Several dependent variables were used in the overall study, which concerned memory for pictures. The study required subjects to indicate whether a picture had or had not been presented previously. None of the sets of data involving accuracy measures showed extreme outliers. Because the subjects presumably were strongly motivated, one would not expect extreme differences between the best subject and the next lower, or between the worst subject and the next higher. The 11 distributions that were extremely skewed due to outliers all involved dependent variables where no ability was required to obtain a high measurement. For example, some of the outlier distributions were distributions of confidence ratings made by subjects to indicate their confidence in the accuracy of their judgments. Because an occasional subject was extremely confident of their accuracy (whether justifiably so or not), such subjects had confidence scores much larger than the rest. Some dependent variables by their very nature are then likely to lead to outliers, and in such cases the use of data permutation may provide much smaller P-values than t tables.

It is not the intent of the preceding discussion to imply that randomization tests always will provide smaller P-values than t tables when there are outliers. Undoubtedly, the relative sensitivity of the two methods of determining the P-value is a function of such considerations as the number of outliers and whether they fall mainly at one end of the distribution or are evenly divided between the two ends. Much needs to be investigated before the conditions under which randomization tests are more powerful for data with outliers can be specified. However, the discussion does show that randomization tests not only control for Type 1 error but, when there are treatment effects, also can show greater power than conventional means of determining the P-value for the same test statistic. Furthermore, it illustrates the fact that conditions that adversely affect P-values determined on the basis of parametric distributions of test statistics need not have a comparable effect on randomization test P-values because of the randomization test's use of the relative rather than the absolute value of a test statistic.

Outliers are sometimes identified and dropped from the data set before performing a statistical test. One means is to compute the mean and standard deviation of the joint distribution of measurements and discard all measurements more than three standard deviations from the mean. The rationale for discarding outliers seems to be that those measurements are in some sense flawed. Whether one should rely on such criteria to decide what data to analyze depends on one's assumptions about the way "correct" measurements should be distributed. If otherwise regarded as appropriate, such a procedure of discarding outliers can be employed with a randomization test by initially discarding outliers and then permuting the remaining data, because under H_0 the same observations would be discarded no matter which data permutation represented the obtained results. However, discarding an outlier on the basis of its extremeness within a data set for an individual treatment, rather than on the basis of the joint distribution of measurements from all treatments, would be unacceptable for a randomization test procedure. On the other hand, permuting all of the data to generate data permutations and excluding an outlier in the computation of a test statistic would be valid, even though the value (or values) excluded could vary over the data permutations. Spino and Pagano (1991) have developed efficient computer algorithms for permutation tests when outliers are excluded in the computation of test statistics.

4.19 Questions and Exercises

1. What is a test statistic that could be used for a one-way ANOVA randomization test for three treatments that would be sensitive to the directional prediction that treatment A will provide larger measurements than treatment B and treatment B will provide larger measurements than treatment C?

2. Use a systematic listing procedure in listing the permutations of these results:

A	5	7	8
B	3	10	6

   Explain your listing procedure in general terms that can be applied to other data.

3. A set of data is given in Section 4.3 to test a systematic randomization test program. Why would the following set not be as useful — A: 2, 2, 2; B: 5, 5, 5?

4. Users of parametric t tests occasionally use a "difference score," which is the post-test measurement minus the pre-test measurement

for each subject as the data for an independent *t* test. Explain how that procedure could serve a similar function to that of analysis of covariance.

5. When is it invalid to divide a randomization test two-tailed P-value by 2 to get a one-tailed P-value for a correctly predicted difference?

6. How many planned comparisons are possible for four treatments?

7. When should a multiple comparison procedure be used instead of planned comparisons?

8. What would be an appropriate test statistic when there is a dropout if we permute all the data, including a marker for each dropout?

9. Dichotomous data can be represented as rank 1 and rank 2. Discuss the appropriateness or inappropriateness of applying the Mann-Whitney U test to dichotomous data.

10. Even when there are no dropouts for the obtained results, randomization tests must make the assumption that a subject that drops out under the assigned treatment would have dropped out under any of the alternative treatments. Is this an unnecessary assumption when there are no dropouts for the obtained results?

REFERENCES

Alt, F.B., Bonferroni inequalities and intervals, in *Encyclopedia of Statistical Sciences*, Vol. 1, Kotz, S. and Johnson, N.L. (Eds.), John Wiley & Sons, New York, 1982, 294–300.

Baker, R.D. and Tilbury, J.B., Rapid computation of the permutation paired and grouped *t* tests, *Appl. Stat.*, 42, 432, 1993.

Edgington, E.S., *Statistical Inference: The Distribution-free Approach*. McGraw-Hill, New York, 1969.

Edgington, E.S. and Ezinga, G., Randomization tests and outlier scores, *J. Psychol.*, 99, 259, 1978.

Ezinga, G., Detection and memory processes in picture recognition, Ph.D. thesis, University of Calgary, Alberta, Canada, 1976.

Fisher, R.A., The coefficient of racial likeness and the future of craniometry, *J. R. Anthropol. Inst.*, 66, 57, 1936.

Green, B.F., *Digital Computers in Research*. McGraw-Hill, New York, 1963.

Holm, S., A simple sequentially rejective multiple test procedure, *Scand. J. Statist.*, 6, 65, 1979.

Keppel, G., *Design and Analysis: A Researcher's Handbook*, Prentice-Hall, Englewood Cliffs, NJ, 1973.

Krauth, J., *Distribution-Free Statistics: An Application-Oriented Approach*, Elsevier, New York, 1988.

Miller, R.G., *Simultaneous Statistical Inference*, McGraw-Hill, New York, 1966.

Myers, J.L., *Fundamentals of Experimental Design*, Allyn & Bacon, Boston, 1966.

Pagano, M. and Tritchler, D., On obtaining permutation distributions in polynomial time, *J. Am. Stat. Assoc.*, 78, 435, 1984.

Petrondas, D.A. and Gabriel, K.R., Multiple comparisons by randomization tests, *J. Am. Stat. Assoc.*, 78, 949, 1983.

Pitman, E.J.G., Significance tests which may be applied to samples from any populations, *J. R. Statist. Soc. Suppl.*, 4, 119, 1937.

Siegel, S., *Nonparametric Statistics for the Behavioral Sciences*, McGraw-Hill, New York, 1956.

Spino, C. and Pagano, M., Efficient calculation of the permutation distribution of trimmed means, *J. Am. Statist. Assn,.* 86, 729, 1991.

Westfall, P.H. and Young, S.S., *Resampling-Based Multiple Testing: Examples and Methods for p-Value Adjustment*, John Wiley & Sons, New York, 1993.

Winer, B.J., *Statistical Principles in Experimental Design* (2nd ed.), McGraw-Hill, New York, 1971.

5

Factorial Designs

Factorial designs are designs for analyzing data where the data are classified according to more than one independent variable. The independent variables are called *factors*, the source of the name *factorial analysis of variance* and the synonym *multifactor analysis of variance*, for a statistical procedure commonly used with such designs. A magnitude or state of an independent variable or factor is a *level* of the factor. For example, if we have an experiment in which plants are assigned to one of three amounts of humidity and to either high or low temperature, it is a factorial experiment with two factors (humidity and temperature) and there are three levels of the first factor and two levels of the second. In this case, both factors have levels that can be expressed quantitatively, but in some factorial experiments the levels of a factor are not magnitudes of the factor. For example, a factor studied in regard to plant growth could be the type of fertilizer, where four different types of fertilizer were compared; in this case, there would be four levels of the fertilizer factor, although the variation over levels represents qualitative rather than quantitative differences.

One purpose of factorial designs is to use the same data to test the effect of more than one factor. If there are *a* levels of one factor and *b* levels of a second factor in a two-factor experiment, the number of cells in the matrix of treatment combinations is *ab* and each cell represents a distinctive treatment. Many tests are possible: The data can be used to test separately the effects of factor A and factor B, or we can test the effect of either factor within any single level of the other factor. Thus, by using the same data for various comparisons, we can make more complete use of our data than in a single-factor design.

To draw inferences from randomization tests concerning mean effects rather than effects on individuals, some statisticians and researchers make the untenable assumption of unit additivity, usually without making that assumption explicit. That assumption, which is the assumption that any treatment effect is identical for all subjects involved in the treatments under test, serves a similar function to homogeneity of variance for parametric tests, and a randomization test involving that assumption should not be regarded as a distribution-free or nonparametric test. The unit additivity assumption will not be a part of the tests in this chapter or in any other chapter in this book.

5.1   Advantages of Randomization Tests for Factorial Designs

The validity of determining the P-value by the use of randomization tests when the parametric assumptions (including that of random sampling) underlying F tables are not tenable is of more practical importance with factorial analysis of variance than with the simpler procedures in Chapter 4. When the simpler procedures are employed with continuous (not categorical) data with relatively large and equal samples, the use of F tables to determine the P-value seldom is called into question, even in the absence of random sampling. Alternatively, under similar conditions the use of factorial ANOVA may be challenged frequently because of the assumptions underlying the F table when used with factorial designs. The assumptions are more complex, and studies of the "robustness" of tests under violations of parametric assumptions have generally concerned the simpler ANOVA procedures. Additionally, there frequently is need for a factorial design when the sample sizes (in terms of the number of subjects per cell or per level of a factor) must be small or unequal, and in those cases the validity is more likely to be questioned than otherwise.

5.2   Factorial Designs for Completely Randomized Experiments

In this chapter, we will consider completely randomized factorial experiments, that is, factorial experiments with random assignment to levels of all factors. Random assignment for completely randomized factorial experiments is of the same form as that for single-factor experiments for one-way ANOVA. In fact, the data layout for a two-factor experiment in r rows and c columns could be subjected to a randomization test for one-way ANOVA by regarding each of the rc cells in the array as a distinctive treatment rather than as a combination of treatments. However, to do so for testing the H_0 of no treatment effect whatsoever would be to lose certain advantages of a multifactor experiment by reducing it to a single-factor (one-way) experiment. Advantages of factorial designs arise from conceptually dividing distinctive treatments (cells) into groups to permit useful comparisons of groups of treatments. When the desired organization of data into factors and levels of factors is known in advance, the experiment can be conducted more efficiently but valid tests of treatment effects can be conducted even when the grouping of treatments for comparison is arranged after the experiment is performed, provided the grouping is done before the data are examined. For example, for practical purposes an experiment can be run to compare the effectiveness of reading a gauge as a function of its location in one of

TABLE 5.1

Alternative Locations of Gauge

		HORIZONTAL	
		LEFT	RIGHT
VERTICAL	TOP		
	BOTTOM		

four quadrants in front of a machine operator, as shown in Table 5.1. A completely randomized design is used with an equal number of subjects assigned to each of the four conditions. One-way ANOVA over all four cells could be employed to provide a single-factor analysis. However, a colleague with somewhat different interests could analyze the experiment as a factorial experiment with the data for the four quadrants placed into the data array in Table 5.1 for two factors with two levels of each factor. The colleague carries out a test to compare column means to test the effect of horizontal placement of the gauge and then compares row means to test the effect of vertical placement. All four cells are involved in each of these two tests, but what is tested is quite different from what is tested with a one-way ANOVA comparison of all four cells.

Null hypotheses for randomization tests for experiments that do not involve random sampling do not refer to population parameters; consequently, main effects and interactions will be defined here somewhat differently for randomization tests than for parametric tests. We will define the test of a main effect of factor J in a completely randomized experiment as a test of the following H_0: the measurement of each subject (or other experimental unit) is independent of the level of J to which the subject is assigned. This H_0 can be used in reference to all subjects assigned to any of the treatments (cells) or it can be restricted to subjects within certain cells. For instance, given an $A \times B$ factorial design we could test two different H_0s for a main effect of factor A. One H_0 could be this: the measurement of each subject within level B_1 of factor B is independent of the level of A to which the subject is assigned. Alternatively, we could test the following H_0 regarding a main effect of factor A: the measurement of each subject, for any level of B, is independent of the level of A to which the subject is assigned.

In the previous chapters, some of the test statistics described for randomization tests were equivalent to conventional F and t statistics, so that the computed P-values were the same as they would have been if conventional test statistics had been incorporated into the randomization tests. However, for randomization tests for factorial designs there will be no attempt to use conventional test statistics or simplified equivalent test statistics because in factorial ANOVA frequently there is disagreement about the appropriate test statistic to use in testing for a certain effect. For example, in a situation where one person uses a test statistic with the "error variance" based on within-cell variability, another person might use the variability associated with interaction as the error variance. As a result of their complexity, factorial designs permit the use of a number of alternative test statistics that might be of interest to experimenters, and so the test statistics to be described in this chapter will be chosen for their apparent appropriateness without consideration of whether they are equivalent to parametric test statistics for such designs. However, these test statistics are simply examples and alternative test statistics, including conventional factorial F statistics, may be more appropriate for certain applications.

5.3   Proportional Cell Frequencies

Throughout this chapter, the tests will employ test statistics appropriate only for proportional cell frequencies, so it is important at this point to clarify the concept of proportionality of cell frequencies. As well as considering proportionality of cell frequencies in general, the following discussion will focus on a particular example illustrated in Table 5.2. That table illustrates a design with proportional cell frequencies that has three levels of factor A and three levels of factor B, providing an A × B matrix of nine cell frequencies.

In a two-factor experiment, the cell frequencies are said to be completely proportional if the sample-size ratio of any two cells within one row is the same as the sample-size ratio for the corresponding cells in any other row. The matrix in Table 5.2 is an example of completely proportional cell frequencies. It will be noted that the sample-size ratio of any two cells within a column is the same as the ratio for the corresponding two cells in any other column, although proportionality was defined in terms of cell frequency ratios within rows. Proportionality for rows automatically determines proportionality for columns.

Within a two-factor design, there can be proportionality for a portion of the AB matrix without complete proportionality. For example, if the frequency for the lower right cell in the matrix in Table 5.2 was 10, no longer would there be a completely proportional matrix but there would be proportionality for that portion of the matrix consisting of the first two columns or, alternatively, the first two rows. In such a case, the test statistics discussed in this chapter would be appropriate for any comparisons not involving the lower right cell, such as a test of the difference in the effects of a_1 and a_2,

TABLE 5.2

An Example of Completely Proportional Cell Frequencies

	a_1	a_2	a_3
b_1	5	5	10
b_2	4	4	8
b_3	7	7	14

based on the first two columns, or a test of the difference in the effects of b_1 and b_2, based on the first two rows. However, tests of hypotheses including the lower right cell should use special test procedures for disproportional cell frequencies provided in Chapter 6.

For a three-factor experiment, cell frequencies for factors A, B, and C are completely proportional if the AB matrix for each level of C is proportional, the AC matrix is proportional for each level of B, and the BC matrix is proportional for each level of A. With three-factor experiments, as with two-factor experiments, there can be proportionality for parts of the complete matrix without complete proportionality.

The simplified test statistics used earlier with single-factor (one-way) designs also will be used here with multifactor designs with proportional cell frequencies: $\Sigma(T^2/n)$ for a nondirectional test of the difference among treatment levels and T_L, the total for the level predicted to provide the larger mean, for a one-tailed test. However, it must be stressed that for many of the tests to be carried out on factorial designs, the one-way ANOVA and *t* test programs are not applicable, despite identity of test statistics, because the data must be permuted differently to generate the distribution of test statistic values.

Sometimes disproportional cell frequency designs are necessary in factorial experiments for various reasons, such as an inadequate number of available subjects or because of dropouts during an experiment. Special test statistics should be employed in those cases to increase the sensitivity of the tests to treatment effects: The validity of randomization tests does not require different test statistics for designs with proportional and disproportional frequencies, but power or sensitivity considerations do call for different test statistics. Those designs and the test statistics appropriate for them are discussed in Chapter 6.

5.4   Program for Tests of Main Effects

Program 5.1 can test the effect of any manipulated factor in a randomized design with two or more factors. For a test of the difference between two levels of a factor, the program provides both a one-tailed and a two-tailed P-value. Only one factor at a time is tested and the input must be organized separately for each test. No matter how many factors are involved in the design, the different cells (combinations of treatments) can always be represented in a two-dimensional array consisting of rc cells, where r is the number of rows and c the number of columns. If we designate the factor to be tested as A, there are c columns standing for the c levels of A and r rows standing for the number of levels of B, the other factor, for a two-factor design. For designs with more than two factors, the rows stand for combinations of the other factors. A three-factor design with c levels of A, m levels

TABLE 5.3

Cells in a Three-Factor Design

$A_1B_1C_1$	$A_2B_1C_1$	...	$A_cB_1C_1$
$A_1B_1C_2$	$A_2B_1C_2$	...	$A_cB_1C_2$
...	...	...	...
$A_1B_mC_n$	$A_2B_mC_n$	...	$A_cB_mC_n$

of B, and n levels of C is represented in Table 5.3. There are c columns and mn rows. This is a complete table of all cells listed in sequential order where, within each column, the subscripts of B and C increment as two-digit numbers. For example, for the first column, after $A_1B_1C_n$, the next cell listed would be $A_1B_2C_1$, which would be followed by $A_1B_2C_2$, and so on.

For a test of the effect of A over all levels of B and C combined, the entire table would be entered into the program, but for other tests a portion of it would be abstracted and entered. For example, for a test of the difference between A_1 and A_2 over all levels of B and C, the first two columns would be used, whereas if H_0 concerned differences among all levels of A within all levels of C and level 1 of B, the first n rows would be entered into the program. When only two levels of A are to be compared, the level of A designated as A_1 should be the level predicted to provide the larger measurements because the one-tailed test statistic T_L is the total of the first column. After determining the portion of the table to be analyzed, the following steps are performed in applying Program 5.1:

Step 1. The user must specify the number of rows (NROWS) and the number of columns (NGRPS) in the array and the number of measurements in each cell, cell by cell, starting with the first cell in the first row, then the second cell in the first row, and so on ending with the last column in the last row. The data are entered in the same sequence.

Step 2. NPERM, the number of permutations upon which to base the test, is specified. One of the data permutations is that associated with the experimental results and the other (NPERM − 1) data permutations are randomly generated.

Step 3. If only two levels of a factor are being tested, perform Step 6 for the obtained data to compute the obtained test statistic value.

Step 4. Perform Step 7 and Step 8 for the obtained data to compute $\Sigma(T^2/n)$, the nondirectional test statistic value.

Step 5. The computer uses a random-number generation procedure to randomly divide the data in the first row among the columns, holding cell frequencies fixed, and does the same for each successive row. A division of the data carried out on all rows is defined as a single permutation. NPERM − 1 permutations are performed and for each permutation, Step 6 to Step 8 are carried out.

Step 6. Compute T_L, the total of the measurements for the first column of the array. Remember that the level of A selected to be the first column should be the level predicted to provide the larger measurements. Step 6 is carried out only when there are just two levels of A being tested.

Step 7. Compute T_1, ... T_c, the totals for each of the columns.

Step 8. Square each T and divide each T^2 by n, the number of measurements in the column, then add the values of T^2/n to get $\Sigma(T^2/n)$, the nondirectional test statistic.

Step 9. Compute the P-values for T_L (when appropriate) and $\Sigma(T^2/n)$ as the proportion of the NPERM test statistic values that are greater than or equal to the obtained values.

5.5 Completely Randomized Two-Factor Experiments

Sometimes experimenters want to investigate the effects of two experimentally manipulated variables (factors) in a single experiment. An example will show some of the randomization tests that can be performed on data from such experiments. The following example is a variation of an earlier example of a two-factor experiment to test the effects of quality of soil and quality of nutrition on plant growth (Edgington, 1969).

Example 5.1. A botanist conducts a greenhouse experiment to study the effects of both temperature and fertilizer on plant growth. This experimenter uses two levels of temperature, called High and Low, and two levels (types) of fertilizer, called R and S. Twenty plants are randomly assigned to the four treatment cells (combinations of temperature and fertilizer levels), with the sample size constraints shown in Table 5.4.

First, a test is conducted to test the following H_0: the growth of each plant is independent of the type of fertilizer applied to it. In testing this H_0, we want to be sure that the test is valid whether or not there is a temperature effect. Consequently, for testing H_0 the only data permutations performed are those where data are permuted within levels of temperature but not within fertilizer levels. H_0 does not refer to temperature effects, and to

TABLE 5.4

Number of Plants Assigned to Each Experimental Condition

		Fertilizer Type	
		R	S
Temperature	High	4	4
	Low	6	6

TABLE 5.5

Experimental Results

	Fertilizer Type	
	R	S
High temperature	15	12
	18	11
	17	14
	18	15
	$\bar{X} = 17$	$\bar{X} = 13$
Low temperature	22	16
	20	17
	17	14
	20	15
	21	18
	20	16
	$\bar{X} = 20$	$\bar{X} = 16$
	$\bar{R} = 18.8$	$\bar{S} = 14.8$

permute data between temperature levels (i.e., within fertilizer levels) as well as between fertilizer levels would provide data permutations associated with the H_0 of no fertilizer effect and no temperature effect. The P-value is determined by reference to that subset of data permutations where the data for the two temperature levels, shown in Table 5.5, are permuted within those two levels. There are $(8!/4!4!) \times (12!/6!6!) = 64{,}680$ data permutations because each of the $8!/4!4! = 70$ divisions of the measurements for high temperature between the two levels of fertilizer can be associated with each of the $12!/6!6! = 924$ divisions of the measurements for low temperature between the two types of fertilizer. As there are 64,680 data permutations in the subset, the experimenter decides to use a random data permutation procedure instead of a systematic procedure to determine the P-value.

Program 5.1 is used by the experimenter to determine the P-value of the effect of fertilizer. For each random data permutation, the program randomly divides the measurements within both levels of temperature between the fertilizer types and computes the appropriate test statistics for those data permutations.

For the obtained results, the value of the two-tailed test statistic, $\Sigma(T^2/n)$, for the effect of fertilizer is computed in this way: $\Sigma(T^2/n) = (188)^2/10 + (148)^2/10 = 5724.8$. In other words, the obtained value of $\Sigma(T^2/n)$ is the type R fertilizer total squared divided by the number of R measurements plus the type S fertilizer total squared divided by the number of S measurements. For each data permutation, $\Sigma(T^2/n)$ is computed and the proportion of data permutations with $\Sigma(T^2/n)$ as large as 5724.8 is the P-value for the fertilizer effect. For a one-tailed test where type R fertilizer was predicted to provide the larger mean, the test statistic used would be the R total, which for the obtained data is 188.

Inasmuch as the plants were assigned randomly to levels of both factors, the data can be permuted to determine the main effect for either factor.

In addition to finding the P-value of the main effect of fertilizer, one also can find the P-value of the main effect of temperature. The null hypothesis for testing the main effect of temperature is: the growth of each plant is independent of the level of temperature. For this test, the only appropriate data permutations are those associated where data are permuted between levels of temperature but not between fertilizer levels. The P-value is based on the subset of data permutations where the data for the two fertilizer types are permuted within those levels. There are $(10!/4!6!) \times (10!/4!6!) = 44,100$ data permutations because each of the divisions of the measurements for type R fertilizer between the two levels of temperature can be associated with each of the divisions of the measurements for type S fertilizer between the two levels of temperature.

Again, we will consider P-value determination by random data permutation using Program 5.1. For each data permutation, there is computation of $\Sigma(T^2/n)$, where T is a total for a level of temperature and n is the number of plants at that level. Thus, for the obtained results $\Sigma(T^2/n) = (120)^2/8 + (216)^2/12 = 5688$. The proportion of the data permutations produced by Program 5.1 with $\Sigma(T^2/n)$ as large as 5688, the obtained value, is the P-value associated with the main effect of temperature. The alternative (complementary) hypothesis that would be accepted if the H_0 was rejected is this: some of the plants would have grown a different amount if they had been assigned to a different level of temperature. For a one-tailed test, the total for the temperature level predicted to provide the larger mean is used as the test statistic in Program 5.1, and the data for those cells should be entered into the program before the data for the other cells.

Suppose there was interest in testing the effect of fertilizer within the high level of the temperature variable. The null hypothesis to test is that the growth of each of the eight plants assigned to the high level of temperature is independent of the assignment to the level of fertilizer. The relevant data permutations for testing H_0 are those associated with assignment of the same eight plants to the high level of temperature. The top part of the data matrix in Table 5.5 is abstracted and analyzed as a separate unit. The systematic data-permuting program for the independent t test, Program 4.4, can be applied to determine one-tailed and two-tailed P-values, as there are only $8!/4!4! = 70$ data permutations. Rejection of H_0 implies acceptance of the following complementary hypothesis: some of the eight plants assigned to the high level of temperature would have grown a different amount if they had been assigned to the other type of fertilizer.

As mentioned in Chapter 3, the speeding up of randomization test programs has been the focus of much investigation and programs for factorial experiments are no exception. However, what may seem to be simply a more efficient program for performing the same test sometimes does not perform the same test but one that is invalid. That is the case with a factorial ANOVA procedure provided by Manly (1990). The above procedure of testing for effects of row or column factors by permuting only within columns or within rows rather than permuting over all cells was criticized by Manly (1990) as

TABLE 5.6

Data for a 2 × 2 Design

	A_1	A_2
B_1	8, 9	3, 4
B_2	6, 7	1, 2
T_A	30	10

being unnecessarily complex, and he suggested a more convenient procedure. He proposed permuting over "all factor combinations" (i.e., all cells)
to test for main effects, stating that it is easier (a point that is indisputable)
and yet will necessarily give the same P-value (which is incorrect) as permuting only within rows (or within columns). The factorial ANOVA programs for permutation tests in his book are based on that assumption.

Example 5.2. Consider the data in the 2 × 2 design in Table 5.6, which will
be used to show that the P-value for the Manly procedure is not necessarily
the same as for the conventional and valid procedure. To test the A effect by
the conventional method we have described, there are (4!/2!2!) × (4!/2!2!) = 36
data permutations, involving switching data within the B_1 row in all possible
ways in conjunction with switching data within the B_2 row in all possible
ways. For a test where we predicted A_1 to provide the larger measurements,
the P-value for a test based on 36 data permutations would be 1/36, or about
0.028, because no other data permutation provides a total of A_1 as large as 30.

However, Manly proposed permuting over all cells, which is not the same
as permuting within rows in conjunction with permuting within columns.
The 1296 data permutations provided by the above procedure is only a subset
of the total of 8!/2!2!2!2! = 2520 data permutations that result from permuting
over all cells, which he recommended and which his factorial ANOVA program does. The subset includes all of the data permutations providing a total
of A_1 as large as 30, and so the P-value by permuting over all cells is 36/
2520, or about 0.014, not 0.028 as provided by the valid method. Permuting
over all cells to test for the effect of a single factor does not give the same
P-value as the computationally more awkward — but valid — procedure
described in Example 5.1. As the factorial ANOVA program in Manly's book
permutes data over all cells to test for effects of individual factors, it should
not be used.

In Example 2.3, there was a description of random assignment in which
nurse + patient was a systematically constructed experimental unit that was
randomly assigned to a drug or placebo condition. If the nurses actually
influence the results substantially, the two-factor experiment described in
the following example might provide a more sensitive test of drug effect,
while at the same time allowing a test of the nurse effect.

Example 5.3. In designing an experiment to compare the effects of a drug
and a placebo, a medical researcher finds it necessary to employ two different
nurses to administer the treatments. It would have been preferable to control
for the nurse variable by using only one nurse but as that was not feasible,

TABLE 5.7

Number of Subjects for Each Treatment
Condition for Nurses A and B

	Drug	Placebo
Nurse A	10	10
Nurse B	10	10

the researcher introduces the nurse as a factor in a two-factor completely randomized design. The design and the number of patients assigned to each condition are shown in Table 5.7. Each of the 40 patients used was allowed to specify a time for receiving a treatment, and the *patient + treatment* time experimental units were randomly assigned to the four cells of the matrix with 10 patients per cell. The nurses as well as the drug conditions are treatments to which patients are randomly assigned, and so this is a completely randomized factorial design that permits a test of either factor.

5.6 Completely Randomized Three-Factor Experiments

Now we will consider an extension of the completely randomized factorial design to more than two factors. An extension of randomization tests to three factors will generalize to completely randomized designs with more than three factors.

Example 5.4. An experiment is conducted where all three factors are experimentally manipulated. Factor A has two levels, factor B has three levels, and factor C has two levels. The various possible treatments are represented as cells in the array shown in Table 5.8.

The random assignment of subjects to cells is performed within the constraint imposed by predetermined cell frequencies. Given 60 subjects to be assigned, 5 per cell, there would be $60!/(5!)^{12}$ possible assignments. The effect of factor A within level B_1 and level C_1 can be tested by computing F for one-way ANOVA for each of the $10!/5!5! = 252$ data permutations, where the 10 obtained measurements in the first and third cells of

TABLE 5.8

Three-Factor Experimental Design

	A_1		A_2	
	C_1	C_2	C_1	C_2
B_1	$n = 5$	$n = 5$	$n = 5$	$n = 5$
B_2	$n = 5$	$n = 5$	$n = 5$	$n = 5$
B_3	$n = 5$	$n = 5$	$n = 5$	$n = 5$

the top row are permuted. Restriction of the analysis to one level of each of two factors reduces the three-factor design to a one-factor design, and so the data permutations and the F for main effects are those for one-way ANOVA. The main effect of factor A within level B_1, over both levels of factor C, can also be tested. For a randomization test based on random data permutation, $\Sigma(T^2/n)$ and T_L are computed for a random sample of the $(10!/5!5!)^2$ data permutations where the 10 measurements obtained in the first and third cells of the top row are divided between those two cells in every possible way and, for each of those divisions, the 10 obtained measurements for the second and fourth cells in the top row are interchanged in every possible way. Program 5.1 could be used to determine the P-values for the two test statistics.

The effect of factor A over all levels of the other two factors can also be tested. The number of data permutations involved would be the number of ways the data can be divided between levels A_1 and A_2 without moving any measurement to a level of factor B or factor C different from the level in which it actually occurred in the experiment. Thus, data transfers could take place only between the first and third cells and between the second and fourth cells within each row. Consequently, the number of data permutations would be $(10!/5!5!)^6$. The P-value can be determined by use of Program 6.1, based on a random sample of the $(10!/5!5!)^6$ data permutations.

5.7   Interactions in Completely Randomized Experiments

In the following discussion of interactions in 2×2 factorial designs, the upper left cell is cell 1, the upper right is cell 2, the lower left is cell 3, and the lower right is cell 4.

The null hypothesis of no interaction for a 2×2 design for the conventional random sampling model can be expressed in terms of population means in the following way: $\mu_1 - \mu_2 = \mu_3 - \mu_4$, where the subscripts denote the different cells. However, for the typical experiment in which there is no random sampling, a null hypothesis expressed in terms of population parameters is not relevant. Nevertheless, we can formulate an analogous H_0 of no interaction, expressed in terms of a sampling distribution, without reference to population parameters: $E(\bar{X}_1 - \bar{X}_2) = E(\bar{X}_3 - \bar{X}_4)$, where E refers to the expectation or average value of the expression following it, the average being based on all possible assignments. Unfortunately, this H_0 does not refer to measurements for individual subjects, and so there is no basis for generating data permutations for the alternative subject assignments, a necessity for determining the P-value by data permutation.

Let us define the H_0 of no interaction in a way that can be used to associate data permutations with alternative subject assignments: for every subject, $X_1 - X_2 = X_3 - X_4$. The null hypothesis thus refers to four measurements for

TABLE 5.9

Data for Test of Interaction

	A_1	A_2
B_1	3, 4	1, 2
B_2	7, 8	5, 6

each subject. For a 2×2 completely randomized design, where each subject takes only one of the four treatments, one of the four terms represents the subject's actual measurement for that treatment and the other three terms refer to potential measurements that the subject would have provided under alternative assignments, according to the null hypothesis.

We will now make explicit some aspects of the H_0 of no interaction. To do this, first we will expand the statement of H_0: for every subject, $X_1 - X_2 = X_3 - X_4 = \theta$, where θ can be 0 or any positive or negative value for a subject and where the value of θ can vary in any way whatsoever over subjects. Thus, H_0 would be true if, for every subject, $X_1 - X_2 = X_3 - X_4 = 0$; but it would be true also if, for example, for some subjects the value of θ was 0, for others it was −4, and for others it was +8.

Example 5.5. Let us now examine a test of H_0 in a completely randomized experiment, an interaction where the measurements in Table 5.9 were obtained from eight subjects.

Inasmuch as $\bar{X}_1 - \bar{X}_2 = \bar{X}_3 - \bar{X}_4$ for these results, the computation of F for an interaction would give an interaction F with a value of 0. Using F for interaction as a test statistic, the obtained test statistic value is thus 0. Now let us consider a data permutation associated with the random assignment where the subject receiving the 7 was assigned to cell 4 and the subject receiving the 5 was assigned to cell 3. Under H_0, the data permutation associated with such a random assignment is represented in Table 5.10.

Under H_0, θ_1 and θ_2 can but need not be equal. Suppose that θ_1 and θ_2 both have values of 0. For the above data permutation, we would then have a value of interaction F greater than 0 because $X_1 - X_2$ would not equal $X_3 - X_4$. Alternatively, suppose that θ_1 and θ_2 are both equal to 2. Then the above data permutation would be the same as the obtained results and would have an interaction F of 0. Thus, the value of interaction F associated with the above data permutation is indeterminate, and this is true of any data permutation except the one associated with the obtained results. Consequently, because the proportion of data permutations with as large a value of interaction F as the obtained value is a function of the value of θ, the H_0 for the

TABLE 5.10

Data Permutation for Test of Interaction

	A_1	A_2
B_1	3, 4	1, 2
B_2	$5 + \theta_1$, 8	$7 - \theta_2$, 6

interaction cannot be tested because it does not specify the value of θ. Therefore, even for the very simplest of completely randomized factorial designs there is no way to test the H_0 of no interaction by data permutation, and the problem remains as more factors are added to the design.

Even though there is no obvious meaningful H_0 of no interaction for a completely randomized design that can be tested by a randomization test procedure, one could use F for the interaction of A and B as a test statistic to test the overall H_0 of no effect of either factor. To test the H_0 of no treatment effect by the use of F for an interaction as a test statistic, the data can be permuted over all cells in the AB matrix with the interaction test statistic computed for each data permutation for a randomization test based on systematic permutation. The total number of data permutations can be readily determined and would equal the number of possible assignments in the completely randomized design. For random data permutation, the data would be permuted randomly throughout the matrix. As this is a test of no effect of either factor, it is not necessary to permute only within rows or only within columns. For each data permutation, F for the interaction would be computed as the test statistic to provide the reference set for determining the P-value of the obtained interaction F.

Tests that have been described in publications as "randomization tests for interaction" for completely randomized factorial designs thus might be useful as tests of no treatment effect that are sensitive to interaction effects. But the proposed randomization tests frequently are faulty as a result of trying to provide a simpler computation of the test statistic than that required for interaction F for each data permutation. For instance, at times it is proposed that the obtained AB data matrix be transformed to adjust the measurements for A and B main effects. Each measurement is transformed into a residual value showing its deviation from the row and column means and an interaction test statistic is computed from the residuals. Unfortunately, after computing the obtained test statistic from the residuals, the residuals are then permuted to generate the other data permutations from which test statistics are computed, instead of permuting the raw data and generating a new set of residuals for each data permutation. Permuting the residuals does not provide a set of residuals that under the H_0 of no treatment effect would have been obtained under alternative assignments, and in fact might provide matrices of residuals that would have been impossible under the H_0, given the obtained data.

Instead of performing a randomization test of the null hypothesis of no treatment effect using a conventional interaction test statistic, a researcher might find it useful to test the H_0 of no effect of a particular factor using a test statistic sensitive to disordinal interactions, as when the effect of factor A at one level of B is opposite to its effect at the other level of B in a 2×2 factorial design. If we had expected the direction of effect of A to be the same for both levels of B, we would have permuted the data within levels of B. When the direction of effect is expected to be opposite, we still permute data only within levels of B. What must be changed in the test is not the method

TABLE 5.11

A 2×2 Factorial Design

	A_1	A_2
B_1	cell 1	cell 2
B_2	cell 3	cell 4

of permuting data but the computation of a test statistic that will reflect the different prediction. Consider the 2×2 factorial design in Table 5.11. If we expected level A_1 to provide larger values within level B_1 and level A_2 to provide larger values within level B_2, we could use the combined total of the scores in cell 1 and cell 4 (i.e., T_L) as our one-tailed test statistic that would be computed over permutations of data between cell 1 and cell 2 and between cell 3 and cell 4. For a two-tailed test, where the direction of difference between means was expected to be opposite within the two levels of B but where there was no basis for specifying which two cells should have the larger means within their row, the data would be permuted in the same way as for the one-tailed test but the test statistic would be $\Sigma(T^2/n)$ instead of T_L. $\Sigma(T^2/n)$ would then be (combined total of cell 2 and cell 3)^2/(total number of measurements in cell 2 and cell 3) + (combined total of cell 1 and cell 4)^2/(total number of measurements in cell 1 and cell 4).

The data can be arranged differently prior to entry into the computer when a disordinal interaction is predicted to see more readily what combinations of cells are to be compared. For example, we could arrange the AB matrix in this way, reversing the order of the two cells in the second row:

A_1B_1	A_2B_1
A_2B_2	A_1B_2

If the data were entered into the program in that form, Program 5.1 would perform the one-tailed test described above by using the total of the measurements in the left column as T_L, and the two-tailed test with the test statistic $\Sigma(T^2/n)$ would be based on the column totals and the total frequency of measurements within the columns. In Example 5.1, expectation of reversal of direction of effect of fertilizers within low and high levels of temperature could lead to a randomization test for column differences where the cells in one of the rows were switched. A reason for expecting a disordinal interaction could be the effect of temperature on the chemicals in the two types of fertilizer.

5.8 Randomization Test Null Hypotheses and Test Statistics

We have seen that randomization tests can be employed to test the null hypothesis of no effect of any factor or a more specific null hypothesis of no effect of a particular factor. The test statistics that were used are not the only ones that could be used to test those hypotheses; the test statistic to be used for a randomization test is one that is likely to be sensitive to the type of effect expected. But there is a fundamental difference between the role of test statistics in parametric test applications and in applications of randomization tests, which should be clarified.

Parametric tests employ test statistics incorporating properties of samples to test null hypotheses about corresponding properties of populations. Consequently, to test for a difference between population means the test statistic, such as t or F, will include the sample mean differences, and a test for heterogeneity of population variances will involve a measure of sample heterogeneity of variances. The same considerations apply to correlation and other test statistics as well. As a consequence, there is a tendency to think of the null hypothesis tested as being reflected in the test statistic employed, which can be quite misleading in the case of randomization tests.

In Chapter 4, the use of randomization test statistics involving differences between means did not imply a test of a null hypothesis of no difference in mean treatment effects. The null hypothesis was no difference in treatment effect for any subject, and that is the same null hypothesis tested if we expect a difference in the variability of treatment effects and use a test statistic sensitive to that property. Whether we employ a test statistic sensitive to mean differences, differences in variability, or differences in skewness of treatment effects depends on our expectation of the nature of the effect that may exist but that choice does not alter the null hypothesis that is tested, which is that of no treatment effect, nor the complementary hypothesis, which is that for at least some subjects there was an effect of a difference between treatments.

The chapters on correlation and trend tests will not test null hypotheses concerning correlation and trend; they will test the null hypothesis of no differential treatment effect but will do it by employing test statistics sensitive to certain expected effects. For example, if a quadratic U-shaped trend of measurement values is expected over a given range of treatment intensities, a test statistic sensitive to such a trend will be employed, but a small P-value simply implies a differential effect of treatment intensities. Of course, if there are sound theoretical reasons for believing that any trend that exists should be U-shaped, the theoretical justification in conjunction with the small P-value might lead an experimenter to the conclusion that there was a U-shaped trend but such an interpretation goes beyond the results of the test itself.

To sum, whether a test statistic is one sensitive to interaction or to some other expected type of treatment effect is guided by one's expectations, but the choice of a randomization test statistic does not transform a randomization test into a test of interaction, a test of difference between means, or some other difference simply because the test statistic is sensitive to that type of difference. This property of randomization tests is easy to overlook because it is at odds with the statistical link between parametric tests and the hypotheses they test.

5.9 Designs with Factor-Specific Dependent Variables

A factorial design can be arranged so that the dependent variable is different for different factors. As in other factorial designs, the experimental units would be assigned randomly to levels of two or more factors but instead of taking measures on a single dependent variable, the experimenter takes measures on several dependent variables, one for each of the factors.

Example 5.6. A neuroscientist is interested in the effect of a lesion in location L_1 in a rat's brain on the rat's brightness discrimination, and is also interested in the effect of a lesion in location L_2 in the rat's brain on the rat's loudness discrimination. To make the optimum use of data from a small number of rats, and to minimize time for performing the operations and allowing the rats to recover, the same rats are used for both experiments and any required lesions are made during the same operation. Five rats are assigned at random to each of the four treatment conditions represented by the four cells in Table 5.12. Five rats have both lesions, five have no lesions, five have a lesion only in L_1, and five have a lesion only in L_2.

After all of the lesioned rats have recovered from the operation, all of the rats are tested on brightness and loudness discrimination tasks, which they have learned previously. Thus, each rat gets a measurement for two dependent variables: brightness discrimination and loudness discrimination.

The experimenter evaluates the results of the lesion in location L_1 with only the brightness discrimination measurement and the results of the lesion in location L_2 with only the loudness discrimination measurement. For the effect of a lesion in L_1, the H_0 would be: whether or not a rat had a lesion in location L_1 had no effect on its brightness discrimination measurement. There would be 63,504 data permutations used to test this H_0, consisting of those associated with each of the $10!/5!5! = 252$ divisions of brightness discrimination data in the top row between the two cells in conjunction with each of the 252 divisions of the brightness discrimination data in the bottom row between the two cells. For this test, the loudness discrimination data would be ignored. The test statistic could be $\Sigma(T^2/n)$ for a nondirectional test or it could be the sum of the brightness discrimination measurements in the column expected to provide the larger measurements for a one-tailed test. A test for the effect of lesion L_2 on loudness discrimination would be conducted in an analogous manner, dividing the loudness discrimination

TABLE 5.12

Assignments to Lesion Conditions

	Lesion in Location L_1?	
Lesion in Location L_2?	**Yes**	**No**
Yes	$n = 5$	$n = 5$
No	$n = 5$	$n = 5$

measurements between cells within columns to generate the set of data permutations. Program 5.1 can be used to carry out these tests.

5.10 Dichotomous and Ranked Data

Throughout this chapter, the issues have concerned all types of quantitative data. Consequently, what has been stated so far is also applicable to cases where the responses fall into one of two categories because the dependent variable "measurement" can then be expressed as 0 or 1 to apply the randomization test procedure.

If the data are in the form of ranks, there are two options. One is to rank within rows separately for testing for column effects and within columns for testing for row effects. The other option is to rank over all cells and to permute those ranks within columns or rows in testing for treatment effects. The latter procedure should be more sensitive to treatment effects as it provides for finer discriminations in much the same way as unranked data permits more subtle discriminations than ranks. Consider the two matrices of ranks in Table 5.13, and consider a test of the effect of factor A. Notice that in the matrix on the left, showing ranks over all cells, switching rank 4 and rank 5 has less effect on a one- or two-tailed test statistic than switching rank 2 and rank 6; yet with reranking within rows, as shown in the right matrix, the corresponding switches, namely of rank 2 and rank 3 within B_1 or within B_2 levels, would have the same effect. But clearly, as rank 4 and rank 5 are between rank 2 and rank 6, they represent effects that are not as different, and thus reversals might more readily occur by chance than for rank 2 and rank 6.

5.11 Fractional Factorial and Response Surface Designs

Webster (1983) noted that there was considerable interest in fractional factorial and response surface designs. Randomization tests are applicable to both types of designs. Federer and Raktoe (1983) defined the fractional factorial design as one in which the number of treatment combinations is

TABLE 5.13

Alternative Ranking Methods

	A_1	A_2		A_1	A_2
B_1	3, 5	4, 8	B_1	1, 3	2, 4
B_2	1, 6	2, 7	B_2	1, 3	2, 4

less than the product of the number of levels of the experimental factors. In terms of the factorial matrix of treatment combinations, some cells are empty. Conducting a valid randomization test with empty cells is no problem. For example, testing the main effect of A when one of the A combinations is missing involves permuting data in the same manner as when all cells contain scores — except, of course, the empty cell is not involved. However, the cell frequencies will be disproportional and so a more sensitive test would result by use of the special methods in Chapter 6 for disproportional designs.

Response surface methodology derives from regression analysis (Biles and Swain, 1989), and Dillon (1977) has described the transition from response curves for single-factor manipulation to response surfaces for multiple-factor manipulation with quantitative factors. As Hinkelmann and Kempthorne (1994) observed, applications of response surface methodology have been made in various fields, including industry, agriculture, and medicine. Where there are two or more quantitative experimental factors, the regression of the data on the quantitative levels of the factors might be of interest, and then response surface procedures can be beneficial. A randomization test in such cases could be conducted as a regular factorial randomization test but with a test statistic reflecting the correlation between treatment magnitude and response magnitude. For example, the reference set for the main effect of A factor could be generated by permuting data over levels of A within levels of the other factors and computing a correlation test statistic from Chapter 8 for each of the data permutations.

5.12 Questions and Exercises

1. Describe a three-factor experiment where it would make sense to have quantitative levels of one factor but qualitative levels of two other factors.

2. How many cells are there for a factorial experimental design involving two factors with two levels of each factor?

3. Why might an experimenter want to employ a factorial design with disproportional cell frequencies?

4. Draw a 2×3 matrix of proportional cell frequencies.

5. How many permutations would be required for a systematic randomization test of the "main effect" of B in the following matrix of cell frequencies?

	A_1	A_2
B_1	2	2
B_2	3	3

6. For a three-factor factorial design, the matrix of results must have at least how many cells?

7. Applying a randomization test to raw data instead of applying it to ranks for the raw data tends to provide a more sensitive test. So why is a randomization test applied to ranks in Table 5.13 in Section 5.10?

8. Why might an experimenter deliberately form a fractional factorial design with an empty cell?

9. For a completely randomized factorial design with two factors, A and B, how would the data be permuted for a test of the randomization test null hypothesis of no effect of factor B within level 1 of factor A?

10. What is the "surface" that would be associated with a "response surface" experimental design? Why must such a design be a factorial design? Why must the factors be quantitative?

REFERENCES

Biles, W.E. and Swain, J.J., *Optimization and Industrial Experimentation*, John Wiley & Sons, New York, 1989.

Dillon, J.L., *The Analysis of Response in Crop and Livestock Production* (2nd ed.), Pergamon Press, New York, 1977.

Edgington, E.S., *Statistical Inference: The Distribution-Free Approach*, McGraw-Hill, New York, 1969.

Federer, W.T. and Raktoe, B.L., Fractional factorial designs, in *Encyclopedia of Statistical Sciences*, Vol. 3, Kotz, S. and Johnson, N.L. (Eds.), John Wiley & Sons, New York, 1983, 189–196.

Hinkelmann, K. and Kempthorne, O., *Design and Analysis of Experiments, Volume I: Introduction to Experimental Design*, John Wiley & Sons, New York, 1994.

Manly, B.F.J., *Randomization and Monte Carlo Methods in Biology*, Chapman & Hall, New York, 1990.

Webster, J.T., Factorial experiments, in *Encyclopedia of Statistical Sciences*, Vol. 3, Kotz, S. and Johnson, N.L. (Eds.), 1983, 13–17.

6

Repeated-Measures and Randomized Block Designs

The use of randomization tests for determining the P-value in repeated-measures (within-subject) and randomized block (within-block) designs is the topic of this chapter. We focus the presentation on repeated-measures designs but many issues are also relevant for randomized block designs. Specific concerns that are related to randomized block designs are discussed from Section 6.19 forward.

All subjects take all treatments in repeated-measures experiments, so there is no need to control for between-subject variability. Random assignment provides statistical control over within-subject variability and is carried out in the following manner. Each of n subjects takes all of k treatments. For each subject, there are k treatment times, each incorporated into a subject-plus-treatment-time experimental unit. Thus, there are nk experimental units and the k experimental units for each subject are randomly assigned to the k treatments, independent of the assignment for other subjects. There are thus $(k!)^n$ possible assignments because the $k!$ possible assignments for each subject can be associated with that many different assignments for every other subject.

6.1 Carry-Over Effects in Repeated-Measures Designs

Carry-over effects, which are influences on measurements from prior treatment administrations, can reduce the power or sensitivity of repeated-measures experiments to show treatment effects. For example, if in an experiment comparing drugs A and B the successive treatment administrations are so close together that the effect of a drug administration has not worn off before the next administration, the resulting mixture of A and B effects on the individual measurements could mask differences between the effects of the two drugs. It is therefore desirable to minimize carry-over effects by whatever procedures are relevant for a particular experiment. For control over persistence of biochemical changes in a subject, an experimenter might concentrate on making sure successive treatment times for each subject are sufficiently far apart. Other types of carry-over effects call for different types of control. For example, in verbal learning experiments, a subject might be required to perform some task designed to interfere with rehearsing a list of words between successive treatment administrations. It is important to control carry-over effects to make a repeated-measures experiment sensitive but carry-over effects do not invalidate a repeated-measures randomization test.

Two types of carry-over effects should be distinguished: carry-over effects that are the same for all treatments and differential carry-over effects, which vary from one treatment to another. It is the differential carry-over effects that have at times been regarded as invalidating a repeated-measures experiment. It is frequently pointed out that if an AB sequence of treatments leads to an inflation of the B treatment measurement when a BA sequence would not lead to an inflation of the A treatment measurement, that could make the B measurements systematically larger than otherwise even with random assignment of treatment times to treatments. And that is true, as is the fact that even if on the first treatment administration the measurement would have been the same for either treatment, a one-tailed repeated-measures t test with a prediction of larger measurements for B would tend to provide small P-values. However, such a differential carry-over effect does not invalidate the test. The reason is that the randomization test null hypothesis of no difference in treatment effect would be false if there was a differential carry-over effect, so the null hypothesis cannot be falsely rejected; in other words, a differential carry-over effect is a treatment effect. If A and B were the same drug, differing simply in the labels A and B, there could be no differential carry-over effect.

A differential carry-over effect may account for a difference that one would like to attribute to a "true" treatment effect, meaning one where for at least one of the treatment times the A or B measurements would have been different in the absence of carry-over effects. That is, it appears that experimenters would like to draw conclusions from repeated-measures experiments about what would have happened in an independent-groups experiment comparing the same treatments. To facilitate this, as well as to sensitize a repeated-measures experiment to treatment effects, the experimenter should strive to make the successive treatment administrations within a subject be affected minimally by previous treatment administrations. The opportunity for differential carry-over effects (which reflect treatment differences but ones not usually of primary concern) is minimized when the opportunity for carry-over effects common to the various treatments is minimized.

6.2   The Power of Repeated-Measures Tests

The repeated-measures test is sensitive to the same type of treatment effects as the independent t test or simple one-way ANOVA: effects wherein some treatments tend to provide larger measurements than others. Thus, an experimenter would not employ a repeated-measures test to look for a different type of effect from that to which a comparison between independent groups (groups employing different subjects for different treatments) is sensitive — unless, of course, there is interest in investigating the influence of prior treatments on the effect of subsequent treatments. The repeated-measures test is simply a more sensitive test than the other type when there is at least a moderate degree of positive correlation between measurements under the various treatments. To see why repeated-measures tests — of which the correlated *t* test is the special type employed for the two-treatment experiment — are likely to be more powerful when there is a positive correlation, let us first examine the question graphically. Figure 6.1 shows two sets of hypothetical data. Set A consists of the same measurements as those shown in set B, but set A measurements are intended to represent the measurements of 10 subjects, five of which were assigned to each of the two treatments, whereas set B measurements represent the measurements of only five subjects, each of which took both treatments. The lines in set B connect the pairs of measurements generated by the same subject. Given this explanation, consider which graph is more suggestive of a treatment effect. Certainly, the overlap of distributions for set A does not very strongly suggest a differential effect of the two treatments. But although the measurements in set B have the same overlap, the lines that connect the measurements of the same subjects indicate that there is a consistent directional difference in measurements of about the same magnitude in all subjects, and this fact leads one to suspect a difference in the effects of the two treatments.

Repeated-measures tests deal with the differences between the effects of treatments within individual subjects, and where those differences tend to be consistent from subject to subject even small differences among treatments can provide significant results. In other words, when the variability within treatment groups is large, a large difference among means is required to provide significant results for one-way ANOVA, whereas with the same between-subject variability the repeated-measures test may find a small difference among means to be significant if there is small variability within the difference scores.

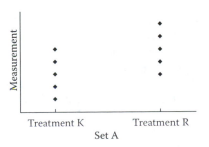 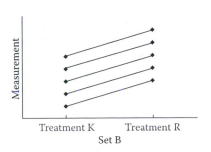

FIGURE 6.1
Graphs of two sets of data.

As with the *t* test for differences between independent groups, the *t* test of the repeated-measures type (which is sometimes called the correlated *t* test) is simply a special case of the corresponding ANOVA test, and the squared *t* value equals the value of *F* that repeated-measures ANOVA would give for the same data. Reference to the published probability tables would provide the same P-values for *t* and *F*. However, it is customary for textbooks to present both the *t* test and ANOVA for the repeated-measures designs because a person may be interested in the one-tailed P-value, and such a P-value sometimes is available only for the *t* test. There are fewer procedures to be considered here in connection with repeated-measures designs because the standard repeated-measures designs use the same subjects for all treatments, providing an equal-sample-size design.

6.3   Systematic Listing of Data Permutations

Because the random assignment procedure is different from that of independent-groups experiments, repeated-measures experiments require a different procedure of listing permutations to determine the P-value by systematic data permutation.

Example 6.1. We have three subjects, each to be tested under two conditions, A and B. We could simply administer treatment A to all three subjects and obtain measurements, then administer B to all subjects and obtain measurements. Then we could pair the measurements subject by subject and, no doubt, we would find that within at least some of the pairs of measurements there was a difference. That is, at least one of the subjects responded differently to one of the treatments than to the other treatment. But that does not answer the question of interest, which concerns a treatment effect. After all, from one treatment administration time to another, a subject may give different responses to the same treatment but we are not interested in the within-subject variability that is simply a function of variation over time. We are interested in the variability produced by the treatment differences. For this reason we should not give one treatment to all subjects first, then the other treatment. Instead, we should randomly assign treatment times to treatments. For each subject, we have two treatment times — times that are convenient to both the experimenter and the subject. The earlier of the times is called t_1 and the later t_2, and two slips of paper are marked with those designations. One of the paper slips is randomly selected for the first subject and that slip indicates the treatment time for treatment A, the other treatment time being associated with treatment B. The slip is replaced, a drawing is made for the second subject to tell which treatment time is associated with treatment A, and the same is done for the third subject. The number of possible assignments is $2^3 = 8$ because there are two possible assignments for the first subject, each of which could be paired with two possible assignments

TABLE 6.1

Data and Treatment Times

Subjects	A	B
a	(t_2) 18	(t_1) 15
b	(t_1) 14	(t_2) 13
c	(t_1) 16	(t_2) 14
	$\bar{X} = 16$	14
	$t = 3.46$	

for the second subject, which in turn could be paired with two possible assignments for the third subject. The experiment is conducted and t for a correlated t test is calculated for the observed results using the classical formula from parametric statistics

$$t = \frac{D}{s_D / \sqrt{n}} \tag{6.1}$$

with D equal to $(\bar{A} - \bar{B})$ or the mean of the difference scores, s_D, equal to the standard deviation of the difference scores and n equal to the number of pairs. The experimental results and the t value are shown in Table 6.1.

To have a systematic procedure for listing data permutations, we first assign index numbers to the data and permute the index numbers systematically. Index numbers 1 and 2 are used to represent the data, where for each subject the measurement obtained under treatment A is given an index of 1, and the measurement obtained under treatment B is given an index of 2. Table 6.2 shows how the obtained results would be indexed.

For purposes of listing a sequence of data permutations, the obtained results can be regarded as a six-digit number where the first two digits represent the index numbers for the first subject, the next two the index numbers for the second subject, and the last two the index numbers for the third subject: 121212. Each data permutation will be represented as a six-digit number where the digits 1 and 2 appear in each of the three groups of two digits in the six-digit number. Each successive data permutation provides a six-digit number that is the next larger than the previous one. For example, the second permutation would be represented by 121221, the order of index numbers for the third subject being changed while the order for the

TABLE 6.2

Systematic Listing of Data

Subjects	A	Index	B	Index
a	18	1	15	2
b	14	1	13	2
c	16	1	14	2

TABLE 6.3

Eight Data Permutations

	Permutation 1		Permutation 2		Permutation 3		Permutation 4	
	A	B	A	B	A	B	A	B
	18	15	18	15	18	15	18	15
	14	13	14	13	13	14	13	14
	16	14	14	16	16	14	14	16
$\bar{X} =$	16	14	15 1/3	14 2/3	15 2/3	14 1/3	15	15
	$t = 3.46$		$t = 0.46$		$t = 1.11$		$t = 0$	

	Permutation 5		Permutation 6		Permutation 7		Permutation 8	
	A	B	A	B	A	B	A	B
	15	18	15	18	15	18	15	18
	14	13	14	13	13	14	13	14
	16	14	14	16	16	14	14	16
$\bar{X} =$	15	15	14 1/3	15 2/3	14 2/3	15 1/3	14	16
	$t = 0$		$t = -1.11$		$t = -0.46$		$t = -3.46$	

first two subjects is held constant. For a complete listing of the data permutations, the six-digit number is incremented until it reaches its maximum value of 212121 on the eighth data permutation. The eight data permutations are listed in sequential order in Table 6.3.

The P-value associated with the obtained results (which is represented by permutation 1) for a one-tailed t test where $\bar{A}$ is expected to be the larger mean is 1/8, or 0.125, because only one of the eight data permutations provides a t value as large as the obtained value, 3.46. If no direction of difference were predicted, the P-value would be determined for a two-tailed t test, and in that case the P-value of the results would be 2/8, or 0.25, because two of the eight data permutations (permutation 1 and permutation 8) provide $|t|$'s as large as the obtained value. If only the subset consisting of the first four listed permutations in Table 6.3 was used to determine P-value, the same two-tailed P-value would be obtained. For a two-tailed test, the last four data permutations are redundant because each is a mirror image of one of the first four. Permutation 1 and permutation 8 are alike except for transposition of the treatment designations, which does not affect the $|t|$ value, and the same is true of permutations 2 and 7, 3 and 6, and 4 and 5.

6.4 A Nonredundant Listing Procedure

When the permutations are listed according to the rule in Section 6.3, it always will be the case for two treatments that the second half of the data permutations will be redundant for determining a two-tailed P-value. The first half of the data permutations are those for which the measurements for the first subject are in the obtained order; with all data permutations in the first half having the same order of measurements for the first subject, none can be a mirror image of another. Obviously, in the set of all data permutations each data permutation can be paired with another that is its mirror image. Consequently, for each data permutation in the first half of the set there is a corresponding mirror-image permutation in the second half that will give the same $|t|$ value, and so the two-tailed P-value given by the subset consisting of the first half of the data permutations is necessarily the same as that based on the entire set.

With three treatments — because there are six ways the treatment designations can be assigned to a particular division of data — the data permutations can be grouped into sixes, where each of the six is a mirror image of the others in that group. Using the same sort of listing procedure for permutations, only the first sixth of the data permutations must be listed to determine the P-value of F because they are the ones that have the measurements for the first subject in the obtained order. In general, for k treatments only the first $1/k!$ of the data permutations then need be computed to determine the P-value of F or $|t|$. (Program 6.1, which will be presented in Section 6.6, stops performing data permutations when the next permutation would permute the scores of the first subject, thereby ensuring that the appropriate subset of $1/k!$ of the data permutations has been used for determining the P-value.)

6.5 ΣT^2 as an Equivalent Test Statistic to F

Table 6.4 is a summary table for repeated-measures ANOVA. In the table, k is the number of treatments and n is the number of subjects. $SS_{B(t)}$ is the SS between treatments, $SS_{B(s)}$ is the SS between subjects, and SS_e is the residual SS. $SS_{B(t)}$ and SS_e are defined as:

$$SS_{B(t)} = \sum \frac{T^2}{n} - \frac{(\Sigma X)^2}{kn} \qquad (6.2)$$

$$SS_e = \Sigma X^2 - \sum \frac{T^2}{n} - \frac{\Sigma S^2}{k} + \frac{(\Sigma X)^2}{kn} \qquad (6.3)$$

where k is the number of treatments, n is the number of subjects, T is the sum of measurements for a treatment, and S is the sum of measurements for a subject.

It will be shown now that ΣT^2 is an equivalent test statistic to F for repeated-measures ANOVA for determining the P-value by data permutation. The formula for F in Table 6.4 is equivalent to $(SS_{B(t)}/SS_e) \times [(n-1)(k-1)/(k-1)]$. Because $(n-1)(k-1)/(k-1) = n-1$ is a constant multiplier over all data permutations, its elimination has no effect on the ordering of the data permutations with respect to the test statistic value; thus, $SS_{B(t)}/SS_e$ is an equivalent test statistic to F. Reference to Equation 6.2 and Equation 6.3 shows that

TABLE 6.4

Summary Table for Repeated-Measures ANOVA

Source of Variation	Sum of Squares	Degrees of Freedom	Mean Square	F
Treatments	$SS_{B(t)}$	$k-1$	$SS_{B(t)}/(k-1)$	$\dfrac{SS_{B(t)}/(k-1)}{SS_e/(k-1)(n-1)}$
Subjects	$SS_{B(s)}$	$n-1$	$SS_{B(s)}/(n-1)$	
Error	SS_e	$(k-1)(n-1)$	$SS_e/(k-1)(n-1)$	
Total	SS_T	$kn-1$		

all of the terms for the numerator and denominator of $SS_{B(t)}/SS_e$ except $\Sigma(T^2/n)$ are constant over all data permutations. An increase in $\Sigma(T^2/n)$ increases the numerator, shown in Equation 6.2, and decreases the denominator, shown in Equation 6.3, thereby increasing the value of $SS_{B(t)}/SS_e$. Therefore, $\Sigma(T^2/n)$ is an equivalent test statistic to $SS_{B(t)}/SS_e$ and thus is an equivalent test statistic to F. Each subject takes all treatments, so $\Sigma(T^2/n)$ is the same as $(\Sigma T^2)/n$. With n constant over all data permutations, this is an equivalent test statistic to ΣT^2, the sum of the squares of the treatment totals. Because ΣT^2 is a simpler test statistic to compute for each data permutation than repeated-measures F and will give the same P-value as F, it will be used in the computer programs to determine the P-value of repeated-measures F.

6.6 Repeated-Measures ANOVA with Systematic Data Permutation

The steps to take in determining the P-value of repeated-measures F by systematic data permutation are as follows:

Step 1. Arrange the data in a table with k columns and n rows, the columns indicating the treatments and the rows indicating the subjects. Assign index numbers 1 to n to the subjects and numbers 1 to k to the treatments so that each measurement has associated a compound index number with it, the first part of which indicates the subject, and the second part of which indicates the treatment. For example, the measurement for the fifth subject under the treatment in column 4 would be given the index number (5, 4).

Step 2. Use a permutation procedure to produce a systematic sequence of permutations of the n measurements, where the measurements for the first subject are not permuted. That is, the measurements for the first subject are always associated with the treatments under which those measurements were obtained in the experiment, whereas measurements for the other subjects are permuted over treatments. Because exactly $1/k!$ of the $(k!)^n$ possible permutations are thereby performed, the number of permutations performed is $(k!)^{(n-1)}$, the first of which is the permutation represented by the observed data. Perform Step 3 and Step 4 for each of the permutations.

Step 3. Compute $T_1, T_2, \ldots T_k$, the total of the measurements associated with the index numbers for all treatments.

Step 4. $T_1^2 + \ldots + T_k^2 = \Sigma T^2$ is the test statistic.

Step 5. Compute the P-value for F as the proportion of the test statistic values that are as large as the obtained test statistic value, the test statistic value for the first permutation.

The computational steps for this test will now be illustrated by a numerical example.

Example 6.2. Hypothetical data from an experiment are presented in Table 6.5 according to the format prescribed in Step 1. The grand total of all the scores in Table 6.5 is $10 + 13 + 6 + 11 + 14 + 8 + 13 + 17 + 9 = 101$. Going through Step 3 and Step 4 for the tabled data provides us with the obtained test statistic value. Because the totals for the first two treatments are 29 and 33, respectively, the total for the last treatment is $101 - (29 + 33) = 39$. Therefore, the observed test statistic value is $(29)^2 + (33)^2 + (39)^2 = 3451$. This value, 3451, is the test statistic value for the first of the $(6)^2 = 36$ permutations that must be performed for the test. The second permutation is obtained by changing the sequence of scores for the third subject from "6 8 9" to "6 9 8." Then the test statistic ΣT^2 becomes $(29)^2 + (34)^2 + (38)^2 = 3441$. There would be $(3!)^2 = 36$ permutations to perform and the proportion of those permutations providing as large a test statistic value as the obtained value, 3451, is the P-value associated with repeated-measures F for this example. The obtained data provides the largest test statistic value, as can be seen in the consistent increase from treatment 1 to treatment 3 within each subject, and so the P-value is $1/36 = 0.028$. If all 216 permutations had been used, the P-value would be the same because there would be six orderings of the same columns of measurements that give the observed test statistic value. Because $6/216$ is equal to $1/36$, the P-value obtained with 36 permutations is that which would be obtained with all 216 permutations.

Program 6.1 on the included CD can be used to determine the P-value of F by systematic data permutation. This program can be tested with the data given in Table 6.3, which should give a P-value of $12/7776 = 0.0015$.

TABLE 6.5

Hypothetical Data

Subject Number	Treatment 1	Treatment 2	Treatment 3
1	10	11	13
2	13	14	17
3	6	8	9

Because the number of permutations depends not only on the number of subjects but also on the number of treatments, the number of permutations increases rapidly with an increase in the number of subjects when there are several treatments. The number of permutations to be computed is $(k!)^{(n-1)}$, and so the addition of a subject multiplies the required number of permutations by $k!$. For three groups, the addition of a subject increases the number of permutations by a factor of 6, for four groups by a factor of 24, and so on. For example, an experiment in which each of three subjects takes four different treatments involves a total of 13,824 permutations, 1/24th of which must be performed for the systematic permutation test, and so it is feasible — but if six subjects were to take each of four treatments, the total number of permutations would be 191,102,976, and 1/24th of that number still is too many permutations for the determination of the P-value by systematic permutation to be feasible. Consequently, although determining the P-value by systematic permutation is useful for small samples, it is not likely to be feasible for relatively large samples. For relatively large samples, one should use the random permutation procedure, which will be discussed in the following section.

6.7 Repeated-Measures ANOVA with Random Data Permutation

The steps necessary in performing repeated-measures ANOVA with random permutation are as follows:

Step 1. As with the systematic permutation procedure, arrange the data in a table with k columns and n rows, where k is the number of treatments and n is the number of subjects. Assign index numbers 1 to n to the subjects and 1 to k to the treatments so that each measurement has a compound index number associated with it, the first part of which indicates the subject, and the second part of which indicates the treatment. For example, the index (3, 4) refers to the measurement for the third person under the fourth treatment.

Step 2. Go through Step 4 and Step 5 for the obtained data to compute the obtained test statistic value.

Step 3. Use a random number generation algorithm that will randomly determine for each subject independently of the other subjects, which of the k measurements is to be "assigned" to the first treatment, which of the remaining $(k-1)$ measurements to the second treatment, and so on. The random determination of order of measurements within each subject performed over all subjects constitutes a single permutation. Perform (NPERM − 1) permutations, and for each permutation go through Step 4 and Step 5.

TABLE 6.6

Test Data

A	B	C
2	1	4
4	6	8
10	12	14
16	18	20
22	24	26

Step 4. Compute $T_1, T_2, \ldots T_k$, the total of the measurements associated with the index numbers for all treatments.

Step 5. $T_1^2 + \ldots + T_k^2 = \Sigma T^2$ is the test statistic value.

Step 6. Compute the P-value for F as the proportion of the test statistic values, including the obtained test statistic value, that are as large as the obtained test statistic value.

Program 6.2 on the included CD is the program for determining the P-value of repeated-measures F by random permutation. This program can be tested with the data in Table 6.6, for which the P-value by systematic data permutation is $12/7776 = 0.0015$.

6.8 Correlated *t* test with Systematic Data Permutation

For repeated-measures tests, as for independent-groups tests, F is equal to t^2 and so F and the two-tailed t test statistic $|t|$ are equivalent test statistics for determining the P-value by means of a randomization test. Thus when only two treatments are employed, the P-value given by Program 6.1 is the value associated with a two-tailed correlated t test. Because of the symmetry of the complete sampling distribution of data permutations, halving the two-tailed P-value gives the one-tailed P-value where the direction of difference has been predicted correctly. However, for the purpose of combining P-values it is also useful to have a one-tailed P-value even when the direction of difference is predicted incorrectly, and that value cannot be derived from the P-value given by Program 6.1. For such one-tailed P-value computation, a special program for correlated t is necessary. Program 6.3 on the included CD can be used to calculate one-tailed P-values of correlated t randomization tests with systematic permutation.

For correlated t as well as independent t tests, the difference between means can be computed in a manner such that a positive t is obtained when the difference between means is in the predicted direction and a negative t is obtained if the difference is in the opposite direction. Therefore, we define the one-tailed P-value for correlated t for a randomization test as the probability, under H_0, of getting such a large value of t (taking into consideration the sign of t) as the obtained value. When the direction of difference between means has been correctly predicted, the one-tailed P-value computed in this manner will be the same as half of the two-tailed value given by Program 6.1. A one-tailed test statistic for correlated t that is equivalent under data permutation to t is T_L, the total of the measurements for the treatment predicted to give the larger mean.

6.9   Fast Alternatives to Systematic Data Permutation for Correlated t Tests

In Section 4.11, it was pointed out that there are fast alternatives to systematic data permutation for independent t tests. This is also true for fast alternatives to systematic data permutation for correlated t tests. In both cases, the procedures rapidly generate frequency distributions of test statistics (treatment totals), reducing the need for random data permutation procedures, which depend on sampling the complete frequency distribution even for large samples. Nevertheless, for both types of t tests random data permutation is better for certain applications, so it will be described below.

6.10   Correlated t Test with Random Data Permutation

With random permutation as with systematic permutation, the P-value for F for repeated-measures ANOVA when there are only two treatments is the two-tailed P-value for correlated t. This P-value can be divided by two to get the one-tailed P-value for one-tailed t whenever the direction of difference has been predicted correctly. The justification for obtaining the one-tailed P-value in this way is the same as in Section 4.12 for random data permutation with independent t tests with equal sample sizes: by virtue of the symmetry of distribution of measurements within the complete sampling distribution, the probability is $1/2$ that in a random sample of data permutations, for any $|t|$ the difference between means will be in the predicted direction when H_0 is true. Thus, the probability of getting a value of $|t|$ (or the equivalent test statistic, ΣT^2) under H_0 as large as an obtained value with the obtained direction of difference between means is half of the probability associated with two-tailed t.

Therefore, Program 6.2 can be used to determine the two-tailed P-value for correlated t, and half of that value is the one-tailed value when the direction of difference between means has been predicted correctly. However, when the direction of difference has been predicted incorrectly the one-tailed P-value cannot be derived from the P-value given by Program 6.2 but must be computed separately. Program 6.4 can be used to calculate one-tailed P-values of correlated t randomization tests with random permutation.

6.11 Correlated t Test and Planned Comparisons

In Section 4.13, there was a discussion of the independent t test and planned comparisons. The same sort of considerations applies to planned comparisons with the correlated t test. With a repeated-measures experiment involving k treatments, there are $(k)(k-1)/2$ pairs of treatments that could be subjected to a correlated t test. When there are a small number of comparisons planned before the experiment, the P-value of the correlated t's for those comparisons can be determined by data permutation for each comparison as if the two treatments in the comparison were the only ones in the experiment. For instance, no matter how many treatments were administered in the experiment, for a planned comparison of treatments B and D the P-value of t based on systematic data permutation would depend only on the 2^n permutations of the data for those two treatments. (Of course, H_0 would refer to treatments B and D alone.) Program 6.1 will provide the P-value for a two-tailed test for systematic data permutation. If the number of data permutations is so large as to necessitate random permutation, Program 6.2 can be used for the two-tailed test. Program 6.3 and Program 6.4 can be used if one-tailed P-values are needed.

6.12 Correlated t Test and Multiple Comparisons

In Section 4.14, the discussion concerning the use of the independent t test with multiple comparisons is in general also applicable to the correlated t test. Before the development of special procedures for multiple comparisons, it was customary first to determine the P-value of the overall F for a repeated measures analysis of variance and then, if F was significant at the 0.05 level, to determine the P-value for each of the possible comparisons by the correlated t test with no adjustment of the P-values to take the number of comparisons into account. Such a treatment of post hoc comparisons following determination of a significant F is still used by some researchers; however, adjustment of the P-values to take into account the number of comparisons is more widely accepted.

The Bonferroni procedure described in Section 4.14 is appropriate for multiple comparisons with repeated-measures as well as independent-groups data. The P-value of correlated t is computed by data permutation for each of the $(k)(k-1)/2$ pairs of treatments, where for each comparison only the measurements for the compared treatments are permuted. Next, those P-values are adjusted by multiplying each value by $(k)(k-1)/2$ to get the final adjusted P-values, or Holm's procedure, as described in Section 4.14.

For repeated-measures experiments as for experiments with independent groups, multiple comparisons procedures should be used only when necessary because of their effect of increasing the P-value. A few thoughtfully planned comparisons might be a better approach. If a multiple comparisons procedure is appropriate, the Bonferroni or Holm's procedure can be used.

6.13 Rank Tests

Data permutation can be used to determine the P-value of correlated t or repeated-measures ANOVA for ranked as well as raw data. Because the use of raw data is as valid as the use of ranks for randomization tests, ranks should not be used when more precise measurements are available.

However, sometimes ranks are the most meaningful observations to analyze. Situations arise in which one is able to rank two responses only within subjects. An example would be where the response was a preference for one of two treatments (stimuli); whereas ranking of responses within subjects could be made, there would be no basis for ranking over subjects. If the responses for each subject were assigned rank 1 and rank 2 and the 1s and 2s were entered as data in Program 6.1, the P-values obtained would be those associated with the sign test. However, the sign test is a special application of the binomial distribution, which is so extensively tabled and so easy to derive that a randomization test program is unnecessary. Lehmacher (1980) discussed multiple comparisons procedures — including Bonferroni's — for sign tests applied to contingency tables.

The Friedman ANOVA-by-ranks test is a multi-treatment extension of the sign test to more than two treatments when responses are ranked within subjects. The Friedman ANOVA-by-ranks test is a nonparametric test that uses ranked data to determine the P-value, and the P-values that are published for the Friedman test statistic are based on permutation of ranks. For each subject, the measurements over treatments are ranked and the rank numbers 1 to k are used as the measurements for the k treatments. The Friedman test statistic is equivalent to repeated-measures F under data permutation, and so the P-value obtained from the Friedman table is that which would be obtained for repeated-measures F under data permutation of the ranks. With ranked data, Program 6.1 gives the same P-value as tables for the Friedman test statistic. Program 6.1 can be useful in extending the utility of Friedman's ANOVA because of limitations of the tables. Like other rank-order tables, the Friedman table is derived from ranks with no ties, and so the P-values are only approximate when there are tied values. Also, some tables (e.g., Siegel and Castellan, 1988) are restricted to 13 subjects or less for three treatments, 8 subjects or less for four treatments, and 5 subjects or less for five treatments. For larger samples, approximate chi-square P-values must be used.

Wilcoxon's matched-pairs signed-ranks test is the rank-order counterpart of the correlated t test. Unlike the sign test — which does not require measurements prior to ranking — the Wilcoxon test assumes raw measurements over both treatments for all subjects. Differences between paired measurements for each subject must be ranked and those ranks are employed in the test. Precise information in the form of raw data then is necessary for the application of the Wilcoxon test, but in ranking to make it a nonparametric test, that precision is degraded. Randomization tests for correlated t are nonparametric and serve the same function as the Wilcoxon test without the loss of information that the Wilcoxon test necessarily entails. Thus, randomization tests should be employed instead of Wilcoxon's matched-pairs signed-ranks test.

Some advocates of rank tests (e.g., Krauth, 1985; Lam and Longnecker, 1983) have proposed using more information than simply ranking within subjects and yet object to Wilcoxon's matched-pairs signed-ranks test because they consider the original measurements involved in that test to have little more than rank importance. They have proposed ranking the nk measurements from n subjects over k treatments to provide ranks from 1 to nk prior to conducting a Wilcoxon test for a difference between paired ranks. Approximate significance testing procedures have been devised but in fact they are unnecessary because, as pointed out by Edgington (1980), Program 6.1 can determine P-values for paired ranks whether the ranks are based on ranking within subjects only or ranking all responses within and between subjects.

6.14   Dichotomous Data

When the dependent variable is dichotomous, 0s and 1s can be assigned to the two categories (as with independent groups) and repeated-measures ANOVA can be used with the 0s and 1s to get a value of F whose P-value is determined by data permutation.

The test statistic for the Cochran Q test (see Siegel and Castellan, 1988) is an equivalent test statistic to F for repeated-measures ANOVA for determining the P-value by data permutation with dichotomous data. The Cochran test statistic refers to chi-square tables to get an approximation to the P-value that would be provided by systematic data permutation — the value provided by Program 6.1.

6.15   Counterbalanced Designs

Let us consider next a procedure called "counterbalancing" that is sometimes used in repeated-measures experiments to control for within-subject variability. A method of *partial counterbalancing* is ensuring that each treatment occurs equally often in each serial position over all subjects. For example, given six subjects who each take three treatments, treatment A would be the first treatment taken by two subjects, treatment B would be the first treatment for two others, and treatment C would be the first treatment for the remaining two subjects. Similarly, each of the treatments would be taken second by two subjects and each of the treatments would be taken third by two subjects. Three sequences of treatments are sufficient for this purpose: ABC, BCA, and CAB. If two subjects take the ABC order, two take the BCA order, and two take the CAB order, each of the treatments will occur twice as the first treatment, each will occur twice as the second treatment, and each will occur twice as the third treatment.

To provide adequate control over within-subject variability, there should be random assignment of the sequences to the subjects. This could be done in the following manner: Specify three treatment times for each subject; then designate the six subjects as a, b, c, d, e, and f on slips of paper that are placed in a container. The first two slips of paper that are randomly drawn from the container show the two subjects that take the treatments in the ABC order, the next two show the subjects that take the treatments in the BCA order, and the last two show the subjects that take the treatments in the CAB order. There are $6!/2!2!2! = 90$ possible assignments of subjects to treatment orders and because the treatment order for a subject assigns a treatment time to each treatment for that subject, there are 90 possible assignments of treatment times to treatments.

Example 6.3. Assume that subjects b and e were assigned to the ABC order, subjects a and f were assigned to the BCA order, and subjects c and d were assigned to the CAB order, and that the experimental results shown in Table 6.7 were obtained. To determine the P-value associated with the F-statistic, the equivalent test statistic ΣT^2 is computed for all 90 possible assignments of treatment times to treatments. The null hypothesis is that each subject gave the same response at each of the treatment times as that subject would have given if the treatments had been given in one of the other two possible orders. So under H_0, whatever the treatment order assigned to a subject, the measurements associated with the individual treat-

TABLE 6.7

Partial Counterbalancing

Subjects	Treatments					
	A		B		C	
a	(t_3)	23	(t_1)	38	(t_2)	17
b	(t_1)	26	(t_2)	31	(t_3)	28
c	(t_2)	15	(t_3)	19	(t_1)	22
d	(t_2)	34	(t_3)	35	(t_1)	29
e	(t_1)	18	(t_2)	27	(t_3)	12
f	(t_3)	24	(t_1)	19	(t_2)	18
$T^2 =$	19,600		28,561		15,876	
			$\Sigma T^2 = 64,037$			

ment times stay with their respective treatment times and that determines the data permutations associated with each of the possible 90 assignments. The P-value is the proportion of the 90 data permutations that provide as large a value of ΣT^2 as 64,037, the obtained value.

The number of orders of treatments necessary for partial counterbalancing is simply the number of treatments because the orders can be generated sequentially by taking the first listed treatment and moving it to last place. For example, with five treatments, five orders that allow each treatment to occupy each serial position once are: ABCDE, BCDEA, CDEAB, DEABC, and EABCD. To have each treatment occur as often as any other in each of the serial positions, it is necessary for the number of subjects to be a multiple of the number of treatments. For example, with partial counterbalancing and five treatments, five or ten subjects could be used in the experiment but not six or seven.

Complete counterbalancing is used by experimenters to provide better experimental control over within-subject variability than can be provided by partial counterbalancing. Whereas partial counterbalancing ensures that each treatment occurs equally often in each serial position, complete counterbalancing ensures that all possible serial orders of treatments occur equally often. Not only would ABC, BCA, and CAB be used in an experiment with three treatments but also ACB, BAC, and CBA. Each of the six possible orders would be used in the experiment the same number of times. Thus, as with random assignment of treatment times to treatments independently for each subject, all possible orders of treatments are considered, but unlike the independent assignment procedure, the complete counterbalancing procedure makes sure that every possible order will be used and that each order will be used as often as any other.

To justify statistical analysis, complete counterbalancing should be done by a random assignment procedure. All $k!$ orders of the k treatments would be listed and then the first n randomly selected subjects (where n is the number of subjects per treatment order) would receive the first order, the second n would receive the second order, and so on. The minimum number of subjects for complete counterbalancing would be $k!$ and, if more than $k!$ subjects are used, the number must be a multiple of $k!$ to ensure complete counterbalancing.

For a given number of subjects, partial counterbalancing, complete counterbalancing, and other procedures involving random assignment then lead to different numbers of possible assignments, and so the P-value determination must follow the particular type of random assignment employed in the repeated-measures experiment. There will always be fewer permutations to consider with partial or complete counterbalancing than without counterbalancing. In unrestricted repeated-measures assignment, where the treatment times are randomly assigned for each subject independently of all other subjects, the distribution of permutations must include all those that would be obtained by complete counterbalancing as well as additional permutations. The additional permutations are for cases where the number of subjects receiving the treatment in one order would not be the same as the number receiving the treatments in a different sequential order. If one counterbalances, then when the sequence in which the treatments are given is likely to be of little consequence, the power of the test has been reduced because the total number of possible permutations is reduced. Alternatively, where it is felt that the sequence in which the treatments are given is of considerable importance, one should counterbalance. If the type of random assignment that is performed is taken into consideration in permuting the data, both counterbalancing and not counterbalancing are valid methods but the relative power of the two methods depends on the importance of the sequence in which the treatments are given.

6.16 Outliers

In Chapter 4, it was shown that data permutation can give substantially smaller P-values than *t* tables for the independent *t* test when there are outliers (extremely large or extremely small measurements). The presence of outliers for the correlated *t* test has a similar depressing effect on the power of the *t* table. The value of *t* for a difference between the means of treatment A and treatment B depends on the magnitude of the difference between means relative to the variability of the difference scores. The addition of a subject with the same direction of difference as the means may increase the variability of the difference scores so much that it more than offsets the effect of the increase in the difference between means, causing the value of *t* to decrease.

Example 6.4. Ezinga's (1976) study was discussed in Chapter 4 in regard to the effect of outlier scores on the independent *t* test. He obtained outlier difference scores for correlated *t* tests also, which have been discussed elsewhere (Edgington and Ezinga, 1978). As might be expected, data permutation provided smaller P-values than the *t* tables for such data. Table 6.8 shows one data set that Ezinga analyzed by use of the correlated *t* test. The data are for 10 subjects evaluated under two alternative sets of instructions. The correlated *t* value is 1.97, which, for a two-tailed test, does not reach the 0.05 level in the *t* table — in fact, the parametric *t* distribution gives a P-value just larger than 0.08 for a *t* value of 1.97 with nine degrees of freedom. However, systematic data permutation gives a two-tailed P-value of 0.016.

Data permutation gave more significant results than the *t* table for the data in Table 6.8, but is the difference a function of the outlier score? To answer this question, the data for subject *d*, who provided the outlier difference score of 2.56, were removed and the remaining data were analyzed by a correlated *t* test. With the outlier removed, the new |*t*| value was 3.05, which, with eight degrees of freedom, has a P-value of 0.016 on the basis of the parametric *t* distribution. The direction of difference in scores for subject *d* is consistent

TABLE 6.8

Experimental Data Analyzed by Use of Correlated t Test

Subjects	New Original Instructions (A)	New Repeated Instructions (B)	Difference Scores (B − A)
a	0.28	0.73	0.45
b	0.80	1.14	0.34
c	3.72	3.72	0.00
d	1.16	3.72	2.56
e	1.00	1.00	0.00
f	0.63	1.00	0.37
g	1.14	1.24	1.10
h	0.33	0.59	0.26
i	0.26	0.26	0.00
j	0.63	1.32	0.69

$t = 1.97$, nine degrees of freedom

Two-tailed P-value by t table: 0.08

Two-tailed P-value by randomization test: 0.016

with the direction for the other subjects but the difference is so great that the inclusion of subject d increases the P-value associated with the results.

6.17 Factorial Experiments with Repeated Measures

Let us consider an example of a two-factor experiment in which each subject is assigned to a level of one factor and each subject takes all levels of the other factor. Such designs can occur because of the desirability (and feasibility) of repeated measures for some factors and not for others.

Example 6.5. An experiment with three levels of factor A and two levels of factor B is conducted. Each subject is assigned to all three levels of factor A and to one of the two levels of factor B. The four subjects are assigned in the following manner. Each subject has three designated treatment times and for each subject, independently, there is random assignment of the three treatment times to level 1, level 2, and level 3 of factor A. There is then random assignment of two of the subjects to level 1 of factor B and the remaining two subjects are assigned to level 2. Table 6.9 shows the results of the experiment, where A and B are the factors and Ss are the subjects.

H_0 for the main effect of factor A is this: at each of the three treatment times, the measurement of a subject is independent of the level of A. There are $3! = 6$ ways in which the three measurements for each subject can be divided among the three levels of A, and each of those divisions can be associated with six divisions for each of the other subjects, making a total of $(3!)^4 = 1296$ data permutations. The generation of data permutations is the

TABLE 6.9

Factorial Design with Repeated Measures

		A_1	A_2	A_3
B_1	S_1	5	7	9
	S_2	7	11	12
B_2	S_3	3	5	7
	S_4	5	9	10

same as for univariate repeated-measures ANOVA and the test statistic, ΣT^2, reflects the property of interest in assessing the effect of A, namely the variation of the A means over levels of A. ΣT^2 for the obtained data, shown in Table 6.9, is $(20)^2 + (32)^2 + (38)^2 = 2868$. The value of ΣT^2 is computed for each of the 1296 data permutations, and the proportion of the 1296 test statistic values that are as large as the obtained value is the P-value for the main effect of factor A. Program 5.1 can be used to determine the P-value.

   H_0 for the main effect of factor B is this: the measurement of each subject at each of the levels of A is independent of the subject's assignment to level of B. For example, if H_0 is true the row of measurements for S_1 would have been a row for level B_2 if S_1 had been assigned to level B_2. One hypothesis that can be tested is that of no effect of B at any level of A. For the test of the null hypothesis, the three measurements for each subject can be added to provide a single sum. The four sums are then divided in all $4!/2!2! = 6$ ways between the two levels of B, and the one-tailed and two-tailed test statistics, T_L and $\Sigma(T^2/n)$, are used for a randomization test. A computer program is unnecessary for this simple example but for larger sample sizes, programs from Chapter 4 can be employed. Three other null hypotheses that could be tested would be those of no effect of B at a particular level of A. For testing each of those three null hypotheses, the four measurements for that level of A could be divided between the two levels of B with two measurements for B_1 and two for B_2 in all possible ways by means of a program from Chapter 4 to determine the P-value.

6.18   Interactions in Repeated-Measures Experiments

In Section 5.7, a randomization test counterpart of the H_0 of no interaction for a population model was described and it was shown that it could not be tested for a completely randomized experiment, although an interaction test statistic could be used for testing other H_0s. Alternatively, it is possible to test a H_0 of no interaction for a repeated-measures design in which there are subjects providing repeated measures over levels of factor A within different levels of factor B. The following example is based on a previously published discussion (Edgington, 1969).

Example 6.6. In a two-factor experiment, factor A and factor B both have two levels. Each of 10 subjects has repeated measures over the two levels of A but is assigned randomly to one of the levels of B. Each subject has two designated treatment times that are independently assigned randomly to levels A_1 and A_2 for each subject. The experimenter then selects five of the 10 subjects at random for assignment to level B_1 and the other five are assigned to level B_2. The obtained measurements for this experiment are given in Table 6.10.

The null hypothesis of no interaction in the experiment is this: for each of the 10 subjects, the difference score $(A_1 - A_2)$ shown in the last column of Table 6.10 is independent of assignment of the subject to a level of the B factor. There are $10!/5!5! = 252$ possible assignments to levels of B and, under H_0, the difference score for a subject is that which that subject would have had under either assignment. Thus, there are 252 data permutations representing all possible divisions of the 10 difference scores into two groups of five each. The test statistic $\Sigma(T^2/n)$, where T is the total of the difference scores for a treatment, can be used to test H_0 for a two-tailed test. For the obtained results, the test statistic value is $8^2/5 + 17^2/5 = 70.6$. The proportion of the data permutations with as large a value of the test statistic as 70.6 is the P-value associated with the interaction. The P-value can be determined by using a one-way ANOVA program, as shown in Chapter 4, with the 10 difference scores constituting the data to be analyzed. Rejection of H_0 implies acceptance of the following complementary hypothesis: for at least one subject, the difference score would have been different for the alternative level of B. For a one-tailed test where the difference scores for level B_2 were predicted to be larger, the P-value would be half of that for the two-tailed test because of the equality of number of difference scores in the two levels of B. For situations with unequal sample sizes for B_1 and B_2, the test statistic T_L (that is, the total of B_2) could be used with the P-value determined by one of the independent t test programs in Chapter 4.

Example 6.6 concerns an interaction for a two-factor design with two levels of each factor, but the one-way ANOVA program in Chapter 4 also could be

TABLE 6.10

Data for Test of Interaction

		A_1	A_2	$A_1 - A_2$
	S_1	8	5	3
	S_2	7	4	3
B_1	S_3	9	9	0
	S_4	5	6	−1
	S_5	10	7	3
	S_6	16	10	6
	S_7	13	9	4
B_2	S_8	12	14	−2
	S_9	19	15	4
	S_{10}	22	17	5

applied to the difference scores if there were more than two levels of the factor with the levels to which the subjects were randomly assigned. Randomization tests for repeated-measures interactions in factorial designs can also be performed when there are more than two treatments per subject. For example, with three treatments per subject the interaction null hypothesis tested could be that the successive difference scores — $(A_2 - A_1)$, $(A_3 - A_2)$, and so on — for any subject are the same as they would have been under assignment of that subject to any level of B. However, none of the programs in this book will test that null hypothesis.

6.19 Randomized Block Designs

Instead of having one subject for all k levels of the independent variable, sometimes k different but similar (e.g., matched) subjects are used. Such a set of k subjects is called a *block* and the random assignment can proceed in the same way as with repeated-measures designs. In the case of a randomized complete blocks design, each of n blocks takes all of k treatments. For each of the n blocks, there are k subjects who must be randomly assigned to the k treatments, and the random assignment is performed independently for each block. There are thus $(k!)^n$ possible assignments because the $k!$ possible assignments for each block can be associated with that many different assignments for every other block.

The variables that determine the similarity of the subjects are called *stratifying* or *blocking variables*. Typical blocking variables include age and sex but usually other subject variables are also involved to make the subjects in a block as homogeneous as possible. It is important to notice that subjects cannot be assigned randomly to different levels of a subject variable. However, stratifying (blocking) subjects according to a subject variable can be useful in increasing the probability of detecting an effect of the manipulated variable.

The purpose of the blocking of subjects is similar to that of repeated-measures designs: to control for between-subject variability and thereby provide a more sensitive test of treatment effects. In fact, repeated-measures designs are sometimes regarded as randomized block designs in which each subject acts as a complete block.

6.20 Randomized Complete Blocks

A *randomized complete block* design is one that is "complete" in the sense that within each block there is an experimental unit for every treatment (Margolin, 1982). Following the discussion of randomized complete blocks, attention will be given to incomplete block designs in Section 6.21, which are ones where any or all of the blocks have fewer subjects (or other experimental units) than the number of treatments, resulting in empty cells. Here we will follow the common conception of a complete block design as one in which for every block, there is one experimental unit for each treatment, even though some writers would include designs with more than one experimental unit per treatment in a block.

A common form of randomized complete block design, when the experimental units are subjects, is known as the *matched-subjects* design and can be analyzed by use of Program 6.1 to Program 6.4. In this design, there is random assignment of k subjects to k treatments, followed by a random assignment of other k subjects to the k treatments, and so on, as explained in Section 6.19. It is analogous to the repeated-measures design, where there is random assignment of k treatment times to k treatments for the first subject, followed by an independent random assignment of k treatment times to the k treatments for the second subject, and so on. The null hypothesis tested is this: the measurement for each subject (or other experimental unit) is independent of the assignment to treatments.

It is called a matched-subjects design because each group of k subjects is composed of subjects matched to be as similar as possible with regard to characteristics thought to be relevant to the treatment effect. Instead of matching a subject-plus-treatment time with the experimental units consisting of the same subject plus other treatment times — as is the case for a repeated-measures design — the k subjects in a group may nevertheless be similar genetically. For instance, we may use pairs of identical twins in a matched-subjects design, or it is common to divide k animals from the same litter randomly among k treatments, even though the animals are not identical genetically.

Because each subject takes only one treatment in a matched-subjects experiment, there is no problem with loss of power due to carryover effects and interpretations of significant results are those associated with independent t or one-way ANOVA randomization tests. Furthermore, matched-subjects designs can be employed when a repeated-measures design is infeasible or even impossible. For instance, a swimming instructor would not use a repeated-measures design to compare various methods of teaching the back-stroke; once it has been learned, it cannot be learned again. Some treatments have irreversible effects, making the random determination of order impossible. In a study of the performance of an animal before and after a brain lesion, random assignment of treatment times to "no lesion" and "lesion" conditions is impossible, and so there could be no statistical determination of treatment effect with a repeated-measures design. For these and similar problems, matched-subjects designs can be effective solutions.

6.21 Incomplete Blocks

When any or all of the blocks have fewer subjects (or other experimental units) than the number of treatments, the design is an *incomplete block design*. An experimenter could use identical twins as complete blocks for comparing two treatments but for the comparison of three treatments, the blocks would be incomplete because only two of the three treatments within a block would have an experimental subject. Identical triplets would provide complete blocks but the subject pool would be restrictive. Similarly, to have complete blocks, an experimenter wanting to block by litters would be restricted to litters as large as the number of treatments.

To compare all treatments with an incomplete block design, the assignment possibilities vary from block to block — for example, allowing in the case of blocks of twins for experiments with three treatments some blocks taking treatments A and B, others A and C, and still others B and C. There is more than one possible random assignment and the randomization test would be conducted accordingly. One possibility would be to fix the two treatments for each block of twins and then randomly assign within those treatments, whereas another possibility would be to randomly assign equal numbers of blocks of twins to the three possible treatment combinations and then randomly assign twins within blocks to the two treatments. Special programs must be written for these cases.

Incomplete blocks may be desirable even when complete blocks would be possible. Margolin (1982) referred to the testing and rating of wine as an example, pointing out that expert judges may develop "palate paralysis" if too many wines are judged in a short period of time, making it desirable to limit the number of wines tasted by a judge. (Presumably, this problem occurs even when the mouth is rinsed after each tasting to reduce carryover effects.) This is the repeated-measures counterpart of the incomplete blocks cases in the preceding paragraph and could employ the same programs.

6.22 Treatments-by-Subjects Designs

We will now consider a randomized block design with more subjects per block than the number of treatments. Program 5.1 from the previous chapter can be used for tests of treatment effects.

Example 6.7. An experimenter uses six males and six females to compare the effectiveness of three treatments, A, B, and C. Two males are randomly assigned to each treatment and then two females are randomly assigned to each treatment. It is a randomized block experiment because within both the male and female blocks, there is random assignment to treatments, but inasmuch as the subjects are not randomly assigned to the subject variable (sex), it is not a completely randomized experiment. Table 6.11 shows the results of the experiment.

First, the experimenter tests this H_0: the measurement of each subject is independent of the level of treatment. A test is carried out to test the main effect of the treatment variable. $\Sigma(T^2/n)$ is a test statistic that can be used for this purpose. The value of $\Sigma(T^2/n)$ for the obtained results shown in

TABLE 6.11

Treatments-by-Subjects Design

	A	B	C	
Males	4	6	8	
	6	8	10	$\bar{X} = 7$
	$\bar{X} = 5$	$\bar{X} = 7$	$\bar{X} = 9$	
Females	6	8	17	
	8	10	17	$\bar{X} = 11$
	$\bar{X} = 7$	$\bar{X} = 9$	$\bar{X} = 17$	
	$\bar{X} = 6$	$\bar{X} = 8$	$\bar{X} = 13$	

Table 6.11 is $(24)^2/4 + (32)^2/4 + (52)^2/4 = 1{,}076$. For a systematic randomization test, $\Sigma(T^2/n)$ is computed for each of the $6!/2!2!2! \times 6!/2!2!2! = 8100$ data permutations, in which the six measurements within each level of the sex variable are divided among the three treatments. The proportion of the 8100 data permutations providing as large a value of $\Sigma(T^2/n)$ as 1076 is the P-value for the main effect of the treatment variable. Program 5.1 can be used to perform this test by use of random data permutation. Rejection of H_0 implies acceptance of the following complementary hypothesis: some of the subjects would have given different measurements under alternative treatments.

The main effect of treatments on the males or females alone can be tested by extracting either the top or the bottom half of the table and computing $\Sigma(T^2/n)$ for that portion of the table for all $6!/2!2!2! = 90$ data permutations. The null hypothesis is that the measurement for each of the six persons of that sex is independent of the level of treatment. The randomization test P-value can be determined by using a program from Chapter 4 because $\Sigma(T^2/n)$ for the males or females alone is $\Sigma(T^2/n)$ for one-way ANOVA, and the data permutation procedure also is that of one-way ANOVA. Rejection of H_0 implies acceptance of the following complementary hypothesis: the measurements of some of the six persons of that sex would have been different under alternative treatments.

If desired, two- or one-tailed tests could be conducted on pairs of treatments by extracting the two relevant columns and using $\Sigma(T^2/n)$ as the test statistic for a two-tailed test or T_L (the total of the treatment predicted to provide the larger mean) as the test statistic for a one-tailed test. For a systematic test, the data would be permuted in $4!/2!2! \times 4!/2!2! = 36$ ways, dividing the data repeatedly within male and female levels. A systematic program would be appropriate for this test but the random permuting program, Program 5.1, is quite valid for this application.

Example 6.8. In Example 5.3, we considered an experiment for determining the effect of a drug with the nurse as a factor, the levels being Nurse A and Nurse B, to have a test of drug effects and nurse effects. Suppose the assignment was conducted as in Example 2.3, with 20 patients associated systematically with either Nurse A or Nurse B, with 10 patients per nurse, and that the *patient + nurse* experimental units were assigned randomly to drug or placebo conditions with five of each nurse's patients being assigned to the drug condition and five to the placebo condition. Although planned as a single-factor study, the study provides results that could be represented in a two-factor layout with four cells for combinations of the two treatment conditions and the two nurses. Nominally, the two nurses are levels of the nurse factor but the measurements in the two levels can differ systematically because of the patients associated with the nurses. Thus, no test of the nurse effect is possible with this design. If it is reasonable to expect the measurements associated with the patients injected by one nurse to differ systematically from those associated with the patients injected by the other nurse, then dividing the measurements within two levels of the nurse factor can provide a more sensitive test than a single-factor test. The cell frequencies are shown in Table 6.12. For a randomization test, the 10 measurements for Nurse A would be divided between Drug and Placebo columns in $10!/5!5!$ = 252 ways, in conjunction with the division of the 10 measurements for Nurse B between Drug and Placebo columns in $10!/5!5!$ = 252 ways for generating the reference set of data permutations. The null hypothesis tested in the randomization test just described would be that each of the 20 patients would have provided the same measurement if the same nurse had provided the alternative injection.

Example 6.9. A three-factor experiment in which two factors are manipulated and one factor is a subject variable is conducted. Factor A has levels A_1 and A_2 and factor B has levels B_1 and B_2. The third factor is a blocking variable (sex) with two levels, Male and Female. Each combination of levels of the A and B factors represents a distinctive treatment. Two males are assigned randomly to each of the four distinctive treatments and two females are assigned randomly to each of the treatments. Thus, there are $8!/2!2!2!2! \times 8!/2!2!2!2!$ = 6,350,400 possible assignments. The results of the experiment are shown in Table 6.13.

A test of the main effect of factor A over both levels of factor B and both levels of the subject variable would be conducted in the following way. For a one-tailed test, the test statistic T_L is computed for the obtained results. If the first level of factor A had been expected to provide the larger measurements, the value of T_L would be 68. For a two-tailed test, we compute $\Sigma(T^2/n)$ for the main effect of factor A for the obtained results, which is $(68)^2/8 + (88)^2/8 = 1546$. To determine the P-value by data permutation, the data are permuted according to the following H_0: the measurement for each subject is independent of assignment to level of factor A. Inasmuch as H_0 does not

TABLE 6.12

Distribution of Cell Frequencies

	Drug	Placebo
Nurse A	5	5
Nurse B	5	5

TABLE 6.13

Three-Factor Experimental Design

	A_1		A_2	
	Males	Females	Males	Females
B_1	4	7	8	10
	7	8	9	11
	$\bar{X} = 5.5$	$\bar{X} = 7.5$	$\bar{X} = 8.5$	$\bar{X} = 10.5$
B_2	9	10	11	12
	12	11	13	14
	$\bar{X} = 10.5$	$\bar{X} = 10.5$	$\bar{X} = 12$	$\bar{X} = 13$
		$\bar{X} = 8.5$		$\bar{X} = 11$

refer to the effect of factor B or the effect of factor C, data must be divided between cells that are at the same level of variable B and at the same level of C, the subject variable. (Of course, we could not permute data over levels of C even to test its "effect" because of the lack of random assignment to levels of that variable.) Designating the top row of cells in Table 6.13 as cell 1 to cell 4 and the bottom row as cell 5 to cell 8, we use the data permutations consisting of all possible combinations of divisions of data within the following pairs of cells: 1 and 3, 2 and 4, 5 and 7, 6 and 8. Because there are $4!/2!2! = 6$ data divisions for each of these pairs of cells and because every data division for a pair of cells can be associated with every data division for each of the other pairs of cells, there are $(4!/2!2!)^4 = 1296$ data permutations. The proportion of data permutations with as large a value of T_L as 68 or as large a value of $\Sigma(T^2/n)$ as 1546, as determined by Program 5.1, is the P-value for the test of the main effect of A.

To test the effect of factor A within level B_1, we extract the first row of data from Table 6.13 and compute $\Sigma(T^2/n)$ for the obtained results as $(26)^2/4 + (38)^2/4 = 530$. T_L for the obtained results would be 26 for a prediction that the first level of A would provide the larger measurements. The eight measurements within level B_1 are permuted in $4!/2!2! \times 4!/2!2! = 36$ ways, consisting of every division of the four measurements for males between the cells for males in conjunction with every division of the four measurements for females between the cells for females. The P-value can be determined by use of Program 5.1.

6.23 Disproportional Cell Frequencies

In Chapter 5, it was stated that cell frequencies are completely proportional in a two-factor experiment if the sample size ratio of any two cells in one row is the same as the sample size ratio for the corresponding cells in any other row. When the sample size ratio varies over the rows for corresponding cells, there are said to be *disproportional* cell frequencies. Table 6.14 is an

TABLE 6.14

Disproportional Cell Frequencies

	a_1	a_2	a_3
b_1	3	4	5
b_2	6	8	10
b_3	9	12	13

example of disproportional cell frequencies of the type that could result from the loss of two subjects from the lower right cell in an experiment that started with proportional frequencies. For analyses involving only the a_1 and a_2 levels of factor A, the cell frequencies are proportional and the same is true for analysis of only the b_1 and b_2 levels of factor B. Consequently, such analyses could be run according to the procedures considered previously, which assume proportional cell frequencies. However, any analysis that involves the lower right cell should not employ the standard procedure but either the procedure that will be described in Section 6.24 or the one in Section 6.25.

What will be proposed is the use of an alternative test statistic for designs with disproportional cell frequencies. The test statistic $\Sigma(T^2/n)$ is valid for disproportional as well as proportional cell frequencies when the P-value is based on data permutation, but the test statistic to be developed here is more powerful when cell frequencies are disproportional.

Before considering the new test statistic, let us see why one might have a factorial design with disproportional cell frequencies. When there are several separate single-factor experiments run at different times to provide data for a treatments-by-replications design, one in which there is blocking by experiment, the cell frequencies might be disproportional for a number of reasons. Each experiment might employ whatever subjects are available at the time, and the number of subjects may not be an exact multiple of the number used in another experiment, making it impossible to maintain proportionality of cell sizes. In some cases, the experimenter may deliberately introduce disproportionality. For example, if the variation within one level of a factor is so great that the experimenter feels there should be more subjects at that level to detect a treatment effect, in later experiments a larger proportion of subjects might be assigned to that level.

Factorial designs based on single experiments can also involve disproportionality of cell size. Obviously, the number of available subjects could at times prevent an experimenter from using all subjects if equal sample sizes for all cells was the experimenter's aim. Furthermore, the number of available subjects might not fit into some other proportional cell size design. Also, in single experiments as well as in a series of experiments, experiments designed with proportional cell sizes can become experiments with disproportional cell sizes because of subject mortality or failure of some assigned subjects to participate in the experiment.

6.24 Test Statistic for Disproportional Cell Frequencies

The following example will show why a special test statistic is needed when cell frequencies are disproportional and will describe a test statistic for disproportional cell frequency designs.

Example 6.10. In a treatments-by-subjects experiment, two males are assigned to treatment A and four to treatment B, and four females are assigned to treatment A and two to treatment B. (Although it represents a possible experiment, this example is not intended to represent a likely design; it was devised to make the need for a special test statistic clear.) The experimental results in Table 6.15 are obtained. For both males and females, the larger scores are associated with the B treatment, yet the A and B means are equal. As the means are equal, $\Sigma(T^2/n)$ has as small a value as could be obtained from any of the data permutations, and so the P-value would be the largest possible, namely 1.

Alternatively, an overall one-tailed test using the sum of the B measurements as the test statistic would give a P-value of $1/225$, or about 0.004, because no other data permutation out of the $(6!/2!4!)^2$ data permutations would give such a large sum of B as 46, the obtained value. The sum of the B measurements is an equivalent test statistic, under data permutation, to the arithmetic difference $(\bar{B} - \bar{A})$, and the value of this test statistic is a maximum for the obtained results, although its value is only 0; all other data permutations would have a negative value for the difference between means.

The one-tailed test then leads to a reasonable P-value in Example 6.10 but the nondirectional test does not. Of course, if there were more than two treatment levels some type of nondirectional test statistic related to a difference between means would be used. Yet as we have seen, such a test might lead to nonsignificant results even when within each type of subject there was an indication of an effect with the same direction of difference between means.

A solution to the problem lies in the development of a more appropriate nondirectional test statistic than $\Sigma(T^2/n)$. The new test statistic is based on a redefinition of SS_B that is mathematically equivalent to the conventional definition when the cell sizes are proportional but which differs when there are disproportional cell sizes: $SS_B = n_A [\bar{X}_A - E(\bar{X}_A)]^2 + n_B [\bar{X}_B - E(\bar{X}_B)]^2 + \ldots,$

TABLE 6.15

Treatments-by-Subjects Data

	A	B	
Males	3	5 6	$\bar{X} = 4.83$
	4	5 6	
Females	8 10	12	$\bar{X} = 10.5$
	10 11	12	
$\bar{X} =$	7 2/3	7 2/3	
	7.67	7.67	

where $E(\bar{X}_A)$ and $E(\bar{X}_B)$ are "expected means," the mean value of those means over all data permutations. When cell sizes are proportional, the expected means are the same for all treatments, each expected mean having the value of the grand mean of all measurements. Thus, the redefined SS_B is a generalization of the conventional SS_B statistic resulting from substituting expected means for the grand mean in the standard formula, a substitution that has no effect when there are proportional cell sizes but that may have considerable effect when the cell sizes are disproportional.

Example 6.11. Let us compute the new test statistic for the data in Example 6.10. $\bar{X}$ for males is 4.83 and $\bar{X}$ for females is 10.5. $E(\Sigma X_A) = 2(4\ 5/6)$ $+ 4(10\ 1/2) = 51.67$, and $E(\Sigma X_B) = 4(4\ 5/6) + 2(10\ 1/2) = 40.33$. $E(\bar{X}_A)$ is then $(51\ 2/3)/6$, or 8.61, and $E(\bar{X}_B)$ is $(40\ 1/3)/6$, or 6.72. We now use these expected means to compute our redefined SS_B test statistic. The obtained test statistic value is computed as $6(7\ 2/3 - 8\ 11/18)^2 + 6(7\ 2/3 - 6\ 13/18)^2$, or about 10.7. Of course, $E(\bar{X}_A)$ and $E(\bar{X}_B)$ are the same for all data permutations but $\bar{X}_A$ and $\bar{X}_B$ vary. However, over all data permutations only the obtained data configuration gives as large a test statistic value as 10.7, and so the P-value of the treatment effect by data permutation, using the new nondirectional test statistic, is $1/225$, or about 0.004.

As stated above, the application of Program 5.1 to data in a design with disproportional cell frequencies is valid but not as sensitive as it would be if the special test statistic described above were computed for each data permutation. If the experimentally obtained data were entered into the computer to determine the P-value with the modified SS_B test statistic, it would be necessary to modify Program 5.1 by changing the test statistic computed for main effects. However, by transforming the data prior to entry into the program it is possible to use Program 5.1 in its present form to determine the P-value based on the modified SS_B test statistic. An appropriate data adjustment is described in the next section.

6.25 Data Adjustment for Disproportional Cell Frequency Designs

We have considered an adjustment in the computation of a test statistic to make a randomization test more sensitive to treatment effects when cell frequencies are disproportional. The superiority of the modified SS_B test statistic over $\Sigma(T^2/n)$, the nondirectional test statistic in Program 5.1, has been shown for a particular situation (Example 6.10), and the rationale given in Section 6.24 suggests that the superiority holds for designs with disproportional cell frequencies in general. However, the rationale concerned an analysis of experimental data that cannot be performed by Program 5.1. But an easy adjustment of data prior to entry into the computer will permit use of the modified SS_B test statistic by Program 5.1 without altering the program.

Example 6.12. The treatments-by-subjects experiment in Example 6.10 gave paradoxical results for a nondirectional test because of the difference in the proportional distribution of males and females between the two treatments plus systematically larger measurements for females under each treatment. If the male and female measurements could be adjusted to make them equal on the average, such an adjustment should serve the same function as adjusting the test statistic.

One way to equate the measurements of males and females is to express all measurements as deviations from the mean for that sex. Thus, because the mean for males was 4 5/6 and the mean for females was 10 1/2, the data in Example 6.10 would be adjusted by subtracting 4 5/6 from each measurement for males and subtracting 10 1/2 from each measurement for females. That adjustment would result in the array of "residuals" shown in Table 6.16. SS_B for the above results, computed in the usual way where deviations are taken from the grand mean, is $6(17/18 - 0)^2 + 6(+17/18 - 0)^2$, exactly the same value as the adjusted SS_B in Section 6.24, the value of which rounds off to 10.7. This numerical identity is not an accident; the two procedures are mathematically equivalent ways of determining the same value of SS_B. Now $\Sigma(T^2/n)$ is an equivalent test statistic to SS_B, and so $\Sigma(T^2/n)$ computed for all permutations of the residuals (adjusted data) will give the same P-value as using SS_B for residuals.

The above procedure of computing the residuals once and then permuting the residuals provides the same distribution of test statistics as the procedure in Section 6.24, which consists in permuting the raw data and computing the residuals for each permutation of the data. The reason is that the residual associated with each measurement is not affected by permuting within rows.

Therefore, Program 5.1 will give the same P-value for a nondirectional test as is given by the use of modified SS_B when the data are adjusted in the above manner before entering the data into the program. Thus, Program 5.1 can serve as a sensitive test of treatment effects for either proportional or disproportional cell frequency designs. Although the discussion of the need to adjust for disproportionality of cell frequencies has been in terms of treatments-by-subjects designs, the same considerations apply to other randomized block designs and to completely randomized designs. Data adjustment of the above type can be helpful for any disproportional cell frequency design. However, Program 5.1 cannot be applied to incomplete block designs

TABLE 6.16

Adjusted Data

	A		B	
Males	−1 5/6		+1/6	+1 1/6
	−5/6		+1/6	+1 1/6
Females	−2 1/2	−1/2	+1 1/2	
	−1/2	+1/2	+1 1/2	
$\bar{X} =$	−17/18		+17/18	

because such designs require a different data permuting procedure from that incorporated into the program.

6.26 Restricted-Alternatives Random Assignment

In Section 6.21, the incomplete blocks were of the conventional type, namely blocks with empty cells, because of the number of treatments being too large for the number of experimental units in a block. Here we will consider another type of incomplete blocks design necessitated not simply by the surplus of treatments but by the inappropriateness of particular treatments within a block for the subject (or subjects) in the block. To block the subjects, the experimenter must determine for each subject what treatment or treatments (if any) are inappropriate for that subject.

There are various reasons why subjects might be assigned to some treatments but not others. In an experiment comparing the effects of tobacco, alcohol, and marijuana, some persons may be willing to submit to two of the treatments but not a third, and which two they would be willing to take could vary from subject to subject. For some experiments, a subject's religious beliefs or moral or ethical attitudes may make certain treatments unacceptable and others acceptable. Just as it is unethical to apply certain treatments to human beings, so also it may be unethical to apply treatments to children that would be satisfactory for adults. A block of adults might be randomly assigned to any one of the treatments, whereas a block of children might be randomly assigned only to a subset of the treatments. Similarly, medical reasons justify assigning a certain subject to some treatments but make other assignments inappropriate. If we restrict our experimental designs to the conventional designs in which all subjects could be assigned to any of the treatments — or, in the case of repeated-measures experiments, designs in which each of the subjects takes all of the treatments — we are restricting our potential subject pool. This might not matter for experiments where it is easy to get enough subjects but it does matter when subjects are difficult to obtain.

Example 6.13. We have an independent-groups experiment with three treatments — tobacco, alcohol, and marijuana — where reaction time is the dependent variable. Three subjects are willing to be assigned to any of the three conditions (T, A, or M), six to be assigned to tobacco or alcohol (T or A), four to tobacco or marijuana (T or M), and two to alcohol or marijuana (A or M). The random assignment was carried out for each of the four groups with the sample size constraints shown in Table 6.17. The subjects are blocked according to their treatment alternatives. Within each subject block, the subjects are assigned randomly to treatments with the sample size and treatment constraints shown. The disproportionality of cell sizes (as is always the case for incomplete block designs) implies that we should use the new SS_B test

TABLE 6.17

Restricted-Alternatives Assignment

	T	A	M
T, A, or M	1	1	1
T or A	3	3	—
T or M	2	—	2
A or M	—	1	1

statistic to test the difference between treatments. The alternative treatment assignments are a function of the individual subjects. If older persons (who might be expected to have longer reaction times) were less willing than younger persons to have marijuana as a possible assignment, younger persons would be more likely to be assigned to marijuana; therefore, if the treatments had identical effects, one would expect the marijuana reaction times to be lower. So as in Section 6.24, equality of column means might well be indicative of a treatment effect.

We will explain the application of the data adjustment procedure described in Section 6.25. The measurements for the 15 subjects are transformed into "residuals" by subtracting the subject's block mean of the measurements from the subject's measurement. Those residuals or deviations from block means then constitute the data to be analyzed by a randomization test. For a test of the difference among all three treatments, the test statistic could be $\Sigma(T^2/n)$. A program would compute $\Sigma(T^2/n)$ for all $3!/1!1!1! \times 6!/3!3! \times 4!/2!2! \times 2!/1!1! = 1440$ permutations of the measurements within blocks to derive the reference set for determining the P-value. Although Program 5.1 computes $\Sigma(T^2/n)$ as a test statistic, it does not permute data for incomplete block designs, where there are empty cells.

For comparisons of only two treatments (tobacco and alcohol, for example), only the subjects who took one of those two treatments and who could have been assigned to the other treatment are considered. For instance, for a comparison of tobacco and alcohol, only the top two blocks are relevant because those are the only blocks containing subjects who could have been assigned either to tobacco or alcohol. The test concerns tobacco and alcohol measurements for those two blocks alone. As there is proportionality for this part of the table, there is no need to compute residuals to compute the special test statistic. For a two-tailed test, we compute $\Sigma(T^2/n)$ for the division of data between T and A for the obtained results and determine the P-value of the result by comparison with a reference set generated by computing the test statistic for all $2!/1!1! \times 6!/3!3! = 40$ permutations of the measurements of the eight subjects in block 1 and block 2 with measurements for tobacco or alcohol, involving every permutation of measurements within block 1 in conjunction with every permutation of measurements within block 2. For a directional test, the total of the treatment expected to provide the larger values, T_L, could be the test statistic to compute over the 40 data permutations. The number of data permutations is obviously small enough for a systematic program to be used but Program 5.1, which is a random data permutation program, can be employed validly if a systematic program is unavailable. Example 6.13 dealt with the analysis of data from an independent-groups design involving restricted-alternatives random assignment. A repeated-measures design could be analyzed in a similar way as follows.

Example 6.14. Three subjects are willing to be assigned to tobacco, alcohol, and marijuana in the following manner: six to tobacco and alcohol, four to tobacco and marijuana, and two to alcohol and marijuana. The experimenter uses a repeated-measures design in which each subject takes all of the treatments he would be willing to take. Table 6.18 shows the treatments the individual subjects would take. For each subject, there would be random assignment of treatment times to the treatments to be taken, so there are 15 blocks within which data are to be permuted for an overall test of differential effects of all three treatments. Of course, the disproportionality of cell frequencies of the 45 cells could again favor one treatment over another because of the type of subject most likely to take it.

For the overall test of a difference among T, A, and M, the data for each subject are converted into residuals by subtracting the mean of the subject's measurements from each measurement. $\Sigma(T^2/n)$ would be computed for all data permutations to determine the proportion that give a test statistic value as large as the obtained value. There would be six possible assignments for each subject taking three treatments and two possible assignments for each subject taking two treatments; so the number of distinctive assignments for the experiment as a whole would be $6^3 \times 2^{12} = 884,736$. If only subjects willing to take all three treatments had been used, there would have been only the first three subjects, providing only 216 possible assignments and, therefore, 216 data permutations instead of 884,736, making it a much less powerful experiment. For comparing only two of the treatments, the data permuted

TABLE 6.18

Repeated-Measures Restricted-Alternatives Assignment

	T	A	M
T, A, and M	S_1	S_1	S_1
	S_2	S_2	S_2
	S_3	S_3	S_3
T and A	S_4	S_4	—
	S_5	S_5	—
	S_6	S_6	—
	S_7	S_7	—
	S_8	S_8	—
	S_9	S_9	—
T and M	S_{10}	—	S_{10}
	S_{11}	—	S_{11}
	S_{12}	—	S_{12}
	S_{13}	—	S_{13}
A and M	—	S_{14}	S_{14}
	—	S_{15}	S_{15}

would be for each subject taking both treatments and just data obtained by those subjects for the two treatments. For example, for a comparison of T and A, there would be 2^9 data permutations because for each of the nine subjects taking both T and A, there are two possible assignments of those treatments to the two treatment times actually used for those treatments by the subject. There is proportionality for this part of the table, so the modified SS_B test statistic is not required. Program 5.1 and Program 5.3 for repeated-measures ANOVA can be applied to determine the proportion of the 2^9 data permutations that provide a one-tailed or two-tailed test statistic value as large as the obtained value.

Both independent-groups and repeated-measures experiments have been discussed as factorial experiments where the subjects are stratified according to treatment preferences. However, there could be restricted-alternatives random assignment where the experimenter decides to restrict the alternatives rather than to leave the choice up to the subject. Of course, the analysis would be the same because it is the restriction on the alternatives for a subject that affects the analysis, not the reason for the restriction. In a treatments-by-subjects design where an experimenter used children, young adults, and older adults as levels of the subject variable, the experimenter might use levels of treatments for young and older adults that would be dangerous or otherwise inappropriate for children, and so the assignments of children would involve greater restriction than the assignments for the other subjects. Another reason for having different restrictions on the random assignment for some levels than for other levels could be the experimenter's greater interest in certain cells (treatment combinations). For instance, in a two-factor design for examining the effects of intensity of sound and intensity of light on work performance, an experimenter might decide that a high intensity of sound combined with a high intensity of light — although not harmful — would be so uncomfortable as to distract subjects. Because it is not the distracting effect in which he is interested, the experimenter decides to use all three levels of intensity for both sound and light but to assign no subjects to the ninth cell; the analyses use only eight of the cells in the 3×3 matrix, the ninth cell being empty.

6.27 Combining P-Values

Example 6.10, which concerned blocking by sex, was presented as if the males and females were participating in a single experiment, but in fact the randomization test as conducted would have been perfectly valid if the males and females were given treatments at different times and places and by different experimenters. It is not surprising that the randomization tests for randomized blocks can be applied when the blocks represent subjects in different experiments; after all, the experimental randomization for any randomized block experiment is precisely the same as in a series of experiments involving the same treatments. (In fact, restricted-alternatives random assignment, like other incomplete block assignments, corresponds to the random assignment for a series of experiments that vary somewhat with respect to the treatments administered.)

An alternative to a randomization test of main effects, which is based on data pooled over blocks, is to combine P-values. In Example 6.10, a procedure that would have been completely valid in the absence of random selection of subjects and which is therefore distribution-free would have been the following: carry out separate randomization tests for the males and females and combine the two P-values by methods to be described in this and subsequent sections.

One of the principal uses of combining P-values is to pool P-values from independent experiments. P-value combining can be employed when the results of different statistical tests are independent. The required independence is this: when H_0 is true for each of the tests for which the P-values are combined, the probability of getting a P-value as small as p for one of the tests must be no greater than p, no matter what P-value is associated with any of the other tests. A valid test of main effects can be carried out by the randomized block tests considered in this chapter or by combining P-values. But why use P-value combining when the randomized block tests can be applied? There are several special situations where the combining of P-values is the preferred approach. Any randomized block design for which there is no appropriate randomization test provides an opportunity for combining P-values. For example, if there were no procedure for the adjustment of test statistics for incomplete block designs or any disproportional cell frequency designs, combining P-values would have been useful. For the experiment in Example 6.10, combining a two-tailed P-value for males with a two-tailed P-value for females would take care of the disproportionality of cell frequencies. Another situation where combining of P-values is favored is when the experiments whose results are to be combined employ different dependent variables. To make a comparison of means rational, it might be necessary to transform the measurements on the different dependent variables by using ranks, z scores, or some other standardizing procedure, whereas it would be simple to combine P-values. Similarly, for some experiments the data may be categorical, for others ranks, and for others continuous data, a circumstance working against pooling data over the separate experiments. Alternatively, there are numerous situations when the data to be pooled are comparable from one block (experiment) to the next, situations favoring randomized block tests over P-value combining. The method of P-value combining to be described in Section 6.28 uses the sum of the P-values to determine the overall P-value of the results of the statistical tests. The general H_0 tested by combining P-values is this: there is no treatment effect for any of the tests that provided P-values to be combined.

One type of factorial design where P-values can be combined is that where the probabilities come from a series of experiments. The separate experiments could be a series of pilot studies and the main experiment; pilot studies are used to determine how to conduct the main experiment. Not only the general conduct of the experiment but also the independent and dependent variables can be based on preliminary pilot work. Observations in a pilot experiment can suggest changes likely to make an experiment more sensitive even when the pilot experiment is not sensitive enough to give significant results. P-value combining permits the experimenter to use pilot studies not only for designing a final experiment but also to provide P-values that can be combined with the P-value from the main experiment.

Example 6.15. A veterinarian at a zoo wants to determine whether certain food being fed to the chimpanzees is making them irritable, so a randomized experiment is devised to compare animals given the food and animals from which the food is withheld. The veterinarian decides to run a pilot study to help in designing a larger experiment and to combine the P-values from the pilot and main experiments. During the course of the pilot study, ways to reduce distractions that adversely affect the eating behavior of the chimpanzees are discovered and other ways of improving experimental control over relevant variables are noted, including systematically associating certain variable values with the individual chimpanzees to form a complex experimental unit that is randomly assigned to a treatment condition. A few months later, a larger group of animals is used in a follow-up experiment incorporating improvements based on knowledge gained from the pilot study. The P-value from the pilot study and the P-value from that experiment are combined to test the overall null hypothesis of no differential effect of the treatment conditions on any of the chimpanzees in either experiment.

A different area of application is the combining of P-values from the same subjects in different experiments. Use of the same subjects is not always appropriate but with independent random assignment, valid and independent statistical tests can be performed on data from the same subjects in different experiments. Randomization test P-values based on the same subjects in different experiments are statistically independent if there is independent random assignment to treatments in the different experiments. Participation in an earlier experiment may very well affect performance in a later experiment, in which case the responses in the second experiment would not be independent of the responses in the first experiment; but it is the independence of the P-values when H_0 is true that is of importance. If the first experiment made the subjects responsive to the difference between treatments in the second experiment, it could affect the P-values in the second experiment. However, this effect would not invalidate P-value combining because in such a case, H_0 for the second experiment would be false and so the probability of rejecting a true H_0 at any level α still would be no greater than α. Separate random assignment ensures the independence of the P-values if the generic null hypothesis of no treatment effect in either (any) of the experiments is true. However, it should be noted that the experience of subjects in a prior experiment of a similar type could reduce the sensitivity of a later experiment.

6.28 Additive Method of Combining P-Values

There are two main procedures for combining P-values, which can be designated as the multiplicative and the additive methods. The *multiplicative* method determines the overall P-value of the product of the separate P-values to be combined, whereas the *additive* method determines the overall P-value of the sum of the P-values. In explaining his multiplicative method, Fisher (1948) stated that "if combination by the product method is intended, it will be worth while to aim at a succession of trials of approximately equal sensitiveness." In fact, to the extent that Fisher's objective of similar P-values is attained, the additive method (Edgington, 1972a) will give a smaller overall combined P-value than the multiplicative method. (The multiplicative method gives smaller combined P-values than the additive method when the individual P-values are quite dissimilar.) Both the additive and multiplicative methods are valid methods for employment with independent P-values but only the additive method will be described below. Both methods are implemented in the COMBINE program on the accompanying CD.

If we take the sum of n independent P-values, then given that H_0 associated with each P-value is true, the probability of getting a sum as small as S is:

$$\frac{C(n,0)(S-0)^n - C(n,1)(S-1)^n + C(n,2)(S-2)^n - \cdots}{n!} \tag{6.4}$$

The minus and plus signs between terms in the numerator alternate, and additional terms are used as long as the number subtracted from S is less than S. The symbol $C(n, r)$ is the same as $n!/r!(n-r)!$ and refers to the number of combinations of n things taken r at a time. Combining as many as seven different probabilities will not be necessary very often, but such an example will be helpful in showing how to use Equation (6.4).

Example 6.16. Suppose that seven experiments provided P-values of 0.20, 0.35, 0.40, 0.40, 0.35, 0.40, and 0.40. The sum is 2.50 and the probability of getting a sum as small as 2.50 from seven independent P-values when each of the seven null hypotheses is true is:

$$\frac{C(7,0)(2.50-0)^7 - C(7,1)(2.50-1)^7 + C(7,2)(2.50-2)^7}{7!} = 0.097$$

This value can easily be verified using the COMBINE program on the accompanying CD.

If the seven values had been somewhat smaller, providing a sum of 1.40, the overall P-value would have been:

$$\frac{C(7,0)(1.40-0)^7 - C(7,1)(1.40-1)^7}{7!} = 0.002$$

Because no single P-value can exceed 1, the maximum possible sum of n P-values is n, and so there never needs to be more than n terms computed. Thus, for three P-values to be summed, no more than three terms would be required, and for four P-values no more than four terms would be required.

<hr />

6.29   Combining One-Tailed and Two-Tailed P-Values

When all of the P-values to be pooled are one-tailed P-values, the matter is simple: add the P-values for the predicted direction of difference; if the obtained results differ in the predicted direction, the P-value may be fairly small, but if the obtained results differ in the opposite direction, the P-value will be large.

When some P-values are one-tailed and some two-tailed, the P-values are added without regard to whether they are one-tailed or two-tailed. A small overall P-value derived from combining a set of P-values from both one-tailed and two-tailed tests is support for the conclusion that at least one of the null hypotheses for the tests providing the individual P-values was false, but the false null hypothesis or hypotheses need not be one-tailed.

When all of the P-values to be combined are two-tailed, there are two alternative procedures that can be followed depending on the predicted effect. First, consider the case where the experimenter has no basis for predicting the direction of difference for any of the experiments and does not even have reason to expect that the unknown direction of difference will be the same over all experiments. In such a case, a simple combination of the two-tailed P-values is appropriate.

In the second situation, suppose that the experimenter has no basis for predicting the direction of difference in any of the experiments but does expect the direction of difference to be the same for all experiments, as the set of experiments are essentially replications of the same experiment. In this case, the experimenter should construct two sets of P-values, one set based on one-tailed P-values for one direction of difference and the other set based on one-tailed P-values for the opposite direction of difference. There are two sums, one from one set and one from the other set. The overall P-value is based on the smaller sum of one-tailed P-values, and the overall P-value for the smaller sum is doubled. The doubled overall P-value is two-tailed, not one-tailed.

Example 6.17. We have three experiments that, although not identical, all involve two treatments, A and B, that are analogous from one experiment to the next. The experimenter expects the treatment that is more effective to be the same one in every experiment but does not know whether the treatment that is more effective would be the A treatment or the B treatment. Based on a prediction that the A treatment measurements would be larger, the one-tailed P-values are 0.10, 0.08, and 0.16; the sum of this set is 0.34.

Of course, the sum of the set of one-tailed P-values based on the opposite prediction is considerably larger. The value $(0.34)^3/6$, which is about 0.0066, is doubled to give 0.013 as the two-tailed P-value of the combined results.

Suppose the two-tailed P-values were twice the one-tailed P-values for the correctly predicted direction. (They need not be for randomization tests, in cases of unequal sample sizes.) If they had been combined without taking into consideration the expectation that the direction of difference would be the same for all three experiments, we would have combined the values 0.20, 0.16, and 0.32 to get a total of 0.68 and would have obtained a P-value of $(0.68)^3/6$, or about 0.052. The difference between 0.052 and 0.013 shows the importance of taking into account a prediction of a consistent direction of difference in combining two-tailed P-values, even when the direction cannot be predicted.

This method of combining two-tailed P-values was recommended by David (1934) and is based on the same principle as the Bonferroni procedure, described in regard to multiple comparisons in Section 4.14 and Section 6.12. The principle is this: when H_0 is true for each of k different tests on the same or different sets of data, the probability of rejecting H_0 at the α level is no greater than $k\alpha$. David's method of combining two-tailed P-values carries out two "tests" on the results from a set of experiments, and the probability of one of the tests rejecting H_0 at the α level is no greater than 2α.

Further discussion of the combining of P-values can be found in reports by Baker (1952), Bancroft (1950), Edgington (1972b), Edgington and Haller (1983; 1984), Gordon, et al. (1952), Guilford and Fruchter (1973), Jones and Fiske (1953), Lancaster (1949), Wallis (1942), and Winer (1971).

6.30 Questions and Exercises

1. (a) What is the smallest number of subjects required to have the possibility of getting a P-value as small as 0.01 for a systematic randomization test when they take three treatments in a repeated-measures design. (b) What would the smallest possible P-value be?

2. Five subjects each take three treatments in a repeated-measures experiment. Would it be possible to have exactly two outliers? Why or why not?

3. Describe a manipulation in the way an experiment is carried out that is designed to minimize carryover effects within individual subjects in an experimental comparison of the effectiveness of drug A and drug B.

4. For between-subjects designs, subjects who drop out can be represented by a marker that has no numerical value, and the marker is permuted over treatments as if it were a measurement. In a repeated-measures experiment, if a subject drops out without taking

any of the treatments, it would be unnecessary to permute markers that represent missing measurements. Why?

5. What adjustment, if any, in the test statistic should be made when a subject in a repeated-measures experiment that has been randomly assigned to treatments does not take any of the treatments?

6. Table 6.16 shows adjusted data for Table 6.15. Construct a table showing adjusted data for the following data:

	A	B
Males	2 3	4
Females	7	5 6

7. Calculate the two-tailed P-value for a randomization test of the difference between A and B for the data in Question 2.

8. Use the additive method for combining P-values from two experiments, one having a P-value of 0.12 and the other a P-value of 0.08.

9. In a between-subjects design, rats are randomly assigned to two treatment conditions to test differences in the amount eaten under the two conditions. Rats with certain abnormalities of the intestinal tract are expected to respond to the treatment differences in a different way from rats without those abnormalities, so a randomized block design could be helpful if those abnormalities could be detected before random assignment. This is not possible, so as an alternative the experimenter does a post-mortem examination of the rats following the experiment and blocks the experimental results on the basis of that examination. Can a randomization test be validly performed on such blocked data? Explain.

10. Under what experimental circumstances would P-value combining over blocks in a randomized block design be preferable to running a single test with combining data over blocks?

REFERENCES

Baker, P.C., Combining tests of significance in cross-validation, *Educ. Psychol. Meas.*, 12, 300, 1952.

Bancroft, T.A., Probability values for the common tests of hypotheses, *J. Am. Statist. Assn.*, 45, 211, 1950.

David, F., On the P_n test for randomness; remarks, further illustrations and table for P_n, *Biometrika*, 26, 1, 1934.

Edgington, E.S., Statistical inference: The distribution-free approach. McGraw-Hill, New York, 1969.

Edgington, E.S., An additive method for combining probability values from independent experiments, *J. Psychol.*, 80, 351, 1972a.

Edgington, E.S., A normal curve method for combining probability values from independent experiments, *J. Psychol.*, 82, 85, 1972b.

Edgington, E.S., *Randomization Tests*, Dekker, New York, 1980.

Edgington, E.S. and Ezinga, G., Randomization tests and outlier scores, *J. Psychol.*, 99, 259, 1978.

Edgington, E.S. and Haller, O., A computer program for combining probabilities, *Educ. Psychol. Meas.*, 43, 835, 1983.

Edgington, E.S. and Haller, O., Combining probabilities from discrete probability distributions, *Educ. Psychol. Meas.*, 44, 265, 1984.

Ezinga, G., Detection and Memory Processes in Picture Recognition, Ph.D. thesis, University of Calgary, Alberta, Canada, 1976.

Fisher, R.A., Combining independent tests of significance, *Am. Statist.*, 30, 1948.

Gordon, M.H., Loveland, E.H., and Cureton, E.E., An extended table of chi-square for two degrees of freedom, for use in combining probabilities from independent samples, *Psychometrika*, 17, 311, 1952.

Guilford, J.P. and Fruchter, B., *Fundamental Statistics in Psychology and Education* (5th ed.), McGraw-Hill, New York, 1973.

Holm, S., A simple sequentially rejective multiple test procedure, *Scand. J. Statist.*, 6, 65, 1979.

Jones, L.V. and Fiske, D.W., Models for testing the significance of combined results, *Psychol. Bull.*, 50, 375, 1953.

Krauth, J., A comparison of tests for marginal homogeneity in square contingency tables, *Biomet. J.*, 27, 3, 1985.

Lam, F.C. and Longnecker, M.T., A modified Wilcoxon rank sum test for paired data, *Biometrika*, 70, 510, 1983.

Lancaster, H.O., The combination of probabilities arising from data in discrete distributions, *Biometrika*, 36, 370, 1949.

Lehmacher, W., Simultaneous sign tests for marginal homogeneity of square contingency tables, *Biomet. J.*, 22, 795, 1980.

Margolin, B.H., Blocks, balanced incomplete, in *Encyclopedia of Statistical Sciences*, Vol. 1, Kotz, S. and Johnson, N.L. (Eds.), John Wiley & Sons, New York, 1982, 284–288.

Siegel, S. and Castellan, N.J., Jr., *Nonparametric Statistics for the Behavioral Sciences* (2nd ed.), McGraw-Hill, New York, 1988.

Wallis, W.A., Compounding probabilities from independent significance tests, *Econometrica*, 10, 229, 1942.

Winer, B.J., *Statistical Principles in Experimental Design* (2nd ed.), McGraw-Hill, New York, 1971.

7

Multivariate Designs

Factorial designs are designs with two or more *independent* variables. Alternatively, multivariate designs are designs with two or more *dependent* variables. Treatment effects in experiments are likely to be manifested in a number of ways, so it frequently is useful to employ more than one dependent variable. For example, an experimenter may expect that under one condition a rat would show greater motivation to solve a task than under another, and that this greater motivation would be reflected in both the speed and the strength of response. If only one of the two responses is measured, the experiment is likely to be less sensitive than if both measures were incorporated into the experiment because one measure might reflect an effect when the other does not. By allowing for the use of two or more dependent variables in a single test, multivariate designs thus can be more sensitive to treatment effects than univariate (single-dependent-variable) designs.

7.1 Importance of Parametric Assumptions Underlying MANOVA

Multivariate analysis of variance (MANOVA) is a common statistical procedure for analyzing data from multivariate experimental designs. There are several conventional test statistics for MANOVA, and these alternative test statistics sometimes provide quite different P-values. Olson (1976) studied the robustness of MANOVA and found that violations of the parametric assumptions have a considerable effect on the probability of incorrectly rejecting the null hypothesis for several of the commonly used test statistics. (This finding supports a widely held view that parametric assumptions are much more important for multivariate than for univariate, or one-variable, ANOVA.) For a critical comment and a rejoinder, see Stevens (1979) and Olson (1979).

Olson constructed three populations of numerical values with identical means on three dependent variables. For one population, the standard deviation for each dependent variable was three times the standard deviation for that variable in the other two populations. A random sample of measurements for five subjects was taken from each population, and each of four MANOVA test statistics was computed with the P-values being determined by parametric statistical tables. The sampling and the P-value determination were done repeatedly, and the proportion of the test statistics of each type significant at the 0.05 level was computed. This proportion ranged from 0.09 to 0.17 for the four test statistics. (Of course, the proportion should be no greater than 0.05 if the method of determining the P-value was valid.) A similar study was performed with six dependent variables where one population had dependent variable standard deviations six times as large as the other two populations. The proportion of the time that a test statistic was judged to be significant at the 0.05 level ranged from 0.09 to 0.62 for the four test statistics. A test statistic known as the Pillai-Bartlett Trace V, along with its conventional procedure for determining the P-value, was the most robust, providing the proportion of 0.09 for both studies. Considered in conjunction with other theoretical and numerical demonstrations of the sensitivity of MANOVA procedures to violations of assumptions, Olson's results tend to make people critical of the validity of MANOVA (e.g., Fouladi and Yockey, 2002). A randomization test determination of the P-value for MANOVA can thus increase the acceptability of results at times.

7.2   Randomization Tests for Conventional MANOVA

Apparently the most common approach in using a multivariate t test or MANOVA is equivalent to combining measurements on different dependent variables into single numbers to which univariate tests are applied. Harris (1975, p. xi) stated: "Each of the commonly employed techniques in multivariate statistics is a straightforward generalization of some univariate statistical tool, with the single variable being replaced by a linear combination of several original variables." The set or "vector" of measurements for a subject is reduced to a single numerical value by determining the "optimum" weight by which to multiply each value and then adding the weighted values. For example, Hotelling's T^2 test is, in effect, a procedure of combining dependent variable measurements into a single measurement for each experimental unit before carrying out a t test on the composite measurements (Harris, 1975).

Consider a randomization test for MANOVA for a single-factor completely randomized (independent-groups) design in which N subjects are assigned randomly to three treatments. Two dependent variable measurements are obtained for each subject. The obtained test statistic value is then computed for the obtained results by using a procedure of deriving a single composite measurement from the vector of measurements for each subject and applying one-way ANOVA to compute F. In permuting the data, the vectors of original measures are then permuted and, for each data permutation, the vectors are reduced to composite measurements, which are used to compute F. Computing the linear combination of values on the different variables for each data permutation is more time-consuming than computing the composite measurements only once and permuting the composite measurements each time, but randomization tests require using the slower procedure. If permuting composite measurements was acceptable, the computation for the obtained results could be performed to provide composite measurements to which an ANOVA program from Chapter 4 could be applied. Unfortunately, permuting the composite measurements is not acceptable. The composite measurements are derived by a combining procedure designed to maximize the separation of groups for the obtained results, and thus the composite measurement values vary as the vectors of the original measurements are permuted among groups. The permuting of composite measurements here is unacceptable for the same reason that the permuting of residuals for analysis of covariance and the permuting of residuals based on column and row means for a test of interaction were unacceptable: the adjustment of the data depends on the particular data permutation that represents the obtained results.

The same considerations apply to other designs where a randomization test counterpart of a conventional multivariate test is desired. A computer program would involve repeated computations of composite measurements and permuting of vectors from which the composite measurements are computed.

7.3   Custom-Made Multivariate Randomization Tests

The randomization tests in Chapter 4 through Chapter 6 were almost exclusively counterparts of widely used parametric tests, employing test statistics sensitive to the same type of treatment effects as the conventional test statistics, and that is true of the correlation randomization tests in the next chapter. Those tests are likely to be useful in their present form for some time. Alternatively, the tests in this chapter were not developed as counterparts to any of the parametric multivariate procedures. If there are parametric counterparts, it is accidental; these randomization tests were developed to meet special experimental needs and thus employ special test statistics. They were custom-made to meet the special objectives of the experimenter rather than being randomization test counterparts of conventional multivariate procedures. As such, any one of the procedures is likely to have occasional but not frequent use. Whatever their frequency of usage, they have value as examples of custom-made randomization test procedures.

Chapter 6 dealt with the pooling of data from separate experiments by means of randomized block tests and the combining of P-values from separate experiments. There was considerable flexibility in both procedures, permitting the experimenter some leeway to take into account special expectations or predictions. Multivariate tests deal with the pooling of data from separate dependent variables from a single experiment and also are readily adaptable to different experimental predictions.

These tests are examples of the possibilities of tailoring multivariate randomization tests to meet the needs of an experimenter. Willmes (1988) stressed the importance of developing permutation test counterparts of conventional multivariate procedures, and has published a FORTRAN program for a multivariate test using Pillai's trace test statistic with the P-value determined by systematic or random data permutation. In Section 7.2, we discussed the general approach for devising randomization test counterparts to conventional parametric multivariate tests, and multivariate experts could undoubtedly suggest modifications to those randomization tests likely to be helpful for special applications.

7.4 Effect of Units of Measurement

As pointed out above, when conventional MANOVA procedures are used, in effect measurements from several dependent variables are combined into a single overall measure of treatment effect and then the single composite measurements are analyzed by ordinary univariate statistical procedures. Alternatives to conventional procedures will be employed in deriving composite measurements but before considering them, let us examine the units-of-measurement problem, a matter of importance for any type of composite measurement.

The combining procedure is more likely to increase the sensitivity of the test if we use a procedure that is not affected by units of measurement. Familiar parametric procedures and all of the statistical tests considered so far in this book are such that the P-value is independent of the unit of measurement. (In fact, the P-value is invariant under any linear transformation of the data.) The independence of the P-value and the unit of measurement must be maintained in using composite measurements because we do not want the P-value to depend on such an arbitrary consideration as whether measurements are expressed in inches or millimeters for one of the dependent variables.

Example 7.1. A herpetologist conducts an experiment to determine the relative effect of two levels of humidity on the size of snakes. Three young snakes are assigned to each level of humidity and at the end of six months, the length and weight of the snakes are measured. The measurements are shown in Table 7.1.

TABLE 7.1

Size of Snakes in Inches and Ounces

	A				B		
	Inches	**Ounces**	**Total**		**Inches**	**Ounces**	**Total**
a	12	16	28	d	36	48	84
b	24	62	86	e	60	80	140
c	48	30	78	f	72	98	170
			$T_A = 192$				$T_B = 394$

TABLE 7.2

Size of Snakes in Feet and Pounds

	A				B		
	Feet	Pounds	Total		Feet	Pounds	Total
a	1	1	2	d	3	3	6
b	2	3.875	5.875	e	5	5	10
c	4	1.875	5.875	f	6	6.125	12.125
			$T_A = 13.750$				$T_B = 28.125$

For the test, the six "totals" or composite measurements would be allocated in all $6!/3!3! = 20$ ways between the two treatments, and T_B would be computed for each data permutation for a one-tailed test where B was expected to provide the larger measurements. Of the 20 data permutations, two provide a T_B as large as 394, the obtained value, so the one-tailed P-value would be 0.10. Table 7.2 shows what would have happened if we had expressed the length in feet and the weight in pounds. In this case, there is only one permutation giving as large a value of T_B as 28.125, the obtained value, and so the P-value would be 0.05. Thus with the same data, using different units of length and weight, different P-values can be obtained when the numerical values are combined to give the composite measurement.

The problem of controlling for units of measurement in conventional MANOVA linear combinations of measurements is solved by the procedure for determining the weights to assign the measurements from different dependent variables. If we have two dependent variables X and Y, we determine weights w_1 and w_2 such that $w_1X + w_2Y$ gives composite measurements that are "best" in the sense of, say, providing the largest F for the obtained results when F is applied to the composite measurements. Suppose it turned out that equal weighting was determined to be best in the snake example for measurements when length is expressed in feet and weight in pounds. Then if length was transformed to inches and weight to ounces, the w_1 and w_2 weighting of the inches and pounds measurements would be in the ratio of 12:16 instead of 1:1, but the MANOVA test statistic for any data permutation would be the same.

7.5 Multivariate Tests Based on Composite z Scores

The unit of measurement problem can be handled in other ways of combining measurements from different dependent variables. The composite z score approach that follows also controls for units of measurement and it does have some advantages over conventional MANOVA test procedures from the standpoint of performing randomization tests.

As mentioned earlier, the weights must be computed separately for each data permutation in performing a randomization test to determine the P-value for a conventional MANOVA procedure. Alternatively, for the composite z-score approach the composite measurements, although dependent on the joint distribution of measurements for the dependent variables, are independent of the way the measurements are divided among treatments. Consequently, the composite measurements can be computed for the obtained results and then be permuted instead of being computed separately for each data permutation.

The composite z score procedure involves transforming the measurements into z scores and the z scores for a subject are added to get a composite z score:

$$z = \frac{X - \bar{X}}{\sqrt{\sum (X - \bar{X})^2 / n}} \tag{7.1}$$

The formula is used to determine the z score for the measurement of a subject on one of the dependent variables. A z score is not affected by the unit of measurement, and so a composite z score — and consequently, the P-value of the sum of the composite z scores — is independent of the unit of measurement. The measurements in Example 7.1 will now be transformed to composite z scores.

Example 7.2. The z score for the length of subject b, based on measurement in feet, is $(2 - 3.5)/(17.5/6)^{1/2} = -0.88$, and the z score for the length of subject b, based on measurement in inches, is $(24 - 42)/(2520/6)^{1/2}$, which also equals -0.88. Table 7.3 shows z-score values for the lengths and weights and the composite z scores, no matter what the units of measurement may be. For determining the P-value by data permutation, the six composite z scores constitute the "data" that are divided between the two treatments. For each of the 20 possible data permutations, the one-tailed test statistic T_B is computed. It can be seen that no other data permutation provides as large a value of T_B as +4.16, the obtained value, and so the one-tailed P-value is 1/20, or 0.05.

TABLE 7.3

Composite z Scores

	Length	Weight	Composite z Scores
		A	
a	−1.46	−1.41	−2.87
b	−0.88	+0.23	−0.65
c	+0.29	−0.92	−0.63
			$T_A = -4.15$
		B	
d	−0.29	−0.27	−0.56
e	+0.88	+0.87	+1.75
f	+1.46	+1.51	+2.97
			$T_B = +4.16$

(Program 4.4 or Program 4.5, for the independent t test, can be used with this design for determining the P-value.)

The P-value happens to be the same as with the raw measurements expressed in terms of feet and pounds but is smaller than with the raw measurements expressed in terms of inches and ounces. Composite z scores are used to prevent an arbitrary choice of units of measurement from influencing the P-value; therefore, they should tend to provide more significant results when there is a treatment effect than if unweighted raw measurements for particular units of measurement were pooled.

By using composite z scores in Example 7.2 to test the effect of the two treatments on length and weight, the variables length and weight are implicitly regarded as equally important; they make equal contributions to the composite measurement. Instead of using z scores, if we expressed each subject's length in inches and its weight in pounds and if we used the total of the two numbers as a subject's composite measurement, almost all of the variation in the composite measurements would be due to the variation in length — the weight variable would have little effect. Similarly, if a composite measurement that was the sum of feet and ounces were used, variation in the composite measurement would be primarily a function of variation in the weight variable and so length would have little influence on the P-value. The transformation of the lengths and weights to z scores before combining to obtain a composite measurement serves the function of equating the variability of the two dependent variables so that they have equal influence on the P-value of the results.

This procedure of generating composite scores by adding z scores presupposes that the relationship between the effects on the different dependent variables is predictable. The above example assumes that the experimenter expected increases in length generally to go with increases in weight; otherwise, it would be inappropriate to add the z scores for the two dimensions. Suppose that the experimenter had three dependent variables, X, Y, and Z, and that he expected X and Y to change in the same direction but expected Z to change in the opposite direction. Z could be the time it takes a rat to run a maze in a situation where the speed was expected to be positively correlated with dependent variables X and Y. In such a case, the signs associated with the z scores for Z would be reversed to make them reflect speed rather than running time. The z scores for a subject would then be added to give the composite z score.

Two or more treatments with any number of dependent variables can be handled in this manner. When only two treatments are involved, a one- or two-tailed test can be employed. To determine the P-value by data permutation for a one-tailed test, the test statistic is the sum of the composite z scores for the treatment predicted to have the larger measurements. In Example 7.2, treatment B was predicted to have the larger measurements, and so T_B, the total of the composite z scores for treatment B, was used for the test. For a two-tailed test, the test statistic would be the same as for F for one-way ANOVA: $\Sigma(T^2/n)$. For Example 7.2, the obtained two-tailed test statistic would be $T_A^2/n_A + T_B^2/n_B$, which would be $(4.15)^2/3 + (+4.16)^2/3$ for the composite z scores.

A between-subjects design was considered in Example 7.2 but the composite z score approach also can be applied to repeated-measures designs when there are several dependent variables. For the between-subjects design, the z score for a subject on a particular dependent variable is based on the mean and standard deviation of the measurements for that variable over all subjects and treatments. The same is true of the z score of a subject on a particular dependent variable for a treatment in a repeated-measures experiment.

Example 7.3. Subjects S_1, S_2, and S_3 each take three treatments, A, B, and C, the treatment time for each treatment being randomly and independently determined for the three subjects. Measurements are made on three dependent variables, X, Y, and Z. The results are shown in Table 7.4.

The measurements for each dependent variable are transformed into z scores. The z scores for all nine X measurements are based on the mean and standard deviation of those nine measurements and, similarly, the z scores of the other two dependent variables are based on the mean and standard deviation of the nine measurements for the dependent variable. For example, the z score for S_1 on dependent variable Z is $(16 - 17.44)/3.69 = -0.39$, where 17.44 is the mean and -3.69 is the standard deviation of the nine Z measurements. A z score is determined for each dependent variable for all subjects and then the z scores for each subject for a single treatment are added to get a composite z score. For instance, subject S_1 would have three composite z scores, one each for treatments A, B, and C. The composite z score for S_1 for treatment A is the sum of -0.39 and the z scores for the measurements on the other two dependent variables. The test statistic is ΣT^2, where T is the total of three composite z scores for a treatment. This test statistic is equivalent to F for repeated-measures ANOVA. There are $(3!)^3 = 216$ possible assignments of treatment times to treatments for the subjects. The nine composite z scores constitute the data that are permuted. The 216 data permutations consist of every allocation of the three composite z scores for S_1 to the three treatments, in conjunction with every allocation of the three composite z scores for each of the other two subjects. The proportion of test statistic values that are as large as the obtained value is the P-value associated with the experimental results. Program 6.1 or Program 6.2 could be used to determine the P-value.

TABLE 7.4

Data from a Repeated-Measures Multivariate Experiment

	Treatment A			Treatment B			Treatment C		
	X	Y	Z	X	Y	Z	X	Y	Z
S_1	5	26	16	8	25	14	12	30	21
S_2	8	23	13	9	24	19	10	25	23
S_3	4	20	12	7	23	18	10	26	21

It would be possible to compute z scores for a subject on the basis of the mean and standard deviation of the three measurements for the subject on a particular dependent variable. To illustrate, we could compute the z score for the measurement of 16 for S_1 on the basis of the mean and standard deviation of the measurements 16, 14, and 21 for S_1. However, such a procedure would be inconsistent with determining the P-value with univariate repeated-measures designs. The test statistics t and F for repeated-measures designs are such that subjects with more variability of measurements over treatments have more effect on variation in the test statistic over data permutations than less variable subjects. Computing composite z scores on the basis of each subject's mean and standard deviation for measurements on a particular dependent variable tends to give subjects the same weight in the determination of the test statistic value.

When the dependent variable values are not based on ordinary measurement but on ranking, there is no need for transformation to z scores because the rank values for the different dependent variables have equal variability and make composite measurements based on a sum of ranks appropriate, equalizing the influence of all of the dependent variables on the composite measurements. The P-value could be determined with Program 4.1 or Program 4.3 for between-subjects designs, or with Program 6.1 or Program 6.2 for repeated-measures designs.

The composite z-score approach requires that a person predict the way the dependent variables co-vary over the treatments and to adjust the sign of z scores for a dependent variable accordingly. Useful though this approach may be when there is a basis for judging how dependent variables will co-vary, a different technique is required when there is no basis for such a judgment. One such technique is the combining of values of t or F over the different dependent variables.

7.6 Combining t or F Values over Dependent Variables

Instead of reducing multivariate data to a univariate form by deriving a single composite measure to replace vectors of measurements, one can compute univariate t or F for each dependent variable and use the sum of ts or Fs for all dependent variables as a compound test statistic. Thus, to carry out a randomization test on data from a repeated-measures experiment with three dependent variables, F for each of the three variables is computed separately and the sum of the three Fs is used as the test statistic. Although Fs are computed separately for the dependent variables, the vectors — not the individual measurements — must be permuted for the test, with a sum of Fs being computed for each data permutation because H_0 concerns the vector of measurements. As the unit of measurement has no effect on t or F, there is no need to transform the data to z scores to control for the unit of measurement.

One feature of this procedure that is unappealing is that with only two treatments, the P-value given by using the sum of Fs as a test statistic may not be the same as that given by using the sum of absolute t values. For instance, with two dependent variables, with absolute ts of 1 and 4, the sum is 5, the same as for a data permutation with ts of 2 and 3, but the sum of the Fs (i.e., the square of the t values) for those data permutations would be 17 and 13. One solution to this problem would be to use the sum of logarithms of F or t as the test statistic. The test statistic for F would then always be twice that for absolute t for all data permutations, ensuring the same P-value for the two test statistics. This modification would have the effect of changing the procedure to one based on the product, rather than the sum, of test statistics.

In a nonexperimental study, Chung and Fraser (1958) also proposed the use of a test statistic for MANOVA that was the sum of test statistics for each separate dependent variable and the determination of the P-value by data permutation for the comparison of two groups of people. Another interesting approach is to use a combination of the P-values for each separate dependent variable as a test statistic and to determine the overall P-value by permuting the multivariate data vectors. See Pesarin (2001) for an elaboration of this approach.

7.7   A Geometrical Model

MANOVA has been approached in two ways in the preceding sections — as a procedure of reducing vectors of measurements to single numbers to which univariate ANOVA can be applied, and as a procedure of applying univariate ANOVA repeatedly to each dependent variable and summing the obtained Fs. A third approach, based on a spatial or geometrical model of MANOVA, follows.

Various distances in a geometrical representation of MANOVA are squared in the computation of sums of squares. Squared distances from centroids are computed. A *centroid* is a multidimensional mean, the vector of the means of all dependent variables. For example, in a scatterplot of measures of subjects on variables X and Y, the point of intersection of a line perpendicular to the X axis at $\bar{X}$ and a line perpendicular to the Y axis at $\bar{Y}$ is the centroid of the values in the bivariate array.

The distances within a p-dimensional space for vectors of measurements from p dependent variables are treated as Euclidean distances, where the squared distance between any two points (vectors) is the sum of the squared distance between them with respect to each of the dependent variables. For instance, the squared "straight-line" distance between one corner of a room and the farthest corner from it is $l^2 + w^2 + h^2$, where l, w, and h are the length, width, and height of the room. Computations based on the model of Euclidean distance treat dependent variables as independent spatial dimensions, like the length, width, and height of a room.

First, we will define F for MANOVA in terms of squared distances in multidimensional space and then show that these squared distances can be determined without direct measurement. This will be done only for a multivariate generalization of one-way ANOVA but analogous geometrical generalizations of repeated-measures and factorial ANOVA also are possible. For standardizing the dependent variables to ensure that scale of measurement alterations do not affect a MANOVA test, each of the dependent variables is transformed to z scores on the basis of the mean and standard deviation of measurements of the dependent variable. The points involved in the following conception of MANOVA are then vectors of z scores.

Unlike the procedure of reducing a number of measurements on an experimental unit to a single number by the derivation of composite measurements, the geometrical approach reduces a number of measurements on an experimental unit to a single point in space. The distances between points in p-dimensional space are all that are involved in this form of MANOVA; the actual numerical values determining the location of a point are not required for the geometrical conception of the determination of F for MANOVA. If we regard a centroid as a multidimensional mean, F for one-way MANOVA can be defined in the same way as for univariate one-way ANOVA:

$$F = \frac{SS_B/df_B}{SS_W/df_W} \qquad (7.2)$$

where sums of squares are sums of squared deviations from means (which are centroids in the multidimensional case), df_B is the number of treatment groups minus one, and df_W is the number of subjects (or other experimental units) minus the number of treatment groups. For MANOVA, let C_A, C_B, and C_{GR} be the centroids of groups A and B and of all of the vectors combined ("the grand centroid"). We can then define SS_B and SS_W for MANOVA as:

$$SS_B = n_A(C_A - C_{GR})^2 + n_B(C_B - C_{GR})^2 + \ldots \qquad (7.3)$$

$$SS_W = \Sigma(X_A - C_A)^2 + \Sigma(X_B - C_B)^2 + \ldots \qquad (7.4)$$

The differences enclosed in parentheses for SS_B are straight-line distances between centroids, and the differences in parentheses for SS_W are straight-line distances between measurement vectors and centroids. The following formulas determine those squared distances by adding the squared deviations from means on each of the dependent variables (X, Y, and so on):

$$SS_B = SS_{B(X)} + SS_{B(Y)} + \ldots \qquad (7.5)$$

$$SS_W = SS_{W(X)} + SS_{W(Y)} + \ldots \qquad (7.6)$$

So for computational purposes, F for one-way MANOVA becomes:

$$\frac{(SS_{B(X)} + SS_{B(Y)} + \cdots)}{(SS_{W(X)} + SS_{W(Y)} + \cdots)} \times \frac{df_W}{df_B} \tag{7.7}$$

To obtain a simpler test statistic that is equivalent for the determination of the P-value by the randomization test procedure, we can drop the constant multiplier df_W/df_B. The numerator plus the denominator of the remaining fraction equals the sum of the total sum of squares for all of the dependent variables: $SS_{T(X)} + SS_{T(Y)} + \ldots$, which is constant over all data permutations. Therefore, the numerator varies inversely with the denominator over all data permutations, making the numerator an equivalent test statistic to the fraction as a whole. In turn, the numerator can be reduced to a simpler test statistic. SS_B for each dependent variable is $n(\bar{X} - \bar{X}_{GR})^2$, and as the grand mean for each dependent variable in z score form is 0, the sum of SS_B for all dependent variables is $\Sigma n(\bar{X})^2$, or $\Sigma(T^2/n)$, where the summation is over all dependent variables and all groups.

The geometrical approach here has been applied in various contexts and its potential applications are extensive enough to merit detailed discussion in the next section. The differences in composition discussed there also can be thought of as differences in proportional distribution or, graphically, as differences in histogram or profile shapes.

7.8   Tests of Differences in Composition

At times, an investigator will be interested in changes in composition rather than changes in amount of some compound. In studies designed to detect changes in composition resulting from experimental manipulations, a useful index of the composition is the proportion of a compound entity that each component comprises.

A distinction must be made between the differences in composition, which contingency chi-square can test, and the differences that call for an alternative procedure, such as the one to be described. For experimental applications, contingency chi-square tests differ between treatments with respect to the proportion of experimental units in various categories. Alternatively, the procedure to be described is concerned with proportional composition within individual experimental units. For instance, if animals were independently assigned to different treatments, contingency chi-square could be used to test for a difference in the proportion of subjects that died under the different treatments. But if a number of subjects are randomly assigned to treatments expected to affect their blood composition, contingency chi-square could not be used to test the difference between treatments with respect to the proportion of white blood cells that were lymphocytes. However, the tests to be described would be applicable. To reiterate, the tests of differences in composition that will be described here concern differences in composition of compounds contained within individual experimental units, not differences in proportion of experimental units in various categories.

Components making proportional contributions to a compound of which they are parts can be of various types. In investigations of body fluids like blood or saliva, the proportions may be the relative volume or weight of various constituents. The measurements also could be given as proportions of a length, for example, as the proportional lengths of the upper and lower leg. Thus, the entity whose composition is expressed as a pattern of proportions may be a particular substance, a span of time, a volume, a length, or any other entity that can be divided into its constituents.

The simplest test of a difference in composition of experimental units is a test of a difference between the mean proportion of some component under two treatment conditions. For instance, a sample of apples from a tree is randomly divided between two storage conditions to see the effect of the storage conditions on the moistness of the apples after several months. The proportion of water by weight in each apple is determined and used as a measurement for the apple. A t test can then be applied to test the difference between the mean proportions for the apples under the two storage conditions. If a randomization test is desired, Program 4.3 or Program 4.4 could be used. Because the proportion of an apple that is not water is unity minus the proportion that is water, the value of t for any data permutation would be the same whether the proportions used in the test were proportions of water or proportions of dehydrated material.

With only two dependent variables, the pattern of the two proportions is reflected in the proportions for only one of the variables, so a randomization test for such cases could be applied to the proportions for either variable using programs from Chapter 4 for between-subjects designs. But that cannot be done when there are more than two dependent variables, as the distribution of proportions for one dependent variable does not completely determine the distribution for one of the other dependent variables. However, an adaptation of the geometric multivariate approach given in Section 7.7 is applicable. The test statistic for the test is SS_B accumulated over all dependent variables, where the dependent variable values are expressed as z scores. The z score transformation prevents the arbitrariness of units of measurement for the different dependent variables from influencing the means and standard deviations, and thus the P-values. The z score measurements on all dependent variables are in the same units, e.g., inches, millimeters, ounces, or grams. Therefore, the test statistic proposed for the test for differences between patterns of proportions is SS_B for the proportions accumulated over all dependent variables. Example 7.4 illustrates the application of a randomization test using this statistic.

Example 7.4. Nine children are randomly assigned to three treatments, with three children per treatment. The treatments are expected to provide different patterns of proportions across three dependent variables, so the test statistic chosen is SS_B added over all dependent variables. The three dependent variables are expressed in the same unit of measurement, time. As in a study by Ewashen (1980), who used a randomization test similar to the one used in this example, the subjects are all given an opportunity to play with various toys and the proportion of time spent with each type of toy is recorded for each child. The results are shown in Table 7.5.

Over all three treatments combined, the mean proportion for X, Y, and Z are the same but within treatment groups, the proportions tend not to be evenly distributed over the three types of toys. It will be observed that the children under treatment A spent the largest proportion of time with type X toys and the least with type Y toys. However, the treatment B children spent the greatest proportion of time with type Y toys and least with type Z, and treatment C children spent the greatest amount of time with type Z and the least with type X toys. There are $9!/3!3!3! = 1680$ data permutations, consisting of all divisions of the nine rows of proportions between the three groups with three rows per group. The test statistic is $SS_{B(X)} + SS_{B(Y)} + SS_{B(Z)}$ and for each of the data permutations, the test statistic is computed and the proportion of the 1680 data permutations with as large a value as the obtained value is the P-value. There is no overlap between treatments of measurements within any of the three columns for the obtained data, so SS_B is a maximum for each of the dependent variables and the obtained data permutation must have the largest value of $SS_{B(X)} + SS_{B(Y)} + SS_{B(Z)}$. As this is an equal-$n$ design, there are $3! = 6$ data permutations with the maximum value of the test statistic. The P-value thus is 6/1680, or about 0.004.

When proportions have been computed for several dependent variables, they should be recomputed for a particular pair of dependent variables when there is interest in treatment effects on the relationship between those two variables alone. For instance, for a test involving only variables X and Y of Example 7.4, the X and Y proportions as given should not be used in a test but should be changed. The X and Y proportions should be transformed by dividing each proportion by the sum of that proportion and the proportion

TABLE 7.5

Proportion of Time Spent with Each Type of Toy

Treatment	Type of Toy		
	X	Y	Z
A	0.59	0.13	0.28
	0.54	0.18	0.28
	0.68	0.12	0.20
B	0.31	0.50	0.19
	0.35	0.57	0.08
	0.20	0.65	0.15
C	0.03	0.19	0.78
	0.11	0.28	0.61
	0.19	0.38	0.43

for the other dependent variable for a subject. For example, the X and Y values for the first subject would be $0.59/(0.59 + 0.13) = 0.819$ and $0.13/(0.59 + 0.13) = 0.181$.

There are two advantages in transforming the dependent variable proportions to make them add up to 1 for the dependent variables involved in the test. One advantage is that the transformation makes the test sensitive primarily to the relative distribution of proportions rather than the differences in general levels of proportions between groups. For instance, if every X_A and Y_A was 0.60 and 0.20, respectively, every X_B and Y_B was 0.33 and 0.11, and every X_C and Y_C was 0.45 and 0.15, we would obtain the maximum test statistic value with a P-value of 6/1680, whereas with each X and Y proportion determined strictly on the basis of those variables and not influenced by the Z variable — i.e., adjusting the total $(X + Y)$ proportion to 1 — all nine values would be 0.75 for X and 0.25 for Y, and would give a test statistic value of 0 and a P-value of 1. Without adjustment to a total of 1, the test, by being affected by absolute magnitudes of proportions and not just relative magnitudes, can provide significant results when the relationship between X and Y is not affected, and can fail to provide significant results when the relationship between X and Y varies over treatments, the variation being masked by effects of differences in magnitude of X and Y. A second reason for using adjusted proportions is that the test then is equivalent to a test comparing three sets of proportions for X (or for Y). When the X proportion + Y proportion = 1, $SS_{B(X)} = SS_{B(Y)}$, so $SS_{B(X)} + SS_{B(Y)}$ is simply twice SS_B for either of the variables. Therefore, we can apply Program 4.1 to the distribution of proportions for either X or Y and get the P-value that would be obtained by using $SS_{B(X)} + SS_{B(Y)}$.

An application of this multivariate procedure to categorical data is presented in the following example.

Example 7.5. Data from a social psychological experiment were analyzed by means of the geometrical approach to MANOVA (Edgington, et al., 1988). Subjects were randomly assigned to one of two treatment conditions differing simply in what the subjects were told about a social situation. Responses were assigned to categories within a classification system consisting of four categories. The prediction was that there would be a difference between treatment groups in the way responses would be distributed across the four categories. Twelve responses from each subject were divided among the categories, so there were four dependent variables and the frequency within each category could be transformed into a proportion by dividing the frequency by 12. The randomization test null hypothesis of no treatment difference was this: for each subject, the histogram or "profile" showing the distribution of responses of a subject over the four categories was the same as it would have been under the alternative treatment. Because the fixing of the number of responses at 12 for each subject prevents one treatment from having more responses in general than the other, differences between the two treatments had to be differences in overall pattern of responses or what in psychometrics could be called differences in profile shape.

7.9   Evaluation of Three MANOVA Tests

Three randomization test procedures for MANOVA have been described in this chapter. One used composite z scores, another involved the summing of Fs over the dependent variables, and the third was a geometrical approach. Although attention was focused on one-way MANOVA, all three approaches can be adapted to repeated-measures or factorial MANOVA. The following evaluation concerns the tests as one-way MANOVA, the principal application for which they were described in this chapter.

All three techniques apparently require less time for the determination of the P-value by a randomization test than would a randomization test with a conventional MANOVA test statistic. The composite z-score approach would be the fastest of the three for a randomization test, the time required after the composite z scores have been computed being the time for univariate ANOVA with observations on only one of the dependent variables. The composite z score procedure has the added advantage of allowing the use of programs in preceding chapters as soon as the composite scores have been derived. The other two approaches would require new programs to be written.

In Section 7.8, we discussed the value of the geometrical approach in testing for differences in composition. A serious limitation of the geometrical approach is its inapplicability to one-tailed t tests. The geometrical approach permits spatial analog of sums of squared deviations to be formed but not spatial analog of the arithmetic difference between means that is the numerator of a one-tailed t. The sum of the squared differences between the means of A and B groups with respect to each dependent variable completely determines the squared distance (and the distance) between the A and B centroids in Euclidean space, but the sum of $(\bar{X}_A - \bar{X}_B)$ over all dependent variables does not determine the distance between the A and B centroids nor their relative positions in space. For instance, when the sum of the $(\bar{X}_A - \bar{X}_B)$ differences is 0, the A and B centroids are not necessarily located at the same point. Univariate sums of squares generalize readily in a geometric manner to multivariate situations but arithmetic differences between means do not.

A major strength of the composite z score procedure is its sensitivity to treatment differences when the method in which treatments co-vary can be predicted. The manipulation of the signs of z scores of a variable to reflect the expected direction of correlation between the variables provides the opportunity to increase the likelihood of detecting treatment effects when more than two treatments are compared, but it is in the application of the t test that the chance to increase the power of the test is clearest. Consider the use of the total of the z scores for one of the treatments as the one-tailed test statistic. The one-tailed test can be based on any of the following predictions:

1. X and Y values will be positively correlated over treatments.
2. X and Y values will be negatively correlated over treatments.

3. Both X and Y values will be large for treatment A.

4. X values will be large and Y values will be small for treatment A.

There is no specified direction of difference in the first two predictions. For prediction 1, the absolute value of the sum of group A is the appropriate test statistic. When prediction 2 is made, the sign of the z scores is reversed for either of the dependent variables and the absolute value of either the sum of A or, equivalently, the sum of B is the test statistic to use. The sum of A is the test statistic to use for prediction 3, and the same test statistic is used for prediction 4 after the signs of the z scores for Y are reversed. With more than two dependent variables, the predictions can be even more specific.

The test procedure based on the sum of test statistics can use the sum of one-tailed ts for the dependent variables for group A to test prediction 3 and prediction 4. To test prediction 1, ts with $(\bar{X}_A - \bar{X}_B)$ and $(\bar{Y}_A - \bar{Y}_B)$ as numerators are added and the absolute sum is used as the test statistic. To test prediction 2, ts with $(\bar{X}_A - \bar{X}_B)$ and $(\bar{Y}_B - \bar{Y}_A)$ are added and the absolute sum is used as the test statistic.

Tests that cannot be carried out satisfactorily with the composite z-score approach are those involving two or more groups where no assumption about the relationship between dependent variables is made. The adjustment of z scores to reflect the way in which the dependent variables are expected to co-vary is both a strength and a weakness of the procedure. Clearly, it is a strength when there is a good basis for prediction of the nature of covariation of dependent variable values over the treatments but to use the procedure without such a prediction is inappropriate. Failing to change the sign of z scores for any of the dependent variables is implicitly to predict that all dependent variables change in the same direction from treatment to treatment, and the P-value is affected by the correctness of the implicit prediction. To carry out MANOVA or multivariate t tests without expectations about the way the dependent variables co-vary, one should then employ either the sum of statistics procedure or the geometrical procedure.

7.10 Multivariate Factorial Designs

Chapter 5 and Chapter 6 dealt with the application of randomization tests to univariate data from completely randomized factorial and randomized block designs. Multivariate tests using the test statistics described earlier in this chapter or employing conventional test statistics, such as Wilk's lambda or Hotelling's T^2, can readily be applied to those designs. The power of multivariate tests with disproportional cell frequencies can be improved by the use of the adjustment of measurements described in Section 6.25. The adjustment would be performed on each dependent variable separately.

In Example 6.7, a factorial experiment with two factors and two dependent variables was described, but there each dependent variable was associated with only one of the factors. In the following example, we will consider the same experimental design, analyzed as it would be if both dependent variables were regarded as measures of effect for both factors. Of course, as with any multivariate design, in addition to multivariate tests, univariate tests using values from only one of the dependent variables can be conducted if desired.

Example 7.6. Let us reconsider Example 5.6: A neuroscientist is interested in the effect of minor lesions in the brains of rats on their ability to discriminate among intensities of stimuli. In addition to investigating the relative performance of lesioned and non-lesioned rats, the experimenter also wants to compare the effect of lesions in different locations in the brain. Brightness and loudness discrimination tasks are convenient ways to assess certain types of perceptual deterioration, and the experimenter decides to measure both brightness and loudness discrimination. To make effective use of a small number of rats, the experimenter uses a factorial design in which rats are assigned to levels of two factors, and measurements of brightness and loudness discrimination are made on all rats. A total of 20 rats are assigned at random to each of four treatment combinations, with the assignment constraint that five rats will have no lesions, five will have lesions in both L_1 and L_2, five will have a lesion only in L_1, and five will have a lesion only in L_2, as was shown in Table 5.12.

After all of the lesioned rats have recovered from the operation, all lesioned and non-lesioned rats are tested on brightness and loudness discrimination tasks. Each rat provides a measurement on two dependent variables: brightness discrimination and loudness discrimination. The experimenter expects both brightness discrimination and loudness discrimination to be adversely affected by the presence of a lesion, and for all of the randomization t tests the composite z score approach is used to determine the P-value. Various tests are possible.

To see whether the presence of a lesion reduces the ability to discriminate intensities of stimuli, z scores for the five animals receiving no lesions and the z scores for the five animals receiving both lesions could be determined on the basis of the scores for those 10 animals prior to being converted to composite z scores. Program 4.4, a randomization test program for the independent t test, then could be applied to the composite z scores using either the two-tailed or one-tailed test statistic, depending on the prediction. However, Program 9.2, for a correlation trend randomization test, would probably be more sensitive because it could make effective use of data from all of the animals, not just the 10 at the extremes of the treatment dimension.

To test the main effect of the lesion in L_1, z scores for the animals in the first and second rows would be computed separately on the basis of the scores in the respective rows prior to converting all of the z scores to composite z scores. Program 5.1 could then be used to test the main effect of the lesion in L_1 by permuting the composite z scores within rows using either a one- or two-tailed test statistic. A one- or two-tailed test of the effect of the lesion in L_2 could be performed similarly with Program 5.1 by permuting composite z scores within columns. Those composite z scores would be derived from z scores computed separately on the basis of the scores within the respective columns.

The relative effect of lesions in L_1 and L_2 could be tested by the use of Program 4.4 applied to the 10 composite z scores of the rats in the lower left and upper right cells, i.e., those rats that had only one lesion. The z scores underlying the composite z scores would be based on the scores of the 10 rats in the two cells under test.

7.11   Combining Univariate and Multivariate P-Values

Let us conclude this chapter by pointing out that the procedure of combining P-values that was presented in Chapter 6 is a very general one and can be employed for P-values resulting from independent experiments with any type of design. In the following example, a P-value from an experiment with a univariate design is combined with a P-value from an experiment with a multivariate design.

Example 7.7. A marine biologist conducted a pilot study preliminary to a major experiment investigating the effect of sunlight intensity on the eating behavior of a species of tropical fish. The pilot study used the amount of food eaten as the dependent variable for a statistical test because that variable was to be used in the subsequent, main experiment. The P-value of 0.06 for that test was recorded for combining with the results of tests in the main experiment. But the biologist also was on the lookout for other aspects of eating behavior that seemed to be affected by the manipulation of light intensity, to find other dependent variables to be measured in the main experiment. This biologist noticed that light intensity variation seemed to affect the variety of foods eaten but did not observe any other dependent variables that offered promise as supplements to the food quantity-dependent variable originally planned. Because of the findings in the pilot study, in the main experiment fish were randomly assigned to intensity levels of simulated sunlight and two dependent variables were measured: quantity and variety of food eaten. A multivariate randomization test P-value was 0.08. The pilot study univariate P-value of 0.06 was added to the main study multivariate P-value of 0.08 to give 0.14, and the overall P-value was $(0.14)^2/2$, which is just below 0.01.

7.12   Questions and Exercises

1. What advantages do multivariate tests have over applying univariate tests to each of the dependent variables?

2. Olson's study of the effect of the robustness of parametric MANOVA under violation of parametric assumptions dealt primarily with the violation of which assumption?

3. For a randomization test on the effect of levels of humidity on the development of oak tree seedings, an experimenter wants to use these dependent variables: height, maximum trunk diameter, and number of leaves at six months, with the sum of the measurements on the three dependent variables providing a single numerical value for each plant. Can a valid randomization test be conducted? Explain.

4. Suppose that in the experiment described in Question 3, there were two levels of humidity and three plants for each level. What is the smallest possible P-value?

5. A multivariate design has two dependent variables. Would a perfect positive correlation between the measurements for the two dependent variables be desirable or undesirable for detecting a treatment effect? Explain.

6. Harris said that Hotelling's T^2 test could be conducted as a t test carried out on what?

7. A multivariate randomization test can be conducted without using any of the parametric test statistics. How might an experimenter planning a randomization test benefit from examining those parametric test statistics?

8. In the example of a multivariate test applied to test differences in the proportion of time spent by children on three types of toys, the test statistic was $SS_{B(X)} + SS_{B(Y)} + SS_{B(Z)}$. If the test statistic is not the sum of F for the three dependent variables, of what is it the sum?

9. In the example of a pilot study on the effect of sunlight intensity on the eating behavior of tropical fish, what probability combining was performed to get the overall P-value?

10. For the geometric model, what was the test statistic that is analogous to that used in the example discussed in Question 8?

REFERENCES

Chung, J.H. and Fraser, D.A.S., Randomization tests for a multivariate two-sample problem, *J. Am. Statist. Assn.*, 53, 729, 1958.

Edgington, E.S., Harrod, W.J., Haller, O., Hong, O.P., and Sapp, S.G., A nonparametric test for reward distribution strategies in the minimal group paradigm, *Eur. J. Social Psychol.*, 18, 527, 1988.

Ewashen, I.E., Effects of hospitalization and fantasy predispositions on children's play with stress-related toys, M.Sc. thesis, Department of Psychology, University of Calgary, Alberta, Canada, 1980.

Fouladi, R.T. and Yockey, R.D., Type I error control of two-group multivariate tests on means under conditions of heterogeneous correlation structure and varied multivariate distributions, *Comm. Statist. Simulation Comput.*, 31, 375, 2002.

Harris, R.J., *A Primer of Multivariate Statistics*, Academic Press, New York, 1975.

Olson, C.L., On choosing a test statistic in multivariate analysis of variance, *Psychol. Bull.*, 83, 579, 1976.

Olson, C.L., Practical considerations in choosing a MANOVA test statistic: A rejoinder to Stevens, *Psychol. Bull.*, 86, 1350, 1979.

Pesarin, F., *Multivariate Permutation Tests with Applications in Biostatistics*, John Wiley & Sons, New York, 2001.

Stevens, J., Choosing a test statistic in multivariate analysis of variance: Comment, *Psychol. Bull.*, 86, 355, 1979.

Willmes, K., Beiträge zu Theorie und Andwendung von Permutationstests in der Uni- und Multivariaten Datenanalyse [Contributions to the theory and development of permutation tests in univariate and multivariate data analysis], Ph.D. thesis, University of Trier, Trier, Germany, 1988.

8

Correlation

8.1 Determining P-Values by Data Permutation

In previous chapters, we focused on different experimental designs and we derived the corresponding randomization tests. In Chapter 8, we will focus on specific test statistics that can be used if the independent variable is quantitative. We will provide the rationale for determining P-values by systematic and random data permutation for both univariate and multivariate designs, and we will distinguish the situations in which the dependent variable is measured quantitatively and the situations in which the dependent variable is dichotomous or ordinal. It will be demonstrated that the test statistics and procedure presented in this chapter can also be used if the independent variable is dichotomous or ordinal.

The correlation between a quantitative treatment variable and measurements can constitute a more powerful test than the division of subjects into two or three treatment groups. For example, if we have only five subjects and one measurement from each subject, with random assignment to two treatments the smallest P-value that can be obtained by a randomization test is 0.10 because there are only 10 data permutations associated with the sample sizes providing the most data permutations: two subjects for one treatment and three for the other. Alternatively, if five treatments consisting of five magnitudes of the independent variable were used instead of two treatments with random assignment of one subject per treatment, the P-value for the correlation between the independent variable values and the response measurements could be small. The P-value could be as low as 1/120 because there are 5! = 120 data permutations associated with the 5! ways five subjects can be assigned to five treatments with one subject per treatment. Thus, an experimenter can have a relatively sensitive experiment with very few subjects provided there are several magnitudes of treatment and a correlation test statistic is used.

Correlation can be used to detect treatment effects in ways other than by correlating treatment levels and data. In fact, the treatment levels need not even be quantitatively ordered. Pitman (1937) described a randomization test for correlation for a hypothetical agricultural experiment involving n treatments applied to blocks A and B, each of which was subdivided into n plots. With independent random assignment of the n treatments to the n

plots within the blocks, there are $(n!)^2$ possible assignments, but for pairing A and B plots administered the same treatment there are only $n!$ distinctive pairings. (Each of the $n!$ distinctive pairings could be manifested in $n!$ ways of permuting treatment designations among the pairs.) Under H_0, the measurements for the ordered first, second, ... nth plots within A and also within B were fixed regardless of the assignment, so getting a high correlation between A and B plots administered identical treatments would be the result of happening to assign the treatments in such a way as to pair similar yields.

Example 8.1. We want to test the effect of a certain drug on the reaction time of subjects. One method would be to compare the drug with a placebo and to use the *t* test; this is somewhat like comparing the effect of zero amount of a drug with X amount. To test whether three different amounts of a drug have the same effect is not the same as testing whether presence or absence of a drug has the same effect but sometimes it is just as important. We take three drug dosages consisting of 5, 10, and 15 units of the drug and randomly assign three subjects to the dosages with one subject per dosage. To illustrate how the P-value of the treatment effect is determined by a randomization test, we will use the following hypothesized results and assume that a positive correlation was predicted:

Drug Dosage	5	10	15
Response	20 (1)	22 (2)	25 (3)

The index numbers are assigned to the response measurements according to the rank order of the associated treatments. A standard formula for the product-moment correlation coefficient r (Equation 8.1 given in Section 8.2) is applied to the results and r is found to be 0.99. The P-value of r based on data permutation is determined from the six possible assignments of subjects to treatments. Given that the assignment had no effect on the measurement of any subject, the six possible assignments would provide the six data permutations in Table 8.1, where the numbers in parentheses are indexes that determine the listing of the data permutations.

The indexes for each permutation can be regarded as forming a three-digit number that increases in magnitude from 123 for the first permutation to 321 for the sixth permutation. For correlation, the index numbers 1 to N are assigned to the N measurements in the order in which they are associated with the ordered treatments for the obtained results. The listing procedure is then to have the sequence of indexes for each permutation treated as an N-digit number and to permute the sequence of index numbers in such a way as to make each successive N-digit number the next larger value that can be obtained by permuting the order. If the prediction was a strong positive correlation — not just a strong correlation — between the dosage and response magnitudes, the P-value of the obtained r is the proportion of the six permutations that provide such a large value of r as the obtained value, 0.99. Only the first permutation, which represents the obtained results, shows such a large r and so the P-value for the obtained results is 1/6, or about 0.167. If it had been anticipated that

TABLE 8.1

List of Data Permutations

	Permutation 1		
Drug Dosage	5	10	15
Response	20 (1)	22 (2)	25 (3)
		$r = 0.99$	

	Permutation 2		
Drug Dosage	5	10	15
Response	20 (1)	25 (3)	22 (2)
		$r = 0.40$	

	Permutation 3		
Drug Dosage	5	10	15
Response	22 (2)	20 (1)	25 (3)
		$r = 0.60$	

	Permutation 4		
Drug Dosage	5	10	15
Response	22 (2)	25 (3)	20 (1)
		$r = -0.40$	

	Permutation 5		
Drug Dosage	5	10	15
Response	25 (3)	20 (1)	22 (2)
		$r = -0.60$	

	Permutation 6		
Drug Dosage	5	10	15
Response	25 (3)	22 (2)	20 (1)
		$r = -0.99$	

the correlation between drug dosage and response would be strong but the direction of correlation was not predicted, the absolute value of the correlation coefficient $|r|$ would be an appropriate test statistic. Permutation 1 and permutation 6 both provide values of $|r|$ as large as the obtained value and so the two-tailed P-value would be 2/6, or about 0.333.

For any situation where, as in Example 8.1, the independent variable values are uniformly spaced or are otherwise symmetrically distributed about the middle of the distribution (e.g., 5, 20, 40, 55), the two-tailed P-value for r will be twice the one-tailed P-value for a correctly predicted direction of correlation. However, this is not necessarily the case when the independent variable values are not symmetrically distributed.

Example 8.2. Suppose that the highest drug dosage had been 20 units and that the same responses had been obtained. Then the first data permutation would have been:

Drug Dosage	5	10	20
Response	20 (1)	22 (2)	25 (3)
		$r = 0.997$	

This would be the data permutation providing the largest positive correlation coefficient, so that if a positive correlation was predicted the P-value would be 1/6, or about 0.167. However, the two-tailed P-value would not be double that value but would be the same value because the first data permutation provided the largest $|r|$ value; the largest negative r value would occur for the sixth data permutation, where the sequence of responses was reversed, and it would have a smaller absolute value ($r = -0.95$) than the obtained positive correlation.

Thus, for asymmetrical distributions of the independent variable the two-tailed P-value need not be twice the one-tailed P-value, and so the one-tailed and two-tailed P-values must be derived separately.

8.2   Computer Program for Systematic Data Permutation

In using a randomization test for determining the P-value by either systematic or random data permutation, a simpler but equivalent test statistic to r can be used. A common formula for determining r by use of a desk calculator is:

$$r = \frac{N\Sigma XY - \Sigma X \Sigma Y}{\sqrt{[N\Sigma X^2 - (\Sigma X)^2][N\Sigma Y^2 - (\Sigma Y)^2]}} \qquad (8.1)$$

It gives the same value of r as other formulas but it is especially useful for deriving simpler equivalent test statistics for data permutation.

Over the data permutations for correlation, the series of X values is paired with the series of Y values in every possible way. The denominator of Equation 8.1 is unaffected by the way in which the values are paired, so it is a constant over all data permutations. As a constant divisor it can be eliminated, leaving the numerator as an equivalent test statistic to r. The second term of the numerator is also constant over all data permutations and so it too can be eliminated, leaving $N\Sigma XY$ as an equivalent test statistic to r. As N is a constant multiplier, it too can be eliminated, leaving ΣXY as an equivalent test statistic to r. We have found then that the proportion of data permutations providing as large a value of ΣXY as an obtained value is the proportion providing as large a value of r as the obtained value. Therefore, the minimum possible ΣXY for a set of data is associated with the strongest negative correlation, and the maximum possible ΣXY is associated with the strongest positive correlation. Thus, to find the P-value for a one-tailed test we find the proportion of data permutations providing a value of ΣXY as large as the obtained value for a predicted positive correlation and the proportion with a ΣXY as small as the obtained value for a predicted negative correlation.

In Section 10.2, the programs for correlation in this chapter will be recommended for use in determining the P-value of the proximity of paired values expressed in the same units, as for example where both X and Y are measures of weight expressed in grams. The programs are useful for that purpose because r is an equivalent test statistic to ΣD^2, the sum of the squared differences between paired X and Y values. A simple way to show this equivalence is to show that ΣD^2 is equivalent to ΣXY, which has just been shown to be equivalent to r. $\Sigma D^2 = \Sigma(X - Y)^2 = \Sigma X^2 - 2\Sigma XY + \Sigma Y^2$, and only the middle term, $2\Sigma XY$, varies over the data permutations, making it and therefore ΣXY an equivalent test statistic to ΣD^2.

The absolute value $|r|$ is the test statistic for a two-tailed test, where the direction of correlation has not been predicted. Because the denominator of Equation 8.1 remains constant over all data permutations, the absolute value of the numerator is an equivalent test statistic to $|r|$. Thus, the two-tailed test statistic used in the computer programs that follow is $|N\Sigma XY - \Sigma X\Sigma Y|$.

Now we will consider the principal operations that the program must perform to determine the P-value by systematic data permutation when Program 8.1 is used.

Step 1. Arrange the N independent variable (X) values in order from low to high and allocate index numbers 1 to N to the measurement (Y) values according to the independent variable values with which they are associated, 1 being assigned to the measurement for the lowest independent variable value and N being assigned to the measurement for the highest independent variable value.

Step 2. Use the procedure described in Section 8.1 for generating a systematic sequence of permutations of the index numbers of the Y values, the first sequence of index numbers being 1, 2, ... (N – 1), N, representing the obtained results, the second sequence being 1, 2, ... N, (N – 1), and so on, the last sequence being N, (N – 1), ... 2, 1. The corresponding measurements for each of the sequences are paired with the X values for the same sequential position, and Step 3 and Step 4 are performed for each sequence. The first sequence of index numbers represents the obtained results, so the performance of Step 3 and Step 4 on the first sequence gives the obtained test statistic value.

Step 3. Compute ΣXY, the sum of the products of the paired X and Y values, as the one-tailed test statistic.

Step 4. Compute $|N\Sigma XY - \Sigma X\Sigma Y|$ as the two-tailed test statistic.

Step 5. For a one-tailed test where a positive correlation is predicted, compute the one-tailed P-value as the proportion of the N! permutations, including the one representing the obtained results, that provide a value of ΣXY greater than or equal to the obtained ΣXY. For a one-tailed test where a negative correlation is predicted, the P-value is the proportion of the permutations with a value of ΣXY less than or equal to the obtained ΣXY.

Step 6. For a two-tailed test, the P-value is the proportion of the permutations that provide a value of $|N\Sigma XY - \Sigma X\Sigma Y|$ greater than or equal to the obtained $|N\Sigma XY - \Sigma X\Sigma Y|$.

Program 8.1 goes through these steps in determining the P-value of r by systematic data permutation. However, the P-value is for a one-tailed test where a positive correlation is predicted. A one-tailed test where a negative correlation is predicted can be conducted with Program 8.1 by changing the signs of the measurement (Y) values.

Program 8.1 can be tested with the following data, for which the one-tailed P-value for a predicted positive correlation is 2/120, or about 0.0167, and the two-tailed P-value is 4/120, or about 0.0333:

X	2	4	6	8	10
Y	1	3	5	8	7

8.3   Correlation with Random Data Permutation

The principal operations that the program must perform to determine the P-value of the product-moment correlation coefficient by random data permutation with Program 8.2 are as follows:

Step 1. Arrange the N independent variable (X) values in order from low to high and allocate index numbers 1 to N to the measurement (Y) values according to the independent variable values with which they are associated, 1 being assigned to the measurement for the lowest independent variable value and N being assigned to the measurement for the highest independent variable value.

Step 2. Compute ΣXY, the sum of the product of the paired X and Y values, as the obtained one-tailed test statistic.

Step 3. Compute $|N\Sigma XY - \Sigma X\Sigma Y|$ as the obtained two-tailed test statistic.

Step 4. Use a random number generation algorithm that will provide a permutation of the index numbers 1 to N; that is, an algorithm that will randomly order the index numbers 1 to N. Where NPERM is the requested number of data permutations, perform (NPERM − 1) data permutations and compute ΣXY and $|N\Sigma XY - \Sigma X\Sigma Y|$ for each.

Step 5. For a one-tailed test where a positive correlation is predicted, the P-value of r is the proportion of the NPERM data permutations — the (NPERM − 1) generated by the random number generation algorithm plus the obtained results — that provide a value of ΣXY greater than or equal to the obtained value. Where a negative correlation is predicted, the P-value is the proportion that provide a value of ΣXY less than or equal to the obtained value.

Step 6. For a two-tailed test, the P-value of $|r|$ is the proportion of the data permutations that provide a value of the test statistic $|N\Sigma XY - \Sigma X \Sigma Y|$ greater than or equal to the obtained value.

Program 8.2 goes through these steps in determining the P-value of r by random data permutation. As with Program 8.1, the P-value given for a one-tailed test is for a prediction of a positive correlation. If a negative correlation is predicted, then you just have to change the signs of the measurement (Y) values and input these values in Program 8.2.

Program 8.2 can be tested with the following data, for which systematic data permutation would give a one-tailed P-value of 2/120, or about 0.0167, for a predicted positive correlation and a two-tailed P-value of 4/120, or about 0.0333:

X	2	4	6	8	10
Y	1	3	5	8	7

8.4   Multivariate Correlation

Multivariate correlation for experiments is that for which there are two or more dependent variables. We will consider two types of situations: those situations in which all of the dependent variables are expected to be correlated with the treatment values in the same way, that is, all positively or all negatively correlated; and those situations in which the direction of correlation can vary over the dependent variables. For both types of multivariate correlation, the first step is to transform the values for each dependent variable into z scores. When all dependent variables are expected to show the same direction of correlation with the independent (treatment) variable, a composite z score, which is the sum of the z scores for the dependent variables, is used as a composite measurement for a subject for a particular independent variable value. (See Section 7.4 regarding construction of composite measurements.) The N! data permutations used to determine the P-value consist of every possible pairing of the N composite measurements with the N independent variable values, and for each data permutation a one- or two-tailed test statistic is computed.

When the nature of the variation in the direction of correlation between dependent variables and the independent variable is unpredictable, it is not appropriate to add the z scores for each subject. Instead, the one-tailed test statistic ΣXY or the two-tailed test statistic $|N\Sigma XY - \Sigma X \Sigma Y|$ should be computed for the z scores for each dependent variable separately. For each data permutation, the multivariate test statistic is the sum of the univariate test statistics. For example, for a one-tailed test the multivariate test statistic is ΣXY for the first dependent variable plus ΣXY for the second, and so on, where the Y values are z scores.

8.5   Point-Biserial Correlation

A *point-biserial correlation coefficient* is a product-moment correlation coefficient based on the correlation between a dichotomous and a continuous variable. For experimental applications, the dichotomous variable can be either the independent or the dependent variable.

When the independent variable for a completely randomized experiment is dichotomous and the dependent variable is continuous, Program 4.4 gives P-values for t and $|t|$ that are the same as would be given by Program 8.1 for r and $|r|$, using 0s and 1s as the X values to be paired with the continuous Y values. The two programs give the same P-values for this type of application because, as will be shown below, the test statistics for Program 4.4, which are T_L and $\Sigma(T^2/n)$, are equivalent to the test statistics for Program 8.1, which are ΣXY and $|N\Sigma XY - \Sigma X\Sigma Y|$.

To see the equivalence of the test statistics in the two programs, consider the following X and Y values where a 1 indicates treatment A (the treatment predicted to provide the larger measurements) and a 0 indicates treatment B (predicted to provide the smaller measurements). $\Sigma X = n_A$ because the only nonzero X values are the 1s denoting applications of treatment A. Each Y value associated with treatment A yields an XY value that is the same as the Y value, and each Y value associated with treatment B yields an XY value of 0; thus, $\Sigma XY = T_A$, the total of the measurements for treatment A.

X	Y	XY
1	Y_1	Y_1
1	Y_2	Y_2
.	.	.
.	.	.
.	.	.
0	Y_{N-1}	0
0	Y_N	0
$\Sigma = n_A$	ΣY	T_A

The numerical identity of ΣXY and T_A shows the equivalence of the one-tailed test statistics r and t. The equivalence of the two-tailed test statistics $|r|$ and $|t|$ can be determined by examining a formula (Nunnally, 1967) for point-biserial r:

$$r_{pb} = \frac{\bar{X}_A - \bar{X}_B}{\sigma}\sqrt{pq} \qquad (8.2)$$

where $\bar{X}_A$ and $\bar{X}_B$ are the means of the continuous variable associated with the two levels of the dichotomous variable, σ is the standard deviation of the continuous variable for both groups combined, and p and q are proportions of subjects for the two levels of the dichotomous variable. Over all data permutations, only $\bar{X}_A - \bar{X}_B$ varies, so $|r|$ is an equivalent test statistic to $|\bar{X}_A - \bar{X}_B|$ and thus to $|t|$. To understand the following demonstration of the equivalence of $|N\Sigma XY - \Sigma X\Sigma Y|$ and $|\bar{X}_A - \bar{X}_B|$, it will be necessary to keep in mind that $\Sigma X = n_A$ and $\Sigma XY = T_A$.

The two-tailed correlation test statistic in Program 8.1, $|N\Sigma XY - \Sigma X\Sigma Y|$, can be expanded to $|(n_A + n_B)(n_A \bar{X}_A) - (n_A)(n_A \bar{X}_A + n_B \bar{X}_B)|$, which is algebraically equal to $|(n_A n_B)(\bar{X}_A - \bar{X}_B)|$. Because $n_A n_B$ is a constant multiplier over all data permutations, it can be dropped, leaving $|\bar{X}_A - \bar{X}_B|$ and thus $|t|$ as an equivalent test statistic to $|r|$.

Program 8.1 is not as useful as Program 4.4 for point-biserial correlation because it performs $n_A! n_B!$ times as many data permutations as Program 4.4 to get the same P-values. This redundancy arises from the fact that for each division of Y values between A and B (treatments represented by 1 and 0), there are $n_A!$ pairings of the values for A with the 1s in the X column and $n_B!$ pairings of the values for B with the 0s in the X column.

Next, consider point-biserial correlation when we have a continuous independent variable and a dichotomous dependent variable. The output of Program 8.1 is unaffected by which variable is called the X variable and which the Y variable. The sequential order of values for one variable is held fixed while the sequence of values for the other variable is permuted. Which variable is called X and has its sequence fixed while the other variable's sequence is permuted does not affect the distribution of $(n_A + n_B)!$ pairings and so we can, as in the preceding proof of the equivalence of the test statistics for Program 4.4 and Program 8.1, list the continuous variable under Y and the dichotomous variable under X. The preceding demonstration of equivalence of the test statistics for the two programs now can be seen to apply also to situations where the independent variable is continuous and the dependent variable is dichotomous. So whether the independent variable is dichotomous and the dependent variable continuous or vice versa, Program 4.4 can be used instead of Program 8.1. Furthermore, because Program 4.4 performs the same test with fewer data permutations, it is the preferred program. In using Program 4.4 when the independent variable is continuous, the values of the independent variable are treated as if they were the data, and the dichotomous results of the study are treated as if they represented two treatments.

Example 8.3. An experimenter assigns five dosages of a drug randomly to five animals, with one dosage per animal, and records the presence (1) or absence (0) of adverse side effects. The following results are obtained:

X	5	10	15	20	25
Y	0	0	1	1	1

To statistically test the effect of dosage level, the experimenter uses Program 4.4 as if the five animals had been assigned randomly to treatments A and B, with three animals for A (designated as treatment 1) and two animals for B (designated as treatment 0). Within this framework, the obtained "data" for treatment A are 15, 20, and 25, and the obtained "data" for treatment B are 5 and 10. These five numbers are divided in $5!/2!3! = 10$ ways by Program 4.4 to determine the one-tailed and two-tailed P-values.

When one variable is dichotomous and the other variable consists of ranks, of course Program 4.4 still can be used whether the ranks are ranks of treatments or ranks of data. As explained in Section 4.16, in this case Program 4.4 gives the same P-value as the Mann-Whitney U test if there are no tied ranks. However, the program is more general, providing a valid test even in the presence of tied ranks.

8.6   Correlation between Dichotomous Variables

Next, consider applications of correlation to experiments with dichotomous independent and dependent variables. Application of Program 8.1 to the data, where the 0s and 1s representing the treatments are paired with 0s and 1s representing the experimental results, gives the same P-value for r and $|r|$ as Program 4.4 gives for t and $|t|$ for the same data. (The equivalence of the test statistics for the two programs described in Section 8.5 obviously holds when both variables are dichotomous as well as when one is dichotomous and the other continuous.)

As discussed in Section 4.17, the P-value for t given by Program 4.4 is that for Fisher's exact test for a 2×2 contingency table, and the P-value for $|t|$ given by Program 4.4 is the P-value for contingency chi-square when both the independent and the dependent variables are dichotomous. Thus, the application of Program 4.4 or Program 8.1 to dichotomous data from two treatments provides the P-value for Fisher's exact test (one-tailed test) and the randomization test P-value for the (two-tailed) contingency chi-square test.

8.7   Spearman's Rank Correlation Procedure

Spearman's rank correlation procedure is a commonly used nonparametric procedure for which published significance tables are available. Spearman's rank correlation coefficient, ρ, is a product-moment correlation coefficient where the correlation is based on ranks instead of raw measurements. The independent and dependent variable values are ranked separately from 1 to N, and ρ is computed from the paired ranks by the following formula:

$$\rho = 1 - \frac{6\Sigma D^2}{N(N^2 - 1)} \tag{8.3}$$

where D is the absolute difference between a pair of ranks and N is the number of pairs of ranks. The minimum value of ρ is −1 and the maximum value is +1. ΣD^2 is the only part of Equation 8.3 that varies over data permutations, so it is an equivalent test statistic to ρ and some significance tables for ρ show the P-values for this simpler test statistic. When data for both X and Y variables are expressed as ranks with no tied ranks, the P-value given by Program 8.1 for systematic data permutation is the same as the value given by the significance table for Spearman's ρ because the table is based on data permutation.

8.8 Kendall's Rank Correlation Procedure

Probably Spearman's is the best-known rank correlation procedure, but *Kendall's procedure* is a close second, and as Bradley (1968) observed, it has a high degree of versatility and utility. Siegel (1956) and Gibbons (1971) discuss the extension of the procedure to partial rank correlation. An example will illustrate the computation of Kendall's τ, as the correlation coefficient is known.

Example 8.4. Suppose five treatments are ranked from low to high and that the five dependent variable values also are ranked from low to high. The treatment ranks are arranged in sequential order and the associated dependent variable ranks are listed under the treatment ranks, as follows:

Treatment Ranks	1	2	3	4	5
Dependent Variable Ranks	2	1	5	3	4

Each dependent variable rank is paired with the ranks that follow it, the 2 being paired with the four ranks that follow it, the 1 with three ranks, the 5 with two ranks are in ascending order within a pair (i.e., where the rank on the right is greater than the rank on the left) is $3 + 3 + 0 + 1 = 7$, and the number of pairs in which the ranks are in descending order is $1 + 0 + 2 + 0 = 3$. Kendall's τ is $(7 - 3)/10$, or 0.40. The general formula is:

$$\tau = \frac{S}{n(n-1)/(n/2)} \tag{8.4}$$

where S is the number of pairs where the ranks are in ascending order minus the number of pairs where they are in descending order, and the denominator is the maximum value of the numerator, the total number of pairs for n ranks. The lower limit of τ is −1 and the upper limit is +1.

Significance tables for Kendall's τ, based on the reference set of all $n!$ data permutations, are standard inclusions in books on nonparametric statistics. The $n!$ permutations of the data would be performed on the ranks entered into Program 8.1 but the test statistic portion would have to be revised to compute the above test statistic; otherwise, the P-value would be for Spearman's ρ. The statement by Siegel (1956) that Spearman's ρ and Kendall's τ will give the same P-value is incorrect because they are not equivalent test statistics; if they were, Program 8.1 could be used to test the significance of τ as well as ρ.

8.9   Questions and Exercises

1. What is a point-biserial correlation?

2. A formula for Spearman's rank correlation coefficient is $1 - 6\Sigma D^2/[N(N^2 - 1)]$. What is a simple equivalent test statistic for a randomization test that does not contain D or N?

3. Suppose that instead of using the formula for Spearman's rank correlation given in Question 2 as our randomization test statistic, we used only $6\Sigma D^2/[N(N^2 - 1)]$ without subtracting it from 1. What adjustment should we make in computing the P-value to get the same P-value as would be obtained with the formula in Question 2?

4. In Section 4.15, two methods were given for conducting a randomization test when one or more subjects dropped out after being assigned to a treatment. What are analogous approaches that can be made for dropouts when an experimenter employs a randomization test of treatment effect based on correlation?

5. Suppose that in the experiment described in Question 4, five subjects have been assigned to five treatment levels with one subject per treatment, and one of the five subjects does not show up to take the assigned treatment. How could you still have a randomization test with 120 data permutations?

6. The general formula for Kendall's rank correlation coefficient is given in Equation 8.4. What would be a much simpler equivalent test statistic?

7. What is the general null hypothesis for a randomization test using any of the correlation procedures in this chapter?

8. An experimenter is interested in a randomization test using a correlational procedure with five drug dosages but only one subject. Describe the random assignment. Why would this experimenter use only one subject if it were easy to obtain five subjects for the experiment?

9. Suppose the following X values were used, and Y-values observed:

X	2	4	6	8	10	12
Y	9	7	8	6	5	4

Calculate the one-tailed P-value for a predicted negative correlation.

10. A psychopharmacologist wants to test the side-effects of a medical compound on six psychiatric patients. He randomly assigns three patients to the treatment condition and the other three take a placebo, and after the administration of the pills he assesses the extent to which these patients are experiencing side-effects. The three patients in the treatment condition report 4, 8, and 3 side-effects, respectively. In the placebo condition, only one of the patients reports one side-effect (dry mouth) whereas the others do not report any side-effects. Use a binary indicator for treatment (1) versus placebo (0) and use Program 8.1 to calculate the one-tailed P-value for a predicted positive correlation between treatment and the number of side-effects. What is your conclusion? Do you arrive at the same results with Program 4.4?

REFERENCES

Bradley, J.V., *Distribution-Free Statistical Tests*, Prentice Hall, Englewood Cliffs, NJ, 1968.

Gibbons, J.D., *Nonparametric Statistical Inference*, McGraw-Hill, New York, 1971.

Nunnally, J.C., *Psychometric Theory*, McGraw-Hill, New York, 1967.

Pitman, E.J.G., Significance tests which may be applied to samples from any populations. II. The correlation coefficient test, *J. R. Statist. Soc. B.*, 4, 1937, 225–232.

Siegel, S., *Nonparametric Statistics for the Behavioral Sciences*, McGraw-Hill, New York, 1956.

9

Trend Tests

The principal function served by parametric trend analysis is to determine the relative contribution of linear, quadratic, and higher-order polynomial functions to the variation of measurements over a quantitatively ordered series of treatments. The trend tests in this chapter have a different function, which is to test H_0 of no treatment effect by using a test statistic sensitive to a certain type of trend that is expected. If the predicted trend is realized in the data, the trend test will be more sensitive than without a trend prediction, but it must be noted that randomization trend tests do not test hypotheses about trends; they simply use test statistics sensitive to trend which test the null hypothesis of no differential treatment effect.

The one-tailed t test can be considered a simple trend test where there is a prediction of an upward or downward trend in measurements from treatment A to treatment B. General upward or downward ("linear") trends over several treatment levels frequently are expected when the treatments differ quantitatively, for example, as with variation in drug dosage. But trends of other types also are commonly expected. For example, an inverted U-shaped ("low-high-low") trend in behavioral response rate in rats over doses of cadmium has been found (Newland, et al., 1986), and those investigators stated that the same type of trend "has also been found with chronic exposure to lead (Cory-Slechta and Thompson, 1979; Cory-Slechta, et al., 1983) as well as many drugs (Seiden and Dykstra, 1977)."

It is necessary to stress that in this chapter, as in previous chapters, no attempt at comprehensiveness is made and the tests discussed are not presented as superior to other randomization tests that serve a similar function. The primary objective of this chapter is to make randomization trend tests understandable, thus facilitating the use of existing tests and the development of new tests.

9.1 Goodness-of-Fit Trend Test

The goodness-of-fit trend test (Edgington, 1975) was developed to accommodate very specific predictions of experimental results for Price (Price and Cooper, 1975). Price replicated and extended work done by Huppert and Deutsch (1969) on the retention of an avoidance learning response as a function of the temporal length of the interval between training and testing. Price used training-testing intervals of various lengths, ranging from 30 minutes to 17 days, as the levels of the treatment variable. She wanted a test that would be quite powerful if the experimental data closely matched her prediction: small measurements for the four-day training-testing interval and increasingly larger measurements for shorter and longer intervals. That is, she predicted a U-shaped trend with the low point of the "U" located at the four-day training-testing interval. This prediction was based in part on the results of the study by Huppert and Deutsch, and in part on pilot work by Price. Huppert and Deutsch had obtained a U-shaped distribution of the mean number of trials taken by rats to relearn a task plotted against the length of the training-testing interval. Therefore, Price expected to get U-shaped distributions from her experiments. She used a different level of shock and expected that difference to alter the location of the low point of the U. Consequently, the low point was predicted on the basis of a pilot study by Price instead of on the results of Huppert and Deutsch.

Price had a specific prediction and wanted a procedure that would take into consideration its specificity. The first procedure considered was orthogonal polynomial trend analysis with the P-value determined in the usual way, by reference to F tables. However, there was the question of whether the use of such a complex and little-used technique would induce critical examination of the tenability of the parametric assumptions. Comparative and physiological psychology journals, the appropriate media for her study, tend to use nonparametric tests rather frequently and that might reflect a critical attitude regarding parametric assumptions. (The use of one-way ANOVA, with the P-value determined by F tables, might go unquestioned by those journals but the test would not be sensitive to the predicted trend.) Consequently, parametric polynomial trend analysis was judged to be inappropriate. The second possibility considered was the use of parametric polynomial trend analysis with the P-value determined by a randomization test. Several difficulties arose in trying to do this analysis.

First, the books at hand that discussed trend analysis restricted consideration to cases with equal sample sizes and equally spaced magnitudes of the independent variable, and neither of these restrictions was met by the experiment. Admittedly, there are journal articles that describe parametric trend analysis for unequal sample sizes and unequally spaced magnitudes of the independent variable, but researching such articles would have been worthwhile only if the trend analysis procedure was otherwise useful for the application.

However, even superficial consideration of the standard trend analysis procedure suggested its inadequacy for the desired analysis of the data. The standard test for quadratic trend, the test to use for prediction of a U-shaped data distribution, could not use the specificity of the prediction. The parametric trend test is two-tailed in the sense of not permitting differential predictions for U-shaped and inverted U-shaped trends, whereas Price's prediction was not a general prediction of that type but of a U-shaped trend in most cases. Second, the standard parametric trend procedure uses published tables of polynomial coefficients that do not permit specifying a predicted point of symmetry (low point for a U-shaped curve) near the lower end of the range of the independent variable magnitudes, and so the four-day low-point prediction could not be used.

Because of these considerations, the use of standard parametric trend analysis with the P-value determined by means of a randomization test was ruled out in favor of developing a new test with a test statistic that could use highly specific predictions of trend. The randomization trend test that was developed for this purpose is called the *goodness-of-fit* trend test because of the similarity between its test statistic and that of the chi-square goodness-of-fit test. The goodness-of-fit trend test can be used with various types of predicted trend in addition to the U-shaped trend for which the test was originally developed.

9.2 Power of the Goodness-of-Fit Trend Test

After developing the goodness-of-fit test, it was decided to compare the P-value given by that test with that given by the alternative test that the experimenter would have used, one-way ANOVA with the P-value based on F tables. As mentioned earlier, the complex assumptions underlying orthogonal polynomial trend analysis made it unacceptable because of anticipated objections by journal editors. Alternatively, there probably would have been little objection to using simple one-way ANOVA with the P-value determined by F tables, and so it was a possible alternative to the goodness-of-fit test. Consequently, each set of data was analyzed by both the goodness-of-fit test and simple one-way ANOVA (with the P-values from the F table) to permit a comparison of the P-values given by the two procedures.

Example 9.1. First, consider some training-testing interval data for which simple one-way ANOVA gave a P-value greater than 0.20, whereas the goodness-of-fit test for U-shaped trends gave a P-value of 0.007:

Time	30 min	1 day	3 days	5 days	7 days	10 days
Means	45.5	40.4	33.1	31.9	36.2	43.1

The data conform well to a U-shaped distribution with a low point at 4 days, and so the randomization test P-value is small.

Example 9.2. Next, consider training-testing interval data where ANOVA gave a P-value between 0.05 and 0.10, whereas the goodness-of-fit test gave a P-value of 0.004:

Time	30 min	1 day	3 days	5 days	7 days
Means	12.6	9.9	6.2	6.4	9.6
Time	9 days	10 days	11 days	14 days	17 days
Means	13.6	14.5	13.7	14.1	18.7

Although the trend is not as consistent as in the previous example, the means do increase in value above and below the predicted low point of 4 days and the inconsistency in the upward trend from 10 to 14 days is so slight that the trend over the entire 17 days matches the predicted trend very closely. Therefore, the P-value for the randomization test is considerably lower than that for one-way ANOVA, where no predicted trend of any type is involved.

Example 9.3. The final example based on Price's training-testing interval data is one where she predicted an inverted U-shaped trend. ANOVA provided a P-value of 0.01 whereas the goodness-of-fit test gave a much larger P-value, a value of approximately 1:

Time	30 min	1 day	3 days	10 days
Means	0.46	0.43	0.43	0.36

In this case, the goodness-of-fit randomization test gave a large P-value because the data deviate considerably from the predicted inverted U-shaped trend for this particular experiment in the study. The test determines the probability, under H_0, of getting a distribution of means that so closely matches the expected trend, and in this case the goodness-of-fit P-value is about 1 because the obtained trend deviates so much from the predicted trend.

The goodness-of-fit test for trend is quite powerful when the trend is rather accurately predicted but, as illustrated in Example 9.3, it is very weak when the observed trend differs considerably from the predicted trend. Therefore, the test is very powerful only if it is used with discretion. As some critics of one-tailed tests might say, a directional test should be used only if any treatment differences in the opposite direction are regarded as either irrational or irrelevant.

9.3   Test Statistic for the Goodness-of-Fit Trend Test

The goodness-of-fit trend test can be used to test the null hypothesis of no differential treatment effect by comparing experimental results with a predicted trend. The predicted trend is represented by a distribution of "trend means" for the levels of the treatment. The trend means are computed in various ways, depending on the type of trend that is predicted. The goodness-of-fit trend test statistic is $n_1(\bar{X}_1 - TM_1)^2 + \cdots + n_k(\bar{X}_k - TM_k)^2$, where n_k is the number of subjects in the kth treatment group, $\bar{X}_k$ is the mean for the kth group, and TM_k is the trend mean for the kth group. In other words, the test statistic is the weighted sum of squared deviations of the measurement means from the corresponding trend means. The closer the means to the trend means, the smaller is the test statistic value, and so the P-value associated with the experimental results is the proportion of the test statistic values in the reference set that are as small as the obtained test statistic value.

9.4 Computation of Trend Means

We will now consider the derivation of distributions of trend means. The goodness-of-fit test does not dictate the manner in which trend means are computed for the test, but a particular procedure that has been found useful for computing trend means will be the focus of attention in this section. That procedure is the one associated with the trend tests employed by Price (Price and Cooper, 1975), whose study and experimental results were described earlier in this chapter.

The method for determining the trend means depends on the type of trend expected. Let us start with the method to be used when the experimenter expects a linear upward trend in the means. The first step is to get "coefficients" for the independent variable (treatment) values by subtracting the smallest value from each of the values. For example, if the independent variable values were 20, 25, 50, 75, and 80, the corresponding coefficients would be 0, 5, 30, 55, and 60. We then solve the following equation for a multiplier m: $m(0n_1 + 5n_2 + 30n_3 + 55n_4 + 60n_5) = GT$, where GT is the grand total of the obtained measurements for the five groups. Thus, multiplier m can be computed by dividing the grand total of the obtained measurements for the five groups by the sum of the product of the sample sizes and the coefficients. When the value of m is thus obtained, it is then multiplied by the individual coefficients to obtain the trend means. The trend mean of the first group then is $0m$, the trend mean of the second is $5m$, and so on. The trend means derived by this procedure have the following two properties: first, because of the way the multiplier m is determined the sum of the product of the trend means and their corresponding sample sizes equals the grand total; and second, by setting the smallest coefficient and consequently the smallest trend mean equal to 0, the trend means plotted on a graph form the steepest upward sloping line possible for nonnegative trend means. (If the data contains negative values, the data can be adjusted before computing coefficients by adding a constant to eliminate negative values. Like other linear transformations of the data, this does not affect the goodness-of-fit P-values.) The importance of this second property can be seen in the following example.

Example 9.4. A biologist randomly assigns six rats to three treatment conditions, with two rats per treatment. The treatments consist of exposing an image on a screen in front of a rat for either 5, 10, or 15 seconds. The biologist expects the response magnitudes to increase with an increase in the duration of the projected image. The trend coefficients are obtained by subtracting 5 (the smallest exposure time) from 5, 10, and 15 to provide 0, 5, and 10. These trend coefficients are then multiplied by $90/[(2)(0) + 2(5) + 2(10)] = 3$ to get 0, 15, and 30 as trend means. Consider the computation of the test statistic value for the obtained results, using these trend means:

Exposure time (seconds)	5	10	15
Trend coefficients	0	5	10
Trend means	0	15	30
Measurements	0, 7.5	15, 15	22.5, 30
Test statistic value: $2(3.75 - 0)^2 + 2(15 - 15)^2 + 2(26.25 - 30)^2 = 56.25$			

There are $6!/2!2!2! = 90$ permutations of the measurements over the three treatments and the above data permutation for the observed results is the only one with such a small test statistic value as 56.25, and so the P-value is $1/90$, or about 0.011.

The derivation of trend means for expected downward linear trends can be performed in an analogous manner, as also for quadratic trends. For each trend, the trend means multiplied by the corresponding sample sizes would equal the grand total of the measurements and the trend means would form the steepest trend of the specified type for nonnegative trend means.

If the experimenter predicts a linear downward trend, the coefficients can be determined by subtracting each independent variable value from the largest value. For example, if the independent variable values were 10, 12, 18, and 25, the coefficients would be 15, 13, 7, and 0, respectively. These coefficients would be used to determine the trend means in the manner described earlier for a predicted upward trend.

Now consider the method for determining the trend means when the experimenter expects a U-shaped distribution of means that is symmetrical about a predetermined point of symmetry. Obtaining coefficients for the independent variable values for this type of distribution is done by squaring the deviation of the independent variable values from the predetermined value expected to be the low point of the distribution.

Example 9.5. In Example 9.1, the predicted low point was 4 days, so the coefficient for a training-testing interval is the square of the deviation of the training-testing interval from 4 days:

Time	30 min	1 day	3 days	5 days	7 days	10 days
Coefficients	16	9	1	1	9	36

(Actually, 30 minutes is not 4 days deviation from the predicted low point of 4 days, but it is very close.) Using these coefficients, the trend means were determined by solving for the multiplier m, as shown in the second paragraph of this section. The sample sizes for the six intervals were 11, 10, 15, 12, 8, and 10, respectively, and the grand total was 2503.3. (Multiplying each mean in Example 9.1 by the respective sample size and summing gives a total of 2504.4 instead of 2503.3 because of rounding errors for the means in this example.) Consequently, the following equation is solved for m: $m[(16)(11) + (9)(10) + (1)(15) + (1)(12) + (9)(8) + (36)(10)] = 2503.3$. The value of m is then $2503.3/725$, or about 3.45. Therefore, in order of training-testing interval length, the trend means were: $(3.45)(16), (3.45)(9), (3.45)(1), (3.45)(1), (3.45)(9),$ and $(3.45)(36)$; that is, the trend means were: 55.20, 31.05, 3.45, 3.45, 31.05, and 124.20.

If these means were plotted against their corresponding training-testing intervals, they would form a U-shaped quadratic curve with a low point of 0 at 4 days. The obtained test statistic value then was: $11(45.5 - 55.20)^2 + 10(40.4 - 31.05)^2 + 15(33.1 - 3.45)^2 + 12(31.9 - 3.45)^2 + 8(36.2 - 31.05)^2 + 10(43.1 - 124.20)^2 = 90,793.2$. At this point, it should be mentioned that there was no basis for predicting a particular U-shaped quadratic function for this study; the quadratic function produced by the above procedure was derived because it constituted an objective, quantitative definition of a steep, symmetrical U-shaped distribution centered at 4 days, with which the observed distribution and the distributions under all permutations of the data could be compared.

The method for determining trend means for an inverted U-shaped distribution about a predetermined high point is similar to that for the U-shaped distribution. The coefficients are computed as for a U-shaped distribution and then subtracted from the largest coefficient.

Example 9.6. For the data in Example 9.5, if the experimenter had predicted an inverted U-shaped trend centered on the independent variable value of 4 days, the coefficients would be:

Time	30 min	1 day	3 days	5 days	7 days	10 days
Coefficients	20	27	35	35	27	0

Each of the coefficients for a U-shaped trend test was subtracted from 36, the coefficient for 10 days. Using these new coefficients, the following equation is solved for m: $m[(20)(11) + \ldots + (0)(10)] = 2503.3$. Each of the coefficients is then multiplied by the computed value of m to get the trend means.

In addition to linear and quadratic functions, other polynomial functions like cubic, quartic, and quintic can be used as predicted trends. When the independent variable values are uniformly spaced, published tables of polynomial coefficients can be used to generate trend means for polynomial functions.

Example 9.7. Assuming that there are no negative measurement values, if we have uniformly spaced independent variable values and expect a cubic trend we could use cubic polynomial coefficients from a table of orthogonal polynomial coefficients in Winer (1971) to generate the following goodness-of-fit coefficients:

Independent variable values	15	20	25	30	35	40	45	50	55
Polynomial coefficients	−14	7	13	9	0	−9	−13	−7	14
Goodness-of-fit coefficients	0	21	27	23	14	5	1	7	28

The distribution of goodness-of-fit coefficients is obtained from the polynomial coefficients by adding a constant, 14, that will make the lowest value 0. It can be seen that the derived distribution predicts an upward trend followed by a downward trend and, finally, another upward trend.

Example 9.8. Alternatively, if it is more reasonable to expect the trend to be "down-up-down," one should not use the derived coefficients but should subtract each goodness-of-fit coefficient in Example 9.7 from 28, the largest coefficient. This provides the following sequence of coefficients:

Independent variable values	15	20	25	30	35	40	45	50	55
Coefficients		28	7	1	5	14	23	27	21	0

The new distribution of coefficients is appropriate for the "down-up-down" trend.

9.5   Computer Program for Goodness-of-Fit Trend Test

Program 9.1, which is part of the RT4Win package on the included CD, is satisfactory for any type of predicted trend provided that the distribution of trend means is computed separately and entered into the program along with the data. Program 9.1 can be tested with the following data, for which the trend means for an upward linear trend have been computed and which should give a P-value of 2/90, or about 0.0222, by systematic data permutation:

Independent variable values	100	150	200
Trend means	0	5	10
Measurements	1, 3	2, 4	9, 11

9.6   Modification of the Goodness-of-Fit Trend Test Statistic

The goodness-of-fit trend test permits very precise predictions of trend and can considerably increase the likelihood of detecting a treatment effect. The computation of the trend means complicates the procedure but that computation could readily be incorporated into Program 9.1. However, it became evident that the test statistic given in Section 9.3 does not adequately reflect the predicted trend. For a linear trend, the trend means represent a straight line on a graph and we are interested in the variability of the measurements about the straight line, small variability being indicative of close agreement between the predicted trend and the data. Instead, the test statistic involves the variability of treatment means about the predicted straight line, making it equivalent for a randomization test to the variance of the treatment means about the line. A slight change in the test statistic in Section 9.3 would transform it into one equivalent to the variance of the measurements about the trend line: change $n_1 (\bar{X}_1 - TM_1)^2 + \cdots + n_k (\bar{X}_k - TM_k)^2$ to $\Sigma(X - TM)^2$.

9.7 Correlation Trend Test

The modified goodness-of-fit trend test statistic, $\Sigma(X - TM)^2$, in Section 9.6 is obtained by pairing each measurement with a trend mean, as for determining the correlation between them. In fact, the product-moment correlation between the paired X and TM values would be an equivalent test statistic to $\Sigma(X - TM)^2$ because, as was shown in Section 8.2, the sum of squared differences between paired values is an equivalent (but inversely related) test statistic to the product-moment correlation coefficient when data permutations are generated by pairing n pairs of values in all $n!$ ways.

We call this test a *correlation trend test* because the test statistic is the product-moment correlation coefficient. With the correlation trend test, one can consider two-tailed tests as well as one-tailed tests, with test statistics equivalent to r and $|r|$, where the correlation is between the measurements and trend means that represent the expected trend. And because the trend means computed for the goodness-of-fit trend test are linearly related to the coefficients, trend means need not be computed for the correlation trend test because the same P-value would be obtained by correlating the measurements with the coefficients. The correlation trend test does not require the trend coefficients to be computed in some particular way; it is a valid randomization test procedure for any set of coefficients that is independent of the data. Instead of discussing the correlation trend test as a modified goodness-of-fit trend test, it will be presented as a different, independently developed test with its own computer program.

This test is both computationally and conceptually simpler than the goodness-of-fit trend test, yet it can accommodate quite specific predictions. There is no need to compute trend means; only the coefficients that define the shape of the trend are required. The logical link between this test and conventional statistical tests is clear. First, this test is simply a generalization of the correlation test considered in Chapter 8. When there is only one subject assigned to each level of the independent variable, the correlation trend test is reduced to the ordinary correlation test. Second, the correlation trend test is related to the parametric trend test. The numerator of the orthogonal polynomial trend test F is equivalent under data permutation to r^2, the square of the correlation between the trend polynomial coefficients and the measurements. For the special case where the coefficients are the same as a set of polynomial coefficients (or where the two sets are linearly related), the two-tailed correlation trend test statistic is an equivalent test statistic to the numerator of the parametric trend F. An example will illustrate the employment of the correlation trend test where a linear upward trend is predicted.

Example 9.9. Suppose that we expected a linear upward trend over five levels of an independent variable, the five levels being 10, 15, 20, 25, and 30. We use the goodness-of-fit procedure to determine the following coefficients:

Treatment magnitude	10	15	20	25	30
Coefficients	0	5	10	15	20

Fifteen subjects randomly assigned to the treatment magnitudes provided these results: (23, 21, 20), (18, 25, 16), (12, 10, 13), (17, 10, 8), and (12, 16, 4). The measurements would be paired with the independent variable coefficients in the way shown in Table 9.1. The obtained test statistic value for our one-tailed test is 1810. Thus, we could determine the proportion of the $15!/(3!)^5$ data permutations that give a value of 1810 or larger for ΣXY, and that would be the P-value for the treatment effect.

With correlation trend tests, we are concerned with the set of $N!/n_1!n_2!...n_k!$ data permutations associated with random assignment to independent groups. The permutation of the data for the correlation trend test is the same as for one-way ANOVA because the random assignment procedure is the same.

For a two-tailed test where a linear trend is expected but the direction is not predicted, the test statistic used is $|N\Sigma XY - \Sigma X\Sigma Y|$, the two-tailed test statistic for correlation. ΣX is the sum of the N coefficients for all N subjects, not the sum of the k coefficients for the k treatment levels. It does not matter whether the coefficients used are those for upward or for downward linear trend; the two-tailed test statistic value for any data permutation is the same for both sets of coefficients and, consequently, the P-value is the same.

Example 9.10. We have the same independent variable values as in Example 9.9 but predict a U-shaped trend, with the low value of the U at the independent variable value of 15. As described in Section 9.4, we compute our coefficients for the goodness-of-fit test to get the following values:

TABLE 9.1

Correlation Trend Test Data

Coefficients (X)	Measurements (Y)	XY
0	23	0
0	21	0
0	20	0
5	18	90
5	25	125
5	16	80
10	12	120
10	10	100
10	13	130
15	17	255
15	10	150
15	8	120
20	12	240
20·	16	320
20	4	80
$\Sigma X = 150$	$\Sigma Y = 225$	$\Sigma XY = 1810$

Treatment magnitude	10	15	20	25	30
Coefficients	25	0	25	100	225

These coefficients are paired with the measurements to determine how strongly they are correlated. A strong product-moment correlation between the measurements and the coefficients implies that the magnitudes of the measurements tend to be distributed over the independent variable levels in the same way as the coefficients.

The two-tailed test statistic $|N\Sigma XY - \Sigma X\Sigma Y|$ would have been used in Example 9.10 if no prediction was made as to whether the trend would be U-shaped or inverted U-shaped. The two-tailed prediction then would be that of a large absolute value of the correlation between the coefficients and the measurements, and that prediction would be confirmed by either a strong positive or a strong negative correlation between the coefficients and the measurements. As with any predicted trend, it makes no difference for the correlation trend test which of the two sets of coefficients is used to determine the two-tailed P-value.

Program 9.2 from the included CD can be used to determine one- and two-tailed P-values by random data permutation for predicted trends using the correlation trend test. It presupposes independent-groups random assignment of subjects to treatments with the treatment sample sizes fixed in advance, the type of random assignment used with one-way ANOVA. Program 9.2 therefore was written by combining the data permuting portion of Program 4.3 (one-way ANOVA: random permutation) with the test statistic computing portion of Program 8.2 (product-moment correlation: random permutation). Whereas the program for the goodness-of-fit trend test required trend means to be specified, this program requires specifying the coefficients. Program 9.2 can be tested with the following data, where the coefficients are based on a predicted downward linear trend. The one-tailed P-value for systematic data permutation is 2/90, or about 0.0222, and the two-tailed P-value is 4/90, or about 0.0444.

Independent variable values	30	35	40
Coefficients	3	2	1
Measurements	16, 19	14, 18	8, 9

9.8 Correlation Trend Test for Factorial Designs

Tests of main effects can be carried out on data from factorial designs when a trend is predicted for some or all of the factors, for either completely randomized or randomized block experiments. When a factor under test does not have a predicted trend, Program 5.1 can be used in its present form

TABLE 9.2

Treatments-by-Subjects Design

	Coefficients		
	0	5	10
Males	2	5	8
	4	5	9
Females	6	10	12
	7	11	14

to test the effect of the factor, but to test effects of factors with predicted trends, extensive modifications would be necessary. However, it would appear that Program 9.2 could be employed with only slight modifications, those modifications necessary to permute the data independently for separate blocks (or levels) of a factor. We will now describe a two-factor experiment with a blocking factor (sex) in which there is the same predicted trend within each block.

Example 9.11. We have a treatments-by-sex experimental design where we expect a linear upward trend over levels of the manipulated factor for both sexes. There are three levels of the independent variable and the coefficients are determined to be 0, 5, and 10. There are six males and six females, randomly assigned to the three treatment levels with the restriction that two males and two females take each treatment. The results are shown in Table 9.2. The obtained ΣXY is $(0)(2) + (0)(4) + \ldots + (10)(12) + (10)(14) = 585$. The P-value given by systematic data permutation would be the proportion of the $(6!/2!2!2!)^2 = 8100$ data permutations that provide as large a value of ΣXY as 585. No other data permutation would provide as large a value as the obtained results because within both sexes there is a consistent increase in size of the scores from the lowest to the highest levels of the treatments. Thus, the P-value would be 1/8100, or about 0.00012.

9.9 Disproportional Cell Frequencies

Disproportional cell frequencies in factorial designs were discussed in Chapter 6 in connection with tests not involving a predicted trend. Here we will consider disproportional cell frequencies in relation to the correlation trend test. The basic difference between this and the earlier discussion is in the test statistic. Earlier, a special SS_B test statistic based on the deviation of a treatment mean from the expected value of the treatment mean over all data permutations was proposed for use with disproportional cell frequencies. A similar modification of the two-tailed correlation test statistic will be proposed to provide a more powerful test when cell frequencies are disproportional. For a one-tailed correlation trend test, the test statistic ΣXY is appropriate, even when cell frequencies are disproportional.

When cell frequencies are disproportional, the two-tailed correlation trend test statistic $|N\Sigma XY - \Sigma X\Sigma Y|$ is not as powerful a test statistic as when cell frequencies are proportional. When cell frequencies are proportional, $\Sigma X\Sigma Y$ is the expected value of $N\Sigma XY$ over all data permutations, so that if we use the expression $E(N\Sigma XY)$ to refer to the average or expected value of $N\Sigma XY$ over all data permutations, a general formula for the two-tailed correlation trend (and ordinary correlation) test statistic is $|N\Sigma XY - E(N\Sigma XY)|$. In computing the two-tailed test statistic for factorial designs with disproportional cell frequencies, we will use the same general formula but we will not use $\Sigma X\Sigma Y$ as $E(N\Sigma XY)$ in the formula because then $E(N\Sigma XY)$ over all data permutations is not necessarily $\Sigma X\Sigma Y$. Instead, we compute $E(N\Sigma XY)$ directly from the data.

As with simple independent-groups or repeated-measures designs, with the correlation trend test it sometimes is useful to restrict the random assignment procedure according to the type of subject assigned, which is likely to result in disproportional cell frequencies. For religious, medical, or other reasons, it might be appropriate to restrict the random assignment to the lower levels of the independent variable, for example, whereas other subjects might be assigned to any of the levels. In the following example of restricted-alternatives random assignment, the relevance of the general two-tailed test statistic $|N\Sigma XY - E(N\Sigma XY)|$ is discussed.

Example 9.12. We have four young subjects to be assigned to level 1 or level 2 of four treatment levels, with two subjects per level, and four older subjects to be assigned to any one of the four levels, with one subject per level. There are then $4!/2!2! \times 4! = 144$ possible assignments. A linear upward trend is predicted and the results in Table 9.3 are obtained. We compute the means of the two age groups to get 15 for the young subjects and 9 for the old. $E(\Sigma Y)$ for the first treatment level would be the number of subjects in the top cell within that level multiplied by 15, plus the number of subjects in the bottom cell multiplied by 9, which is $(2)(15) + (1)(9) = 39$. $E(\Sigma Y)$ for the second level also is 39, computed in the same way. The $E(\Sigma Y)$ value for each of the remaining treatment levels is 9. $E(\Sigma XY)$ is the sum of the products of the coefficients and $E(\Sigma Y)$ values for the respective levels. Thus, we have $(0)(39) + (10)(39) + (20)(9) + (30)(9) = 840$ for $E(\Sigma XY)$. $E(N\Sigma XY)$ is 8×840, or 6720. The two-tailed test statistic value for the obtained results is $|N\Sigma XY - 6720|$ which is $|7600 - 6720| = 880$. There is one other data permutation that would give such a large value as 880: that where the order of the measurements

TABLE 9.3

Restricted-Alternatives Random Assignment

	Trend Coefficients			
	0	10	20	30
Young	12	16		
	14	18		
Old	7	8	10	11

over treatment levels is reversed within each age group and so the two-tailed P-value is 2/144, or about 0.014.

Example 9.13. Suppose the data in Example 9.12 were to be analyzed to obtain a two-tailed P-value, using $|N\Sigma XY - \Sigma X\Sigma Y|$ as the two-tailed test statistic instead of the one based on the expected value of ΣXY over the data permutations. Using $|N\Sigma XY - \Sigma X\Sigma Y|$ would be perfectly valid because it could be computed over all data permutations so that if there was no treatment effect, only chance would tend to make the value for the obtained results large relative to the other data permutations. However, there would have been a loss of power. Although the lowest measurements within each row are separately associated with the lowest coefficients, the highest measurements for both groups combined are associated with the lowest coefficients so that, overall, there tends to be little correlation between the coefficients and the measurements. Put differently, by restricting the random assignment of the high-scoring (i.e., young) subjects to the lower levels of treatment, the strong positive correlation between coefficients and measurements within each row separately is masked when the regular two-tailed test statistic is used. For the obtained results, $|N\Sigma XY - \Sigma X\Sigma Y|$ would be $|7600 - 7680| = 80$. $\Sigma X\Sigma Y$ is constant and so no other data permutation could provide as large a value of $N\Sigma XY$ as 7600; consequently, every data permutation would have a value of $|N\Sigma XY - \Sigma X\Sigma Y|$ as large as the obtained results. It follows that the two-tailed P-value would have been 1 instead of 0.014, the two-tailed P-value associated with use of the test statistic that takes into account the disproportional cell frequencies.

It can be seen that when there are disproportional cell frequencies, the general test statistic $|N\Sigma XY - E(N\Sigma XY)|$ can be much more powerful than $|N\Sigma XY - \Sigma X\Sigma Y|$. For a randomization test, the general test statistic is equivalent to the simpler test statistic $|\Sigma XY - E(\Sigma XY)|$, the absolute difference between ΣXY for a data permutation and the expected or average value of ΣXY over all data permutations. This in turn is equivalent to $|r - E(r)|$, the absolute difference between the product-moment correlation coefficient for a data permutation and the mean correlation coefficient over all data permutations. In other words, for determining the P-value by the randomization test procedure the general two-tailed test statistic is equivalent to the absolute difference between a value of r and the value that would be expected by "chance." When cell frequencies are proportional, $E(r)$ becomes 0, making the special test statistic then equivalent to $|r|$.

9.10 Data Adjustment for Disproportional Cell Frequency Designs

In Section 6.23, it was shown that an adjustment of data could serve the same function as modifying the two-tailed test statistic in making a randomization test more sensitive to treatment effects when cell frequencies are disproportional. The same data adjustment can be used before computing $|N\Sigma XY - \Sigma X\Sigma Y|$ for a correlation trend test to achieve the same P-value as that of the special test statistic $|N\Sigma XY - E(\Sigma XY)|$.

Example 9.14. To provide a way of determining the P-value for a correlation trend test applied to factorial designs, we modify Program 5.1 by adding trend coefficients as inputs and changing the test statistics from T_L and $\Sigma(T^2/n)$ to ΣXY and $|N\Sigma XY - \Sigma X\Sigma Y|$. We use these test statistics whether the designs have proportional or disproportional cell frequencies but when the designs have disproportional cell frequencies, the data are adjusted before being subjected to the statistical test. The adjustment carried out on the data in Table 9.3 (in Example 9.12) would be to express measurements for young subjects as $(Y - \bar{Y})$ and measurements for old subjects as $(O - \bar{O})$, where $\bar{Y}$ and $\bar{O}$ are the means for the two age groups. Table 9.4 shows the results of the study in terms of adjusted data.

First, let us examine the relationship between one-tailed P-values for r for the unadjusted measurements (Table 9.3) and the adjusted measurements (Table 9.4). The adjusted measurements are linearly related to the unadjusted measurements for the young subjects, and the adjusted measurements are linearly related to the unadjusted measurements for the old subjects. However, one should not expect r to be the same for the adjusted and unadjusted data because the linear relationship between adjusted and unadjusted measurements is different for the old and young subjects. Nevertheless, it can be shown that r for adjusted data is an equivalent test statistic to r for unadjusted data, providing the same P-value.

For either adjusted or unadjusted data, ΣXY is an equivalent test statistic to r, so to show the equivalence of rs for adjusted and unadjusted data, it is sufficient to demonstrate that ΣXY for unadjusted data (ΣXY_{un}) and ΣXY for adjusted data (ΣXY_{adj}) are equivalent test statistics. The demonstration of this equivalence will be accomplished by showing that $\Sigma XY_{un} - \Sigma XY_{adj}$ is constant over all data permutations.

The following argument depends on the fact that all $N!/n_1!n_2!...n_k!$ data permutations associated with dividing N things into k groups with fixed sample sizes can be derived from the obtained data permutation by successive switching of two data points between groups. For example, the set of six data permutations associated with the young group consist of the observed results in Table 9.4 plus five other data permutations that can be produced by successively switching a pair of measurements between

TABLE 9.4

Adjusted Data

		Trend Coefficients		
	0	10	20	30
Young	−3	+1		
	−1	+3		
Old	−2	−1	+1	+2

columns and the same is true for the set of 24 data permutations associated with the old group. The reference set for all eight subjects jointly, consisting of $6 \times 24 = 144$ data permutations, then could be listed in the order in which they were produced, each data permutation being the same as the one preceding it except for the location of two measurements.

The amount of change in ΣXY resulting from exchanging two measurements, a and b, in the same block between columns for a higher coefficient (HC) and a lower coefficient (LC) is $|HC - LC| \times |a - b|$. ΣXY is reduced if a and b are negatively correlated with the coefficients after switching, and ΣXY is increased if a and b are positively correlated with the coefficients after switching. Now the difference $(a - b)$ is the same for two measurements whether a and b refer to unadjusted measurements or the corresponding adjusted measurements. Suppose two series of 144 data permutations have been produced by successive switching of two measurements for the unadjusted data, starting with the obtained results, and simultaneously switching the corresponding measurements for the adjusted data, starting with the obtained adjusted data. Then because the difference $(a - b)$ is the same for two measurements, whether the measurements are adjusted or unadjusted, ΣXY_{adj} and ΣXY_{un} must increase or decrease by the same amount between the first data permutation and the second, increase or decrease equally between the second and third data permutations, and so on for all successive data permutations. Thus, $\Sigma XY_{un} - \Sigma XY_{adj}$ is constant over all data permutations, so the two one-tailed test statistics are equivalent, giving the same P-values.

Let us turn to consideration of the two-tailed test statistic $|N\Sigma XY - \Sigma X\Sigma Y|$, the test statistic whose apparent inappropriateness for factorial designs with disproportional cell frequencies led to the modified two-tailed test statistic, $|N\Sigma XY - E(N\Sigma XY)|$. We will show in six steps that even though $|r|$ for unadjusted data may be different from $|r|$ for adjusted data, the test statistics $|N\Sigma XY_{adj} - \Sigma X\Sigma Y_{adj}|$ and $|N\Sigma XY_{un} - E(N\Sigma XY_{un})|$ are numerically equal. We start with the relationship between the one-tailed test statistics,

$$\Sigma XY_{un} - \Sigma XY_{adj} = c \qquad (9.1)$$

where c is a constant over all data permutations. From Equation 9.1, we derive

$$N\Sigma XY_{adj} = N\Sigma XY_{un} - Nc \qquad (9.2)$$

From Equation 9.2, we get the following equation of expectations:

$$E(N\Sigma XY_{adj}) = E(N\Sigma XY_{un}) - Nc \qquad (9.3)$$

Because the sum of adjusted measurements must equal 0, $E(N\Sigma XY_{adj}) = 0$, so

$$E(N\Sigma XY_{un}) = Nc \qquad (9.4)$$

Substituting $E(N\Sigma XY_{un})$ for Nc in Equation 9.2, we get

$$N\Sigma XY_{adj} = N\Sigma XY_{un} - E(N\Sigma XY_{un}) \tag{9.5}$$

Because the sum of the adjusted measurements is 0, $\Sigma X\Sigma Y_{adj} = 0$, so $\Sigma X\Sigma Y_{adj}$ can be subtracted from the left side of Equation 9.5, leading to the following equation, which is the expression of the numerical identity of the two-tailed test statistics for adjusted and unadjusted data:

$$|N\Sigma XY_{adj} - \Sigma X\Sigma Y_{adj}| = |N\Sigma XY_{un} - E(N\Sigma XY_{un})| \tag{9.6}$$

Thus, applying Program 9.2 to the adjusted data will provide the same two-tailed P-value as modifying the program and applying it to the raw data. Both test statistics are equivalent to $|r - E(r)|$ for the respective data types, and as $E(r_{adj})$ equals 0, both test statistics are equivalent to $|r_{adj}|$.

9.11 Combining of P-Values for Trend Tests for Factorial Experiments

The application of trend tests to randomized blocks data has been treated in the preceding sections of this chapter. Trend tests also are applicable to data from completely randomized experiments. The permuting of data for a test of the effect of factor A in a two-factor completely randomized experiment must be within each level of factor B to test the effect of A without assumptions about factor B. (To permute data over all levels of A and B would be appropriate for testing no effect of either factor, but would be inappropriate in testing for the effect of only one factor.) Consequently, Program 9.2 can be used as a trend test for completely randomized factorial experiments in the same way as for the treatments-by-subjects experiments discussed earlier. Similarly, adjustments for disproportional cell frequencies would be the same.

We have seen that correlation trend randomization tests and programs become complicated for factorial experiments, especially when there is adjustment for disproportional cell frequencies. However, if instead of developing special procedures for factorial experiments one uses combining of P-values, there is conceptual simplicity as well as practical advantage in many cases.

Example 9.15. In Example 9.11, the P-value determination did not require use of a special program but in cases where the results are not ones that give the smallest possible P-value, the data permuting modification of Program 9.2 described in the first paragraph of Section 9.8 may be required. However, as mentioned in Chapter 6 for completely randomized factorial designs as well as for randomized block designs, such as the one in this example, combining of P-values can be employed when a special program is not available. To use combining of P-values with the results in Example 9.11, Program 9.2 could be applied to the data for males and would give a P-value close to 1/90, or about 0.011, and the application of Program 9.2 to the data for the females similarly would give a P-value of about 0.011. (Program 9.2 is based on random data permutation, so the P-values to be combined are likely to deviate slightly from 0.011, even with a large number of data permutations, but the individual P-values for males and females nevertheless would be valid and independent, and thus satisfactory for combining of P-values.) For the sake of illustrating the combining of P-values, suppose the P-values supplied by Program 9.2 — a random program — were 0.011 and 0.011. Using the additive method of combining P-values given in Section 6.26, we have $(0.011 + 0.011)^2/2 = 0.00024$ as the overall combined P-value, whereas a modified Program 9.2 would give approximately 0.00012. The combining of P-values would test the same null hypothesis as would be tested by Program 9.2: for none of the 12 subjects was there a differential effect of the three treatments. Combining P-values is a substitute for modifying Program 9.2 because the experimental random assignment was within blocks, making blocks the equivalent of independent single-factor experiments, each of which is appropriate for Program 9.2.

For factorial designs with proportional cell frequencies, as in the example just cited, the combining of P-values from Program 9.2 can make it unnecessary to modify the data permuting procedure in the program. When there are disproportional cell frequencies, as in Example 9.12, the combining of P-values from applications of Program 9.2 can be performed in exactly the same way, with no necessity of either a special test statistic or data adjustment, because disproportionality of cell frequencies is relevant to the combining of the data over the cells where there is disproportionality, not combining of P-values over what are in effect independent experiments.

Using combining of P-values as a substitute for data combining for tests of what are independent experiments (and even in completely randomized factorial experiments, the levels of a factor can be regarded in that manner for testing the effects of the other factor) can be useful in other ways for trend tests for factorial experiments. By computing P-values separately for different levels of a factorial design, the experimenter can use Program 9.2 as it stands and simply combine the P-values. P-values can be for different trend predictions, such as the prediction of an upward trend for one level with increasing values of the independent variable and a downward trend for a second level, with or without disproportionality of cell frequencies. Another potential use of the combining of P-values for these designs would be when there is a one-tailed prediction at one level and a two-tailed prediction at the other level.

9.12 Repeated-Measures Trend Tests

Repeated-measures trend tests are for experiments where there is an expected trend in the repeated measures for each subject. Each subject in a repeated-measures design has k treatment times for k treatments, and the association of treatment times with treatments is randomly determined separately for each subject. Of course, the introduction of a predicted trend does not affect the data permutations, which are dependent on the random assignment procedure. What is different is the test statistic because it should be sensitive to the particular trend that is predicted. Thus, Program 6.1 and Program 6.2 for repeated-measures ANOVA need little modification to become programs for repeated-measures trend tests. The portion of Program 9.2 that computes the trend test statistics ΣXY and $|N\Sigma XY - \Sigma X\Sigma Y|$ can replace the repeated-measures test statistic computation portion of Program 6.1 or Program 6.2.

Repeated-measures experiments can be regarded as randomized block experiments where each block is a separate subject. The discussion of correlation trend tests for factorial designs in preceding sections of this chapter, such as restricted-alternatives random assignment and the need for special computational procedures when cell frequencies are disproportional, thus applies to repeated-measures trend tests as well as to trend tests for other factorial designs.

9.13 Differences in Trends

Willmes (1982) described a randomization test for a difference in trend, which he referred to as a test of a difference between response curves. Response curves are curves showing trends in measurements over quantitative levels of another variable; thus, tests of differences between response curves can be testable by tests for differences in trend. We have been considering trend tests in which the trend is over different levels of a manipulated treatment. In the example given by Willmes, the trend is over fixed points in time at which measurements are taken. There were two levels of the manipulated factor — alcohol or no alcohol — and the measurements were heart rates taken at three different times for each subject — before, during, and after "performing a speeded intelligence test." To test the null hypothesis of no differential effect of alcohol on heart rate by means of a randomization test, the data would be permuted as for a multivariate test because there are in fact three different measures of heart rate for each subject. A test of the difference in heart rate overall could readily be tested by, for example, a univariate t test if all three measurements were added together to provide a single value but this or similar pooling procedures ignores the difference in trend that also is expected. What was desired was a test statistic sensitive not just to the difference in level of responses between the alcohol and no alcohol conditions but also to the difference in trend over the three treatment times. Willmes discusses rank tests as well as tests on raw data from this standpoint.

Krauth (1993) also considered tests for a difference in trend, referred to as tests of a difference among response curves. As in the study by Willmes, the trends to be compared are trends over fixed points in time, not over treatment levels. There are three levels of the manipulated factor — placebo, drug 1, and drug 2 — and behavioral measures on each subject are recorded at eight different times after administration of the drug, ranging from 1 hour to 24 days. To test the null hypothesis of no differential effect of the three treatments on the behavioral measurements, the test would be of the same form as that of Willmes, using a test statistic sensitive both to the difference in overall size of the measurements under the three conditions and the difference in trend of measurements over the eight points in time.

The Willmes and Krauth examples concerned differences in trend over levels of a fixed, unmanipulated factor. Now a randomization test for a difference between trends over levels of a manipulated factor will be described. The design is a completely randomized 2×5 two-factor design with equal numbers of subjects in all 10 cells. There are two qualitatively different levels of factor B and five dosages of drug A. Interest is focused on the main effect of factor B, so the null hypothesis to be tested is that the responses of all subjects at their assigned levels of A are independent of assignment to level of B. There is no prediction as to whether one level of B would in general provide larger measurements than the other but theory suggests that as the drug dosage increases, the level B_1 measurements should increase more rapidly than the B_2 measurements. Thus, $(\bar{B}_1 - \bar{B}_2)$ is expected to increase from the smaller to the larger dosages of A. The trend that is predicted from the lowest to the highest levels of A then is an increase in $(\bar{B}_1 - \bar{B}_2)$; thus, the trend coefficients for the five levels of A will be 1, 2, 3, 4, and 5. The test statistic is ΣcD, where c is the trend coefficient and $D = \bar{B}_1 - \bar{B}_2$ for each level of A. To construct the reference set, the test statistic would be computed for each of the data permutations associated with permuting the data between levels of B within each level of A.

9.14 Correlation Trend Test and Simple Correlation

When there is only one subject for each of k levels of a quantitative independent variable, the test statistics are equivalent to the correlation between the k coefficients and the k measurements. Coefficients for linear trend are linearly related to the independent variable values, so the P-values are the same for the correlation trend test for linear trend with the use of the linear trend coefficients as with the independent variable magnitudes used as the coefficients. Thus, simple correlation described in Chapter 8 gives the same P-value as if the data were tested by a correlation trend test for linear trend. Simple correlation is then a special case of the correlation trend test — the case where there is only one subject per level of treatment and a linear trend is predicted.

However, there is no reason why only a linear trend can be tested when there is just one subject per treatment level. The coefficients that are determined are independent of the number of subjects per treatment level, so they are the same for a U-shaped trend with a low point at the independent variable value of 15 whether there are 20 subjects for each level of the independent variable or only one subject. Thus, the use of the correlation trend test with data where there is only one subject per treatment level is effectively an extension of product-moment correlation to situations where any type of trend can be predicted. The same degree of specificity of prediction can be accommodated as if there were several subjects at each treatment level. Therefore, with one subject per treatment the coefficients can be determined for the predicted trends and used as the X values to be paired with measurements in Program 9.1 or Program 9.2.

9.15 Ordered Levels of Treatments

The correlation trend test can be applied even when the levels of the independent variable are not quantitative, provided that they can be ordered with respect to the expected trend in the data.

Example 9.16. For short periods of time, people apparently are able to ignore or block out distractions that otherwise would impair their performance on a skilled task. However, one would expect that over long periods of time the effort to carry out such blocking would be too great to sustain and that as a consequence, the performance would suffer. An experiment is designed to determine whether the noise of other typewriters and movement in the periphery of the field of vision — of a level comparable to that in a typical office — would adversely affect the quality of typing if the noise and the visual distraction were sustained over several hours. The quality of typing is operationally defined as the typing speed in words per minute, adjusted for errors. Three levels of the independent variable are used: N (no visual or auditory distraction), A (auditory distraction), and A + V (auditory and visual distraction). The null hypothesis is this: the typing speed of each typist is independent of the experimental conditions. We expect N to provide the highest measurements, (A + V) to provide the lowest, and A to provide measurements of intermediate size; consequently, the coefficients for the correlation trend test are ranks 1, 2, and 3, assigned in the following way:

Treatment level	A + V	A	N
Coefficients	1	2	3

Six typists with similar experience are selected for the experiment and two are randomly assigned to each treatment level, there being $6!/2!2!2! = 90$ possible assignments. The following typing speeds were obtained: (A + V): 53, 52; A: 56, 55; and N: 59, 60. Program 9.2 can be used to determine the P-value. To compute the one-tailed test statistic ΣXY for the obtained data, the program pairs the measurements with the coefficients as in Table 9.1 for the data in Example 9.10. For the obtained data in the present example, $\Sigma X = 12$, $\Sigma Y = 335$, and $\Sigma XY = 684$. ΣX and ΣY are constant over all 90 data permutations but the pairing of the X and Y values, and therefore ΣXY, varies. No other data permutation gives a value of ΣXY as large as 684, the obtained value, and so the P-value for systematic data permutation would be 1/90, or about 0.011. However, Program 9.2 uses random data permutation to determine the P-value of the results and may not give a P-value that is exactly 1/90. It should be kept in mind that the random data permutation procedure is valid no matter how small the sample size or how few distinctive data permutations are possible.

If a systematic data permutation program for the correlation trend test is desired, it can be produced by combining the part of Program 4.1 that systematically permutes the data for one-way ANOVA with the part of Program 8.1 that computes the test statistics for product-moment correlation. Program 9.2 is composed of a similar combination, using the computation of test statistics for product-moment correlation in conjunction with the random permuting component of Program 4.3, which is for one-way ANOVA with the P-value determined by random permutation.

9.16   Ranked and Dichotomous Data

If the data are in the form of ranks, the correlation trend test can be applied to the ranks as if they were raw measurements.

Example 9.17. As in Example 9.16, two of six typists are randomly assigned to each of the three conditions specified in that example: N (no visual or auditory distraction), A (auditory distraction), and A + V (auditory and visual distraction). The null hypothesis is this: cheerfulness is independent of experimental conditions. An independent observer ranks the typists with respect to cheerfulness after the experiment, assigning a rank of 1 to the least cheerful and 6 to the most cheerful. N is expected to produce the highest ranks, (A + V) to produce the lowest, and A to produce the intermediate ranks. The coefficients for the correlation trend test thus are the same as in Example 9.16:

Treatment level	A + V	A	N
Coefficients	1	2	3

The obtained cheerfulness ranks were: (A + V): 1, 2; A: 3, 4; and N: 5, 6. ΣXY for the obtained results is $(1)(1) + (1)(2) + (2)(3) + (2)(4) + (3)(5) + (3)(6) = 50$. Of the 90 data permutations, only one gives a value of ΣXY as large as 50, and so the P-value is 1/90, or about 0.011.

Although in Example 9.17 the treatment coefficients as well as the data are ranks, the same procedure could be followed if only the data were ranked;

for example, for a quantitative independent variable the coefficients could be 0, 5, and 10, although the data consisted of ranks. Furthermore, tied ranks cause no problem with regard to validity when the P-value is determined by data permutation. It would be relatively easy to construct tables for a rank-order correlation trend test where both the data and the treatment coefficients were ranks, with no tied ranks within either the treatment coefficients or the data, and with equal sample sizes. However, with the availability of Program 9.2 that is unnecessary. Program 9.2 has great versatility in determining the P-value for ranked data: the treatment coefficients need not be ranks — tied ranks for either coefficients or data are acceptable — and the sample sizes can be equal or unequal.

Example 9.18. If the dependent variable is dichotomous, like lived-died or correct-incorrect, the numerical values 0 and 1 can be assigned to designate the categorical response and those numerical values can then be used as the quantitative data for trend analysis. Suppose that we have three levels of dosage (5, 10, and 15 units) of a pesticide and want to test its lethality on six rats. Two rats are randomly assigned to each level, providing $6!/2!2!2! = 90$ possible assignments. A linear upward trend in the proportion of rats killed by the dosages is predicted, and so the coefficients 0, 5, and 10 are associated with the treatments. The following results, where 0 indicates that a rat lived and 1 indicates that it died, were obtained: 5-unit dosage: 0, 0; 10-unit dosage: 0, 1; and 15-unit dosage: 1, 1. The obtained ΣXY is then $(0)(0) + (0)(0) + (5)(0) + (5)(1) + (10)(1) + (10)(1) = 25$. There is no data permutation with a larger value of ΣXY than 25 but there are nine data permutations with that value, and so the P-value is 9/90, or 0.10. None of the 5-unit dosage rats died, half of the 10-unit dosage rats died, and all of the 15-unit dosage rats died, so the results could not have been more consistent with the prediction; yet the P-value associated with the results is not at all small. Although the correlation trend test can be used with dichotomous data, the sensitivity of the test is considerably less than with continuous data.

9.17 Questions and Exercises

1. What is the complement of the null hypothesis for a correlation trend test?

2. What is the basic difference between a correlation trend test and the tests in Chapter 8?

3. To determine whether the strength of a sunscreen lotion affected the growth of hair on the treated skin, an experimenter selects six adults for his randomization correlation trend test. There are three strengths of sunscreen lotion with two persons for each strength. What is the smallest possible P-value?

4. An intermediate level of electrical shock is expected to be more effective in speeding up the learning of a task than either a low level or a high level. Instead of using a correlation trend test with an inverted U-shaped predicted trend, the experimenter decides to use a different randomization test. How could he compute a test statistic that takes into consideration that they have no basis for predicting a direction of difference in effect between the low and high levels of shock?

5. For what type of experimental design for a correlation trend test should disproportionality of cell frequencies be taken into consideration in constructing a test statistic?

6. What is the difference in the meaning of the adjective "expected" in "expected trend" and "expected ΣXY"?

7. Describe a situation where a restricted-alternatives correlation trend test would be appropriate. Explain.

8. In the randomized block design known as a treatments-by-subjects design, what P-values should be combined through the combining of P-values in a test of no treatment effect?

9. In examples in this chapter, the P-value given in an example often was for a systematic randomization test. What effect, if any, on the validity of the randomization test would result from use of a randomization test with random rather than systematic permutation of the data?

10. Which type of randomization test, systematic or random, is most useful for understanding the rationale underlying a test and assessing its validity? Discuss.

REFERENCES

Cory-Slechta, D.A. and Thompson, T., Behavioral toxicity of chronic postweaning lead exposure in the rat, *Toxicol. Appl. Pharmacol.*, 47, 151, 1979.

Cory-Slechta, D.A., Weiss, B., and Cox, C., Delayed behavioral toxicity of lead with increasing exposure concentration, *Toxicol. Appl. Pharmacol.*, 71, 342, 1983.

Edgington, E.S., Randomization tests for predicted trends, *Can. Psychol. Rev.*, 16, 49, 1975.

Huppert, F.A. and Deutsch, J.A., Improvement in memory with time, *Q.J. Exp. Psychol.*, 21, 267, 1969.

Krauth, J., Experimental design and data analysis in behavioral pharmacology, in *Methods in Behavioral Pharmacology*, van Haaren, F. (Ed.), Elsevier Science Publishers B.V., New York, 1993, 623–650.

Newland, M.C., Ng, W.W., Baggs, R.B., Gentry, G.D., Weiss, B., and Miller, R.K., Operant behavior in transition reflects neonatal exposure to cadmium, *Teratology*, 34, 231, 1986.

Price, M.T.C. and Cooper, R.M., U-shaped functions in a shock-escape task, *J. Comp. Physiol. Psychol.*, 89, 600, 1975.

Seiden, L. and Dykstra, L., *Psychopharmacology: A Biochemical and Behavioral Approach*, Van Nostrand Reinhold, New York, 1977.

Willmes, K.A., Comparison between the LEHMACHER & WALL rank tests and Pyhel's permutation test for the analysis of *r* independent samples of response curves, *Biomet. J.*, 24, 717, 1982.

Winer, B.J., *Statistical Principles in Experimental Design* (2nd ed.), McGraw-Hill, New York, 1971.

10

Matching and Proximity Experiments

10.1 Randomization Tests for Matching

When responses are qualitatively different, contingency chi-square tests can be employed to test the null hypothesis of no differential treatment effect. However, a contingency chi-square test is nondirectional, and the results of the test reflect only the consistency — and not the nature — of the association between types of responses and types of treatments. For instance, a contingency chi-square test would not use the specific prediction that treatment A would lead to the death of animals and treatment B to survival but only the prediction that the treatments would have different effects on mortality. In this case, there is a dichotomous dependent variable, so Fisher's exact test can be used as the desired one-tailed test, but there are no standard procedures for accommodating specificity of prediction when there are more than two categories of response. Let us now consider an example in which we predict a specific type of response for each subject and determine the P-value of the number of correct predictions.

Example 10.1. We conduct an experiment to see whether the songs of baby English sparrows are influenced by the songs of their mothers. The null hypothesis is that a bird's song is not influenced by the mother. We expect the bird's song to be like that of its mother (foster mother, in this example). We obtain six newly hatched English sparrows to use as subjects. The "treatments" to which the subjects are assigned are six adult female sparrows to serve as foster mothers to the nestlings. The six adult birds represent six different types of sparrows, shown in column 1 of Table 10.1.

A recording is made of the song of each of the six adult birds at the beginning of the experiment for comparison with the songs of the nestlings at a later date. The six nestlings are assigned randomly to the six foster mothers, with one nestling per adult bird. The first two columns of Table 10.1 show the resulting assignment, the first baby bird being assigned to the grasshopper sparrow, the second to the tree sparrow, and so on. After the young birds have been with their foster mothers for a certain period of time, a recording is made of the song of each baby bird. An ornithologist is given

TABLE 10.1

Results of Experiment on Acquisition of Bird Songs

Type of Mother Sparrow	Assignment of Baby Birds	Baby Bird with Most Similar Song
Cape Sable	3	3
Chipping	5	5
Field	4	6
Grasshopper	1	1
Ipswich	6	4
Tree	2	2

the set of six songs along with the set of six prerecorded songs of the adult birds and is asked to pair the songs of the young birds with those of the foster mothers to provide the closest correspondence between the paired songs. The ornithologist pairs the song of bird 1 with the song of the grasshopper sparrow, that of bird 2 with the song of the tree sparrow, that of bird 3 with the song of the Cape Sable sparrow, and so on as shown in column 1 and column 3. To determine the P-value of the results by the randomization test procedure, we use as a test statistic the number of times the ornithologist pairs a baby bird with the bird that raised it.

The results are shown in Table 10.1. There is agreement between the paired birds in column 2 and column 3 for four subjects: 1, 2, 3, and 5. Thus, the obtained test statistic value is 4. To determine the P-value, we must determine how many of the data permutations provide as many as four matches, which can be done without a computer program. Data permutations are derived from the data in Table 10.1 by permuting the sequence of numbers in column 2 to provide new pairings of column 2 and column 3. (The permutations of numbers in column 2 represent possible assignments of baby birds to the birds in column 1, and under H_0 the pairing of column 1 and column 3 is fixed.) There is only one of the 720 pairings that provides six matches (perfect agreement). There are no data permutations with exactly five matches because if five pairs match, the sixth also would have to match. There are $6!/4!2! = 15$ combinations of four numbers from the six numbers in column 2 that can match four numbers in column 3, and for each of those 15 combinations, there is only one way the remaining two numbers in column 2 could disagree with the remaining two numbers in column 3, so there are 15 pairings providing a test statistic value of 4. Thus, there are $1 + 15 = 16$ data permutations with as large a test statistic value as 4, so the P-value is 16/720, or about 0.022.

Example 10.2. In Example 10.1, the determination of the P-value was not difficult but if there had been, say, 20 possible pairings and 8 matches, it would have been considerably more difficult to determine the P-value. For such cases, a formula may be useful. A discussion by Riordan (1958) of combinatorial aspects of the arrangement of n rooks on an $n \times n$ chessboard with no two rooks in the same column or row has been used by Gillett (1985) to deal with matching problems. Gillett provided the following formula to determine the number of pairings of n things for a one-to-one pairing with n other things:

$$f(j) = \sum_{i=j}^{n} (-1)^{i-j} \binom{i}{j}(r_i)(n-i)! \tag{10.1}$$

where r_i is the coefficient of x_i in the expansion of $(1 + x)^n$. For the situation described in Example 10.1, the coefficients are those of $(1 + x)^6$, which are $r_0 = 1$, $r_1 = 6$, $r_2 = 15$, $r_3 = 20$, $r_4 = 15$, $r_5 = 6$, and $r_6 = 1$. There were four matches in six pairs of numbers, so we would use Equation 10.1 by computing the number of ways of getting four, five, or six matches as below:

$$f_6 = (-1)^0 \binom{6}{6}(1)(6-6)! = 1$$

$$f_5 = (-1)^0 \binom{5}{5}(6)(6-5)! + (-1)^1 \binom{6}{5}(1)(6-6)! = 0$$

$$f_4 = (-1)^0 \binom{4}{4}(15)(6-4)! + (-1)^1 \binom{5}{4}(6)(6-5)! + (-1)^2 \binom{6}{4}(1)(6-6)! = 15$$

The results of the formula thus coincide with those given earlier, also yielding a P-value of 16/720, or about 0.022.

Equation 10.1 is inapplicable when the numbers of things in the two sets to be matched are not equal or when more than one member of one set can be matched with a single member of the other set. In Example 10.1, there were the same number of baby birds as foster mothers and the ornithologist was instructed to pair each baby bird's song with one and only one mother bird's song, so the application of Equation 10.1 was appropriate but experiments frequently would not meet the requirements associated with the formula. Usually, more than one subject is assigned to a treatment and some categories of response occur more frequently than others, as in the following example.

Example 10.3. Expecting that the color of illumination under which an animal is raised will be the animal's preference in adulthood, an experimenter randomly assigns 15 young animals to three cages illuminated by red, green, and blue lights, with five animals per color, and keeps them there until they have matured. Then they are removed from their cages and given a choice of cages with the three colors of light. The color of lighting in the cage the animal chooses is recorded and the results are shown in Table 10.2.

It will be observed that the responses tend to be as predicted. Contingency chi-square for the table is 5.60, and with four degrees of freedom, it has a P-value of 0.23 on the basis of the distribution underlying chi-square tables. However, as the contingency chi-square test is nondirectional, a test statistic using the specificity of the matching prediction should be more powerful when the prediction is borne out. The test statistic is the number of matches

TABLE 10.2

Results of Experiment with Colored Lights

		Treatments		
		Red	Blue	Green
Responses	Red	3	1	1
	Blue	1	4	2
	Green	1	0	2

and the obtained value is 9. The P-value is the proportion of the $15!/5!5!5!$ = 756,756 data permutations with nine or more matches. The generation of all data permutations by a computer program would be impractical but a program based on random permuting of data (Program 10.1) can be used. To use the program, the user first provides the number of treatments (3) as input data, then the number of subjects for the first, the second, and the third treatments (i.e., 5, 5, and 5). "Code values" for the treatment levels to serve as input can be any three distinctive numbers, such as 1, 2, and 3. After reading in the codes, the number of data permutations to be generated is entered. The data are then input using the same code numbers for the data, thus substituting, say, 1 for R, 2 for B, and 3 for G. For each data permutation, the data codes are paired with the treatment codes as with Program 9.2 for the correlation trend test. However, instead of using ΣXY as a test statistic, Program 10.1 on the CD uses the number of matches among the 15 pairs of code values and "observations," represented by 1s, 2s, and 3s. Program 10.1 applied to the data in this example (using 10,000 data permutations) gave a P-value of 0.034. This is about one sixth of the P-value of 0.23 given by the (nondirectional) contingency chi-square test.

Program 10.1 is a random permuting program that can be tested on the following data, for which a test based on systematic data permuting should give a P-value of 1/60:

Treatment Codes	Data Codes
1	1, 1, 2
2	2, 2, 1
3	3, 3, 3

There are seven matches of treatment codes and data codes, and only 28 of all $9!/3!3!3!$ = 1680 data permutations provide as many as seven matches, so the P-value is 28/1680, or 1/60.

In the two preceding examples, the matching of songs with songs and colors with colors may have given the impression that some observable characteristic of a treatment must be paired with a response in a matching test. In fact, there need not be an observable characteristic of a treatment to match with responses for a matching test to be effective. The characteristic of a treatment that is relevant is the *hypothesized response*. An example will help clarify this point. Suppose that on the basis of reactions of humans, it is predicted that Drug 1 given to dogs will produce inactivity, Drug 2 aggressive behavior, and Drug 3 timidity. If the predictions were correct, it would not be because the behavior matched that of the drugs because drugs do not exhibit behavior; it would be because the behavior of the dogs matched the behavior predicted to be produced by the drugs. It is useful to regard matching as the correspondence between a predicted and an observed response. A test of matching is useful whenever we predict a particular type of response to a particular type of treatment because there is then a basis for pairing predicted and obtained responses.

Program 10.1 can be used when there are several treatments, not just two or three, and this versatility permits an experimenter to be more flexible in designing matching experiments. Many treatments should be included in an experiment — when responses to them can be predicted — to strengthen the test. For example, in a study by Zajonc, Wilson, and Rajecki (1975) in which chicks were dyed either red or green to determine whether they would be attracted to chicks of the same color, the use of more colors would have made the experiment more sensitive. Another example of the use of two treatments when more treatments would have increased the power of the study is Burghardt's (1967) experiment in which horsemeat or worms were fed to baby turtles to determine the effect of early feeding on later food preferences.

10.2 Randomization Tests of Proximity

Matching tests are more useful in some situations than tests of proximity or nearness of an obtained response to a predicted response. Such a situation occurs when a subject responds to identity or nonidentity rather than similarity. For example, in predicting that children under stress will go to their mothers instead of their nursemaids, fathers, sisters, or others, there is the possibility that if the mother is not available, any preference for one of the available persons will not be based on similarity to the mother.

In other situations, there may be good reason to measure the quantitative difference between a predicted and an obtained response. To do so requires specification of a quantitative dimension. For instance, if in the experiment in Example 10.1 we expected a particular dimension of the birds' songs, such as length of the song, variation in loudness, or some other property, to be imitated, reliance on an ornithologist to give a judgment on overall similarity might be inappropriate. Suppose we were interested in the proximity of the sound intensity of the songs of the baby birds to those of their mothers. Instead of having an expert make an overall judgment on the similarity of songs, we might measure the sound intensities of the 12 songs and pair the measurements of the baby bird songs with those of the mother bird songs. Program 8.1 is then applied to the paired measurements to determine the

P-value of the correlation between the paired sound intensities. The product-moment correlation coefficient is a measure of the proximity of paired measurements relative to their proximity under alternative data permutations, as is evident from the equivalence of r and ΣD^2, the sum of the squared differences between paired measurements, for a randomization test. This equivalence is easy to demonstrate. For paired values of X and Y, $\Sigma D^2 = \Sigma(X_i - Y_i)^2 = \Sigma X^2 - 2\Sigma XY + \Sigma Y^2$. Over all data permutations (pairings of X and Y values), ΣX^2 and ΣY^2 are constant, so $-2\Sigma XY$ is an equivalent test statistic to ΣD^2. As $-2\Sigma XY$ is a test statistic that is equivalent (although inversely related) to ΣXY, which is equivalent to r, ΣD^2 and r are equivalent test statistics.

Program 8.1 is not appropriate when there is more than one subject per treatment. When there are several subjects for each treatment, the correlation trend test program (Program 9.1) can be used. Even when there can be no paired measurements and consequently no possibility of using r as a test statistic, it still may be possible to use ΣD^2 as a test statistic for a randomization test of proximity. An example of this type will be considered in Example 10.6.

10.3   Matching and Proximity Tests Based on Random Selection of Treatment Levels

Tests of matching and of proximity both use test statistics expressing a relationship between the treatment conditions and dependent variable measurements. As a consequence of this relationship, it is possible to make effective use of a small number of treatments by randomly selecting them from a population of treatments and then randomly assigning subjects to the randomly selected treatments.

In Example 10.1, we considered the pairing of the songs of six baby birds with the songs of six mother birds to determine the P-value of the number of matches. The test was relatively sensitive for that experiment but if there had been only three nestlings and three mothers, the smallest possible P-value would have been 1/6, the value associated with perfect matching. However, by increasing the number of possible assignments with only three nestlings raised by three mothers, we can have a sensitive randomization test.

Example 10.4. We conduct an experiment to see whether the songs of baby English sparrows are influenced by the mother birds. The experiment is conducted in the manner described in Example 10.1 with this exception: only three baby birds and three mother birds are involved, the three mother birds being randomly selected from the six listed in column 1 of Table 10.1. The random assignment is performed in the following way: one of the six mother birds listed in column 1 of Table 10.1 is randomly selected to be the foster mother of baby bird 1, one of the remaining five is selected for baby bird 2, and one of the remaining four for baby bird 3. Thus, there are $6 \times 5 \times 4 = 120$ possible assignments. (This is equivalent to selecting three of the six

TABLE 10.3

Assignment of Baby Birds to Randomly Selected "Mothers"

Type of Mother Sparrow	Assignment of Baby Birds	Baby Bird with Most Similar Song
Cape Sable	3	3
Chipping	—	—
Field	—	—
Grasshopper	1	1
Ipswich	—	—
Tree	2	2

adult birds randomly and then assigning the three nestlings randomly to the three selected adults.) As we are concerned here with only some of the adult birds shown in Table 10.1, the assignments and results of this experiment are shown in a different table, Table 10.3.

The first two columns of Table 10.3 show this assignment: baby bird 1 was assigned to the grasshopper sparrow, bird 2 to the tree sparrow, and bird 3 to the Cape Sable sparrow. After listening to the prerecorded songs of all six mother sparrows and to the songs of the three baby sparrows, the ornithologist pairs the song of bird 1 with that of the grasshopper sparrow, the song of bird 2 with the tree sparrow's song, and the song of bird 3 with the song of the Cape Sable sparrow, as shown in Table 10.3. As the ornithologist paired each young bird with the foster mother that raised it, the number of matches is three. To determine the P-value, the number of matches is computed for all 120 data permutations and the proportion of the data permutations with three or more matches is the P-value associated with the results. Under the null hypothesis, the relationship between column 1 and column 3 of Table 10.3 is constant over all possible assignments. The 120 possible assignments correspond to the ways in which the numbers 1, 2, and 3 can be arranged over the six rows in column 2. Out of the 120 possible assignments, only the actual assignment provides as many as three matches; thus, the P-value is 1/120, or about 0.008.

Example 10.4 shows how a single "measurement" (pairing of a baby bird and an adult bird on the basis of similarity of songs) from each of three subjects assigned to three treatments can be sufficient for a sensitive randomization test. By selecting three treatments randomly from six possible treatments, rather than nonrandomly specifying or fixing three treatments to which three subjects could be assigned randomly, the size of the smallest possible P-value is reduced from 1/6 to 1/120.

To construct a systematic randomization test program for this type of test, one could use the test statistic of Program 10.1 and change the permuting procedure to one that selects combinations of the treatments, then divides the observations among the treatments for each combination of treatments. A random permuting program could randomly select combinations of treatments then randomly divide observations among the treatments.

Sampling of treatment levels functions in the same way for measures of proximity. If we had measures of sound intensity of songs of 10 adult birds from which we randomly select three, the product moment r between the paired intensity levels for baby birds and their foster mothers would be the obtained test statistic and a systematic test would consider all $10 \times 9 \times 8 = 720$ data permutations consisting of pairing the three response intensity levels in every way with any three of the 10 adult bird intensity levels. The proportion of the 720 rs that were as large as the obtained r would be the P-value.

In the case of either matching or proximity test statistics, the sampling procedure can be effective in providing a much smaller randomization test P-value than otherwise would be possible, both for cases where each subject is assigned to a different treatment and where several subjects are assigned to each treatment. Although product-moment r for a randomization test is equivalent to ΣD^2 for a randomization test with subjects assigned to all treatments and thus is a measure of relative proximity for a set of paired numerical values, it is not an equivalent test statistic to r for a randomization test based on random subsets of treatments. When proximity is predicted and a set of treatments is randomly sampled, ΣD^2 instead of r should be computed from the paired measurements. Example 10.5 describes the performance of a randomization test of this type.

Example 10.5. A sociologist interested in imitation in children believes that young children will imitate the rate of performance of a new type of activity performed by a person (model) in front of the child. The sociologist specifies five rates of beating a drum which, in beats per minute, are 20, 30, 45, 70, and 100. A random selection of three of these rates provides 30, 45, and 100 as the rates to be used in the experiment, and one of these rates is selected randomly for each of three subjects. The model performs in front of one child at a time, beating the drum at the rate appropriate for the child. Following exposure to the model's beating of the drum, the child is given an opportunity to beat a drum. The children's rates of drum beating follow:

Model Rate	20	30	45	70	100
Child Rate	—	27	55	—	105

There are $5!/2!3! \times 3! = 60$ data permutations, consisting of all pairings of 27, 55, and 105 with three of the five rates of the model. The value of the test statistic ΣD^2 for the obtained results is $(30 - 27)^2 + (45 - 55)^2 + (100 - 105)^2 = 134$. As none of the other data permutations would provide such a small test statistic value as the obtained results, the P-value is 1/60, or about 0.017.

Random sampling of treatments can be used with proximity measures even when there is simply a measure of difference, as in the following example.

Example 10.6. We want to find out whether birds tend to build nests near the trees in which they were raised. An experiment is carried out within a covered enclosure containing six trees, the most widely separated trees being 120 feet apart. The null hypothesis is that the tree in which a bird builds its nest is independent of the tree in which the bird was raised; that is, the tree in which a bird builds its nest is the tree in which it would have built its nest regardless of the tree in which it was raised. The random assignment of Ss (nestlings) to treatments (trees in which the birds will be raised) is carried out in the following way: one of the six trees is randomly selected for S_1, one of the remaining five trees is selected for S_2, and one of the last four trees is selected for S_3. Thus, there are $6 \times 5 \times 4 = 120$ possible assignments. (This is equivalent to randomly selecting three of the six trees and then randomly assigning one of the Ss to each of the three trees.) The baby bird, along with its nest and mother, is placed in the assigned tree. Later, when the baby bird leaves its nest to build a nest of its own, the tree in which the nest is built is noted. The test statistic is ΣD^2, where D is the distance between the tree in which a bird is raised and the tree in which it builds its nest. Table 10.4 shows the distances of every tree from every other tree and shows where each of the Ss was raised and where each built its nest.

S_1 and S_2 built nests in the trees in which they were raised — namely T_4 and T_2, respectively — and so D_1 and D_2 are both 0. S_3 was raised in T_5 but built its nest in T_4, which was 25 feet away. So $\Sigma D^2 = 0 + 0 + 625 = 625$. Under H_0, S_2 would have built its nest in T_2 and S_1 and S_3 would have built in T_4, no matter where they were raised. In terms of Table 10.3, H_0 is that the only variation over the 120 potential assignments would be in the arrangement of S_1, S_2, and S_3 over the six rows within their respective columns. For example, consider the assignment of S_1 to T_1, S_2 to T_2, and S_3 to T_3. The test statistic value for the data permutation for that assignment is $14,400 + 0 + 1,849 = 16,249$ because under H_0, S_1 would have been 120 feet from where it was going to build its nest, S_2 would have been assigned to the tree in which it was going to build its nest, and S_3 would have been assigned to a tree that was 43 feet from T_4, where it would build its nest. The test statistic ΣD^2 is computed for each of the 120 data permutations. There are only two data permutations that give as small a value of ΣD^2 as the obtained value of 625. That is the data permutation for the actual results and the one for the assignment of S_1 to T_5, S_2 to T_2, and S_3 to T_4, which also gives $\Sigma D^2 = 625$. Thus, the P-value associated with the results is 2/120, or about 0.017.

TABLE 10.4

Distances (in Feet) between Trees Where Birds Were Raised and Trees Where They Built Their Nests

Trees Where Birds Were Raised	Trees Where Nests Were Built					
	T_1	T_2	T_3	T_4	T_5	T_6
T_1	0	60	110	120	110	40
T_2	60	0 (S_2)	60	87	91	68
T_3	110	60	0	43	62	97
T_4	120	87	43	0 (S_1)	25	93
T_5	110	91	62	25 (S_3)	0	77
T_6	40	68	97	93	77	0

10.4 Questions and Exercises

1. In the example of an experiment on the effect of mother bird songs on the song of baby birds, why were the mother bird songs not recorded at the same time as the songs of the baby birds?

2. For the bird song experiment, if there were five mother birds and five nestlings, what would be the smallest possible P-value?

3. Why would a randomization test with the product-moment correlation coefficient as a test statistic be inappropriate for the bird song experiment?

4. In the example to study whether birds tended to build their nests near the nest in which they were raised, the experimenter would have to be very careful not to violate the randomization test assumption of experimental independence. Why must the experimenter be careful in this experiment?

5. Describe an agricultural experiment for which a proximity test statistic would be better than a matching test statistic.

6. In the experiment to investigate the effect of a drummer's rate of drumming on a child's drum rate, the test statistic was the sum of the squared differences between the rate of the drummer and the rate of the child. What test statistic is equivalent?

7. For the experiment on the effect of early light color on an animal's later preference, which type of experiment was it — matching or proximity? Why does it not meet the requirements of the other type?

8. Describe an agricultural experiment that would call for a proximity test statistic instead of a matching test statistic.

9. Suppose that in Example 11.3, the following data were obtained

		Treatments		
		Red	**Blue**	**Green**
Responses	Red	3	0	0
	Blue	1	4	0
	Green	1	0	5

Perform a contingency chi-square test and a matching test on these data and compare the results.

10. Suppose that in the Example 10.5 all five rates were used, and the following data were obtained:

Model Rate	20	30	45	70	100
Child Rate	10	27	55	70	105

Calculate the P-value of the proximity test. What is your conclusion?

REFERENCES

Burghardt, G.M., The primacy effect of the first feeding experience in the snapping turtle, *Psychonomic Sci.*, 7, 1967, 383–384.

Gillett, R., The matching paradigm: An exact test procedure, *Psychol. Bull.*, 97, 1985, 106–118.

Riordan, J., *An Introduction to Combinatorial Analysis*, John Wiley & Sons, New York, 1958.

Zajonc, R.B., Wilson, W.R., and Rajecki, D.W., Affiliation and social discrimination produced by brief exposure in day-old domestic chicks, *Animal Behav.*, 23, 1975, 131–138.

11

N-of-1 Designs

A single-subject randomized experiment is the primary focus of this chapter, although the experimental designs and associated randomization tests can be applied to entities other than subjects. Randomized experiments can be conducted on just one person or other individual entity by randomly assigning treatment times to treatments, as in a repeated-measures experiment. However, subjects in a repeated-measures experiment take each treatment only once, necessitating a number of subjects to permit a sensitive randomization test, whereas just one subject is sufficient to provide sensitivity of a randomization test when random assignment of treatment times to treatments permits a subject to take a treatment more than once.

From the outset of this chapter, it is important to note that confining the discussion of randomized N-of-1 experiments almost exclusively to human subjects is motivated primarily by the fact that randomized N-of-1 experiments have been found extremely useful for experiments on people and by no means suggests that the designs are appropriate only for human participants. The research entity involved in an N-of-1 design might just as well be (and occasionally has been) an animal, a tree, an inorganic entity such as an automobile, a part of an organism (e.g., a leaf or a neuron), or a collection (e.g., a family or a fleet of battleships). See also Chapter 1, especially Section 1.1.

11.1 The Importance of N-of-1 Designs

There are many reasons why N-of-1 experiments are a worthwhile undertaking. There are practical as well as theoretical advantages in conducting N-of-1 rather than multiple-N experiments. In documenting a number of important single-subject studies carried out by early psychologists, Dukes (1965), in his ground-breaking article "N = 1," discussed some of those advantages.

The rareness of certain types of experimental subjects can necessitate conducting an experiment whenever such a subject is available. With luck, an experimenter doing research on pain may recruit a person with congenital absence of the sense of pain but not be able to locate enough such persons at one time for a multiple-subject experiment. Cases of multiple personality also are rare and the psychiatrist who has such a patient probably would be interested in carrying out an intensive study on the patient. In addition to the rareness of certain types of individuals, there are other compelling reasons for the intensive study of a single subject. The difficulty and expense of studying several subjects over a long period of time are good reasons for the intensive study of individual subjects in longitudinal studies. For example, a single ape can be raised with a family and be carefully studied but such an undertaking involving several apes simultaneously is not feasible.

Their relevance to theories that concern individual organisms is by no means a trivial advantage in using N-of-1 experiments. Although social, economic, and political theories frequently concern the aggregate behavior of groups of people, they must take into consideration factors that influence the development of opinions, attitudes, and beliefs within individual persons. Educational, psychological, and biological theories more openly show their primary concern with individual behavior, and those theories explain phenomena in terms of processes within individuals. Therefore, experiments in various fields are intended to further our understanding of intraorganismic processes. Typically, a measurement is taken from each subject rather than there being a single measurement from an entire group of subjects. However, in multiple-subject experiments treatment effects on individuals sometimes are concealed by the averaging of data over a heterogeneous group of subjects. For example, if A is more effective than B for some subjects and is less effective than B for about the same number of subjects, a difference between means might be so small as to obscure treatment effects. When virtually all subjects are affected in the same manner, averaging does not have that depressing effect on statistical significance but it still reflects only a general trend. The problem with averaging over subjects is not so much that of overlooking the existence of subjects with atypical responses — although that is a problem — but the inability to identify the individuals who respond in one way and those who respond differently to explore the basis of the variation in type of response. To attain an understanding of the internal processes of individuals, some experimenters prefer a series of N-of-1 experiments to a single multiple-N experiment, when feasible.

Research performed on one person at a time is important for reasons aside from the scientific importance of the findings. The importance of issues relating to human rights and individual freedom throughout the world today indicates the need for precise determination of the needs and abilities of individual persons because it is only in relation to a person's needs and abilities that it makes sense to refer to that person's rights or freedom. The demand to be treated as a person and not just a number is more than a protest against being treated in an impersonal manner; it is a demand that that person's individual characteristics that distinguish them from other people be taken into consideration. Educational techniques have been custom-made to suit the needs of students with special handicaps. Tests of various types are useful in developing an instructional setting appropriate for an individual student. In the field of medicine, a general practitioner who has known a patient for many years will use the knowledge of the patient to adjust their treatment accordingly. Knowing that what works for one patient may not work for another, a physician will try one medication and another to find out what is best for the particular patient being treated. Each person uses the experience of others as a guide but spends a lifetime gradually discovering what they personally like and what is best in education, medicine, occupation, recreation, and other aspects of life. In light of the importance of taking individual differences into consideration in all aspects of life, we should not rely solely on unsystematic investigation to determine the idiosyncratic "laws" that regulate the life of a particular person. Loosely controlled empirical research and crude trial-and-error approaches ("take this new drug for a week and let me know if it helps") are not enough. Even though some types of information are attainable only by other means, N-of-1 experiments should supplement or replace many of the procedures currently in use for intensive investigation of individuals (see also Onghena, 2005).

11.2 Fisher's Lady-Tasting-Tea Experiment

In 1935, Fisher (1951) discussed a hypothetical N-of-1 randomized experiment, the widely known "lady-tasting-tea" study. The experiment was a test to determine whether a lady could tell by tasting a cup of tea whether tea was poured on top of milk or milk was poured on top of tea before being mixed in the cup. Eight cups were prepared, four in each manner. The cups were presented to the lady in a random order with her knowledge that the order of presentation was random, that four cups had been mixed in each way and that she was to decide which cups had milk put in first and which had tea put in first. Fisher considered various possible outcomes of this hypothetical experiment, including the case where the lady correctly identified all eight cups. There are $8!/4!4! = 70$ ways eight cups can be divided into two groups of four cups each, with one group being designated as the "milk first" group and the other as the "tea first" group. Fisher thus gave a P-value of $1/70$ to the results because that is the probability of correctly identifying all eight cups by chance, that is, by randomly dividing the eight cups into four "milk first" and four "tea first" groups.

Fisher provided no guidance in the conduct of N-of-1 experimentation. The title of the chapter devoted to the topic was "The Principles of Experimentation, Illustrated by a Psycho-physical Experiment," and Fisher used the hypothetical experiment precisely as specified in the chapter title — to illustrate principles of experimental design relevant to experiments in general, not just N-of-1 experiments. In discussing the lady-tasting-tea experiment, Fisher put emphasis on the use of randomization to control for various sources of confounding and on aspects of the experimental design that could be modified to increase the sensitivity of the test to treatment effects.

Also, although the statistical procedure was simple and familiar, Fisher did spend time discussing the null hypothesis that was tested and the interpretation to place on the outcomes of the statistical test.

Despite Fisher's insistence that the random assignment of cups to be tasted was the only point in the experimental procedure in which randomness was involved, he seemed to treat the lady's judgment as being based on a random process. In that respect, Fisher's approach is so divergent from that taken in this chapter that the distinction between the two approaches requires careful examination, which is provided in the following section.

11.3   The Concept of Choosing as a Random Process

An element of the lady-testing-tea example that complicates data analysis is the belief, which Fisher seems to have held, that choosing under uncertainty ("guessing") on the part of the subject introduced a chance element into the experiment in addition to that resulting from random presentation of the cups of tea.

Fisher did not state that his null hypothesis involved a probability of being correct but did state that the lady might be claiming to distinguish milk-first and tea-first cups correctly more often than not rather than correctly every time. He then discussed how the test could be made more sensitive by having more cups to test. He did not state that the test of such a claim by the lady would require a null hypothesis involving a probability, but as Gridgeman (1959) noted, Neyman (1950), Wrighton (1953), and Hogben (1957) were critical of Fisher's implicit introduction of a hypothetical infinite population of responses with a certain proportion of correct responses. Alternatively, Gridgeman defended Fisher by asserting the reasonableness of an assumption of the subject having a probability of being correct, based on the randomness of a neural mechanism associated with the taste discrimination. Whether favoring or opposing the introduction of a random element in addition to the random assignment, a number of writers perceive that random element (a probability of correct identification of the way a cup of tea was prepared) as a component of Fisher's outlook.

Another discussion in Fisher's chapter clearly concerned a random element in the lady's judgment, a random element some people associate with "guessing." He referred to a situation in which all cups made with milk poured first had sugar added but none with tea poured in first had sugar added, noting that this could lead to an easy division by the taster into two groups on the basis of the flavor of sugar alone, and remarked:

> These groups might either be classified all right or all wrong, but in such a case the frequency of the critical event in which all cups are classified correctly would not be 1 in 70, but 35 in 70 trials, and the test of significance would be wholly vitiated. (p. 18)

There is no reference to a random element being employed in determining that all cups with milk poured first would have sugar added and all cups with tea poured first would not have sugar added. Therefore, the probability of 35/70 must be based on the assumption that after distinguishing four cups made with sugar from four made without sugar there is a probability of 1/2 of correctly designating the groups as "milk first" or "tea first." Now if a taster usually drank tea with sugar and with tea poured on top of milk ("milk first"), the response could be "milk first" solely on the basis of the sugar taste on every taste trial, in which case the probability of correct classification would be 1, not 35/70. No matter what Fisher meant, what he wrote represents a point of view that unjustifiably treats responses under the null hypothesis as being randomly selected. As will be shown below, the everyday notion of guessing as being like mentally tossing a coin must be abandoned before a randomization test can properly be applied to data from tests of perceptual discrimination.

Careful consideration of the random assignment and the corresponding data permutation procedure, along with H_0, is especially important in connection with experiments where "guessing" is the hypothesis to be rejected. There is a strong temptation to regard the chance element as residing in the subject rather than in the random procedure introduced by the experimenter. However, bear in mind that in the lady-tasting-tea experiment, if H_0 was true, the lady did not by sheer chance make correct responses; rather, the experimenter happened by chance to assign cups of tea that matched the responses that she would have made at that time, in any case. It is not particularly useful to have evidence that a person is not making responses at random but it is useful to get evidence that a person can make certain types of discriminations. Unjustifiable assumptions of randomness of individual behavior can cause the necessity of random assignment to be overlooked, as in the following example.

Example 11.1. An experiment is conducted to determine whether an alleged mind reader can read the mind of another person, a "sender." Elaborate precautions are taken to ensure that the two people do not know each other and have no prearranged signals or other ways of faking telepathic transmission. The sender is given five cards and asked to concentrate on one of the cards, whichever one he wants; at the same time, the mind reader, who is isolated from the sender, is asked to specify the card on which the sender is concentrating. One trial each is given for geometrical figures, colors, numbers, and words, there being five alternatives for each trial. Suppose that in every one of the four trials, the mind reader selected the card on which the sender was concentrating. Then the probability of this degree of agreement resulting by chance is computed as $(1/5)^4$, which is less than 0.002. Is this not strong evidence that the mind reader is reading the mind of the sender because alternative means of transmission have been rendered impossible? No, it is not. It is strong evidence that the relationship between the card the mind reader picks and the card the sender observes is not chance. However, maybe all the mind reader is demonstrating is that he tends to think like the sender. After all, there have been numerous demonstrations to show the predictability of responses of people in general when they are asked to name a color, to give a number between 1 and 10, or to perform some similar task. Instead of the answers being uniformly distributed over all possible answers, there tends to be a concentration on one or two popular answers. We might obtain a very small P-value simply because of similarities between the sender and the mind reader. The sender and the mind reader probably are not only members of the same species but members of the same society with enough experiences in common to give them similar preferences. An experiment where the sender as well as the mind reader indicates preferences can mislead us into thinking we have evidence of telepathy or mind reading because of the similarity of the preferences of the two people.

To get evidence of mind reading, it is necessary to have the chance element in the experiment be introduced by the experimenter. The experimenter randomly selects one of the five cards each time and then has the sender concentrate on it. Under those conditions, the computation of the probability of such a large number of correct responses by the mind reader would test the null hypothesis that the mind reader's response was the same as it would have been if the sender had concentrated on any of the other cards.

11.4 Limitations of the Random Sampling Model for N-of-1 Experiments

The application of statistical tests to N-of-1 experimental data cannot be justified on the basis of the random sampling model, which is the model underlying the significance tables for parametric tests. Random selection of a single subject from a population of subjects does not permit an estimate of the population variance, and so this type of random sampling obviously provides no basis for application of a parametric test. Another type of random sampling, that of random selection of treatment times from a population of such times, could be performed but not in the manner required by the random sampling model. For example, the experimenter could designate a large number of potential treatment times and then randomly select so many of them for one treatment, so many for another, and so on. The number of potential treatment times in the population would have to be large relative to the number selected for the various treatments because the random sampling model involves an infinite population, not one that is almost exhausted by the sampling. Alternatively, taking small random samples from a very large population of treatment times — although inconvenient because of the likelihood of the sample treatment times being spread over a long interval of time — nevertheless is possible. However, there is an assumption of independence associated with the random sampling model that would make such samples inappropriate for testing hypotheses.

The random sampling model underlying parametric significance tables assumes that the measurement associated with a randomly selected element is independent of other elements in the sample. Thus, in reference to randomly selected treatment times for a subject, the independence assumption implies that a subject's measurement associated with a treatment at a particular time t_k is independent of the number of treatments given prior to t_k. Whether t_k is the earliest time in a sample of treatment times or the latest then is assumed to have no effect on a subject's response at time t_k. The inappropriateness of this assumption because of factors like fatigue and practice indicates that random sampling of treatment times cannot provide a basis for the valid use of parametric statistical tables with N-of-1 experimental data.

Thus, neither random sampling of a population of subjects nor of a population of treatment times would justify the determination of P-values by parametric significance tables for N-of-1 experimental data. Apparently, there is no random sampling that can be performed that would make P-value determination by reference to parametric significance tables valid for N-of-1 experiments.

11.5 Random Assignment Model

Alternatively, randomization tests whose P-value is based on random assignment can be applied to N-of-1 experimental data validly when there has been random assignment of treatment times to treatments. Any randomized multiple-N experimental design can be used as an experimental design for a randomized N-of-1 investigation. In fact, although in previous chapters we have regarded the experimental unit to be a subject except for the subject-plus-treatment time units for repeated-measures experiments, all of the properties of randomization tests are applicable to any experimental unit. And similarly, any randomization test applicable to a certain multiple-N design where subjects are randomly assigned to treatments can be applied to data from an N-of-1 design where treatment times are assigned randomly to treatments in an analogous manner. In the typical multiple-N independent-groups experiment, the assignment procedure is this: the number of subjects per treatment is fixed and within that restriction, there is random assignment of subjects to treatments. The N-of-1 analog is as follows: the number of treatment times for each treatment is fixed and within that restriction, there is random assignment of treatment times to treatments. Given such random assignment in an N-of-1 experiment, a randomization test for determining the P-value can be employed to test the following H_0: for each of the treatment times, the response is independent of the treatment given at that time. See Edgington (1996) for an elaboration.

The randomization tests that have been described for determining the P-value for ANOVA and other techniques based on random assignment of subjects to treatments therefore can be applied to N-of-1 experiments employing analogous assignments of treatment times to treatments. Applications of this type will be examined but first we will return to a topic considered earlier in regard to repeated-measures experiments: carry-over effects.

11.6   Carry-Over Effects

Parametric significance tables assume not only random sampling from a population of sampling units but also that each sampling unit has a measurement value associated with it that is independent of the constitution of the sample in which it appears. As indicated in Section 11.4, the independence assumption would be unrealistic for random samples of treatment times. The measurement value associated with a randomly selected treatment time ordinarily would be dependent on the sample in which it appeared because the sample composition determines how many treatment administrations precede any particular treatment time and thus bears on how practiced, fatigued, or bored the subject will be by that time.

The effect of the number of treatment administrations that precede a particular administration can be regarded as a type of carry-over effect from previous treatments. The validity of determining P-values by means of randomization tests for N-of-1 experiments usually is unaffected by such a carry-over effect because the effect is constant over all data permutations whenever all data permutations involve the same set of treatment times, with the number of treatment administrations preceding any treatment time fixed rather than randomly determined.

The carry-over effect just described should not be considered a treatment effect because its existence does not at all depend on differences between the effects of treatments. The carry-over effect could occur even if the same treatment was given repeatedly. However, there is another type of carry-over effect that should be considered a treatment effect. This effect is the differential carry-over effect, wherein the subject's response to a treatment at time t_k depends not only on how many treatments preceded that treatment but also on the particular treatments that were given at each of the preceding treatment times. For example, if at time t_k the response to treatment B is influenced by whether the immediately preceding treatment was A or another treatment B, there is a differential treatment effect. Why is this type of effect considered a treatment effect? Because there is a differential effect of the treatments on the measurements. If A and B were identical treatments given different labels, such carry-over effects would not occur, and so differential carry-over effects indicate differential effects of the treatments on the measurements.

What we usually mean by a treatment effect is quite different from a differential carry-over effect. If at treatment time t_k the response to A was different from what it would have been to B, there was a treatment effect and if the A or B response was the same as it would have been for any possible pattern of preceding treatment assignments, the treatment effect would not be a differential carry-over effect. It is important to distinguish this type of treatment effect from a differential carry-over effect because both types of effects can influence the P-value and can occur when a subject takes a series of treatments. For N-of-1 experiments, the H_0 of no treatment effect tested by a randomization test is that the association between treatment times and measurements is the same as it would have been for any alternative assignment. If the time-ordered series of measurements associated with the treatment times would have been different for certain possible assignments of treatment times to treatments, H_0 is false — there is a treatment effect. But rejection of H_0 does not imply that there is a treatment effect that is not a differential carry-over effect. The only method of drawing statistical inferences about differences that are not the result of carry-over effects is to use an independent-groups, multiple-N design, where each subject takes only one treatment. Only in that situation can statistical inferences be drawn about difference in effectiveness of treatments that are not a function of carry-over effects.

The fact that the statistical inference does not necessarily imply a treatment effect for temporally isolated or once-given treatments is not as important as it first appears. Useful generalizations from experiments almost always have a nonstatistical basis because the statistical inferences concern specific subjects at specific times in specific situations, and so on, and therefore have little generality. The statistical test for an N-of-1 experiment therefore can be run to determine whether there is a treatment effect and then on the basis of theoretical and other nonstatistical considerations, the experimenter can decide whether there is sufficient evidence for inferring that there was a treatment effect not attributable to differential carry-over alone. If the experimenter is especially concerned about controlling for differential carry-over effects and cannot do so by using a multiple-N, independent-groups experiment, he can exercise various experimental controls. If carry-over can result from the memory of previous tasks, the experimenter can introduce distracting stimuli between treatment sessions to impair that memory. Or if carry-over could occur through physiological changes of short duration, the experimenter may decide to space the treatment sessions far apart to cut off this source of differential carry-over. Whatever the situation, it is up to the experimenter to exercise experimental control if he wants to minimize the influence of differential carry-over effects.

11.7   The N-of-1 Randomization Test: An Early Model

The "Fisher randomization test," which was applied to Darwin's data on descendents of cross-fertilized and self-fertilized plants, was discussed by Fisher in his chapter following the lady-tasting-tea experiment (Fisher, 1951).

That fact raises the question of why Fisher did not discuss the lady-tasting-tea experiment from the randomization test perspective as, for example, by using a quantitative response and explicitly deriving a reference set of test statistic values based on the data. A consideration that may help explain Fisher's failure to discuss the lady-tasting-tea testing procedure as another instance of a randomization test is that the "Fisher randomization test" applied to Darwin's data required random sampling, and random sampling in addition to random assignment could be difficult to introduce in a plausible manner into the lady-tasting-tea experiment. It should be kept in mind that it was Pitman (1937), not Fisher, who demonstrated the applicability of randomization tests in the absence of random sampling. Whatever the cause may be, Fisher did not present the lady-tasting-tea experiment as a new procedure for deriving a reference set of test statistics based on data permutation.

A simple N-of-1 randomization test was published about 40 years ago, along with a discussion of its rationale and the replicating of N-of-1 randomization tests (Edgington, 1967). The general nature of this test, which is an independent *t* test applied to N-of-1 data with the P-value based on data permutation, and the experimental design for which it is appropriate are illustrated in Example 11.2.

Example 11.2. In an experiment, we predict that our subject will provide larger measurements under treatment A than under treatment B. In planning the experiment, we decide to administer each treatment six times. We confer with the subject to determine when the subject could be available for the experimental sessions. An afternoon is selected during which the treatments can be given. Then we select 12 times from that afternoon for treatment administration, ensuring that the treatments are far enough apart to serve our purpose. The 12 times are then assigned randomly to the treatments. The null hypothesis is: the subject's response at any treatment time is independent of the assignment of treatment times to treatments. We obtain the following data, listed in the order in which the treatments were given:

A	B	B	A	B	A	A	B	A	B	A	B
17	16	15	18	14	17	18	14	20	13	19	15

The one-tailed test statistic is T_A, the total of the A measurements, and the obtained test statistic value is 109. To perform our statistical test, we determine the proportion of the $12!/6! \ 6! = 924$ data permutations that have a value of T_A as large as 109 using Program 4.4, the program for the independent *t* test that uses systematic data permutation.

We have considered a hypothetical example of the N-of-1 randomization test in a simple form. Let us now examine an actual application of it that has been published. (An experiment involving replications of the N-of-1 design and test in Example 11.2 will be discussed later.)

Example 11.3. An N-of-1 randomized experiment was conducted on a patient who had been taking metronidazole for three years following an operation for ulcerative colitis (McLeod et al., 1986). The physician and her patient agreed on the desirability of performing an experiment to determine whether the medication was relieving unpleasant symptoms, like nausea and abdominal pain, because if metronidazole was not effective both the cost of the drug and risk of cancer associated with its long-term use dictated discontinuing it.

The experiment lasted 20 weeks. The 20 weeks were divided into 10 blocks of two-week intervals, with five of the intervals were randomly selected as the times when the patient received metronidazole daily and the remaining five two-week intervals assigned to "an identical placebo capsule." The primary source of data was a diary in which the patient was asked to report the presence or absence (or the magnitude) of each of seven symptoms, which included nausea, abdominal pain, abdominal gas, and watery stool. Because of the possibility that the effects of the drug would carry over to days after it was withdrawn, only data from the second week of each two-week session were subjected to a statistical test. Daily observations were combined to provide a single weekly measurement for each different symptom. Therefore, there were 10 measurements for each dependent variable (symptom) and, because there were only $10!/5!5! = 252$ data permutations, the experimenters carried out a systematic randomization test using the data permuting procedure in Program 4.4.

Six of the seven symptoms showed a reduction under metronidazole significant at the 0.05 level: two P-values were 0.024 and four were 0.004. As the results confirmed the effectiveness of the drug, the patient was advised to continue taking it.

11.8 Factorial Experiments

An example will now be provided to show how a factorial experiment involving only one subject can be devised and the data analyzed. In this particular example, only two levels of two factors will be considered but the procedure followed in the analysis can be generalized to any number of levels of any number of factors.

Example 11.4. Two factors — illumination and sound, each with two levels — are investigated to determine their effects on physical strength. Twenty treatment times are used. Ten treatment times are randomly drawn and assigned to the "light" condition, with the remaining 10 going to the "dark" condition. The 10 treatment times for each of these treatments are then randomly divided into five for a "quiet" condition and five for a "noise" condition. The sound and illumination conditions are given in four combinations: quiet and light, quiet and dark, noise and light, and noise and dark, with five measurements for each treatment combination.

The measurements in Table 11.1 show the outcome of a hypothetical experiment where measures of strength are expressed in arbitrary units. First, consider the test for determining whether the sound variable had any effect on strength within the light condition; this requires a comparison of the two

TABLE 11.1

Factorial Experimental Data

	Quiet		Noise	
Light	11	14	16	22
	12	14	18	24
	15		20	
Dark	15	8	12	3
	8	6	4	12
	14		5	

upper cells. The null hypothesis is that the measurement for each of the treatment times in the light condition is independent of the assignment of the treatment time to level of noise. In other words, H_0 is that the presence or absence of noise had no effect on physical strength for the light condition. Thus, we divide the 10 measurements in the upper row in every possible way into five for the upper left cell and five for the upper right cell, and we compute independent t for each division. Program 4.4 can be used to determine the P-value by systematic data permutation. For these particular results, a test is unnecessary because it can be seen that the two cells differ maximally, and so the P-value for the obtained results would be $1/252$ for a one-tailed independent t test, or $2/252$ for a two-tailed t test, because there are $10!/5!5! = 252$ permutations of the measurements to be considered.

Of course, similar comparisons could be made to determine the effect of sound within the dark condition, the effect of illumination within the quiet condition, or the effect of illumination within the noise condition. For each comparison, the P-value could be determined by Program 4.4.

In addition to evaluating the main effects of the illumination and noise factors within particular levels of the other factor, one also can test the main effect of illumination or noise over all levels. Consider how the test would be conducted to test the illumination effect over both quiet and noise conditions. Within each of the two sound conditions, the measurements are divided between the light and dark conditions to test the H_0 of no differential effect of the illumination conditions on strength. The number of permutations within the two levels of sound between the two illumination conditions will be $10!/5!5! \times 10!/5!5! = 63,504$ because each division of the 10 measurements for the quiet condition between the two illumination conditions can be paired with each of the 252 divisions of the 10 measurements for the noise condition. Program 5.1 can be used to test the overall main effect of illumination.

11.9   Randomized Blocks

In multiple-N experiments, the same subjects are sometimes given two or more treatments and the data are analyzed by pairing the measurements by subject. The sensitivity of an N-of-1 test can be increased by a similar use of pairing. In N-of-1 experiments, the variability over the time during which the experiment is conducted can be considerable in some cases, and the pairing or clustering of treatment times can be used to control for the intersessional variability.

Example 11.5. An experimenter has 20 treatment times spread over 10 days and performs the following random assignment and pairing. First, two times for each day are specified, one in the morning and one in the afternoon. Thus, each of the days provides a pair of treatment times. On each of the 10 days both treatments are given, the determination of which of the two treatments is to be given in the morning and which in the afternoon being determined by the toss of a coin. The response measurements are paired within days, so that there are 10 pairs of measurements where one member of each pair is a measurement for treatment A and one member is a measurement for treatment B. Program 6.1 and Program 6.3, the programs for repeated-measures ANOVA and the correlated t test with the P-value determined by systematic data permutation, with ΣT^2 (the sum of the squares of the treatment totals) as a test statistic, can be used for determining the P-value for a two-tailed test. If treatment A is predicted to provide the larger measurements, T_A can be used as the one-tailed test statistic with Program 6.3.

Example 11.6. Smith (1963) provided a good example of a randomized N-of-1 clinical trial. He used a randomized block design with a single narcoleptic patient to compare the effectiveness of three drugs (methamphetamine, dextroamphetamine, and adrenaline methyl ether) on reducing narcoleptic symptoms. The study was a double-blind investigation in which neither the therapist nor the patient knew which drug was administered at a given time. A 15-day span for administering the drugs was divided into five three-day blocks. All three drugs were given in each three-day block with random determination for each block separately how the drugs would be distributed over the three days. The dependent variable was a score on a self-report rating scale on which the subject indicated the strength of various symptoms of narcolepsy each day.

ANOVA was applied to the data, with the P-value based on F tables. Smith did not publish the raw data for the 15 days but to show how a randomization test could have been applied to his data, we will use hypothetical data shown in Table 11.2.

Program 6.1 or Program 6.2 can be applied to the data to determine the P-value. Under the null hypothesis of no differential treatment effect, random

TABLE 11.2

Comparison of Three Drugs

	Three-Day Blocks				
	1	2	3	4	5
Methamphetamine	4	8	7	8	4
Dextroamphetamine	2	2	3	6	1
Adrenaline methyl ether	3	6	8	7	3

assignment of treatment times to treatments within each block randomly assigned the measurements for that block to the treatments. Thus, there are $(3!)^5 = 7776$ data permutations for the systematic test, consisting of those that can be produced by switching the measurements between treatments within blocks in every possible way. Program 6.1 gave a P-value of about 0.0039, as there are 30 of the 7776 data permutations that provide as large a value of ΣT^2 as the obtained results.

Even though Smith did not use a randomization test to analyze the data, his conduct of a rigorous experiment on a single patient with a relatively complex randomized experimental design is of considerable importance to the field of randomization testing. Randomization tests depend on experimental randomization, and so innovative random assignment procedures of demonstrated utility open up new opportunities for the employment of randomization tests.

11.10 Correlation

A test for correlation with the P-value determined by a randomization test can be applied to N-of-1 data whether the independent and the dependent variables are raw measurements, ranks, or categorical variables with levels expressed as numbers, as when 0 and 1 represent two categories of a dichotomy.

Example 11.7. An investigator wants to determine whether a certain drug has an effect on a patient's blood pressure. Six dosage levels and six treatment administration times are selected. The six treatment times are then assigned randomly to the six different levels of drug dosage. The time lapse between drug administration and the measurement of blood pressure is set in advance also and is constant over drug dosages. The drug is then administered on the six occasions and the six measurements of blood pressure are obtained. The experimenter predicts that increases in drug dosage will raise the blood pressure, and so the one-tailed correlation test statistic ΣXY is computed for all $6! = 720$ data permutations to determine the P-value.

11.11 Operant Research and Treatment Blocks

Pavlov repeatedly associated food with the sound of a bell to train a dog to salivate at the sound of the bell. Such conditioning has come to be called *classical conditioning*. Alternatively, *operant conditioning* is conditioning in which reinforcement of a certain type of behavior (e.g., by giving food or shock) leads to an increase (or decrease) in the frequency of that type of response. For example, receiving food for pressing a bar increases the frequency of bar-pressing by a rat, and receiving a shock when stepping onto an electric grid causes a rat to avoid repetition of that behavior. *Behavior modification* is applied operant conditioning that is used mainly for getting rid of "problem behavior" in people. The elimination of bed-wetting, thumb-sucking, and stuttering are examples of problems that have been approached through behavior modification.

Operant research frequently involves the intensive study of the individual animal or person. The dependent variable is often the frequency of response of a certain type within a period of time, during which a certain type of reinforcement is in effect. (We will call the time period for which the number of responses is recorded a *treatment block.*) For example, a child who is shy is observed over 20 five-minute periods, during each of which the number of times she speaks to another child is recorded. During some of the periods, she is given a coin whenever she talks to another child, and in some periods there is no such reinforcement. There would be 20 treatment blocks, and for each block there would be a measurement consisting of the number of times the child conversed with other children during that time interval. The use of treatment blocks in which a particular treatment is continuously in effect and in which responses of a certain type are counted to provide a single measurement for that block does not necessitate special procedures of statistical analysis. The problems of analysis of operant data are not fundamentally different from those of the analysis of N-of-1 data where there are no treatment blocks, as will be shown in the following discussion of operant research designs. Randomization tests for operant experiments have been discussed elsewhere by Edgington (1984), Kazdin (1976), and Levin et al. (1978).

11.12   ABAB Design

Operant research with two treatment conditions, A and B, is sometimes carried out with a series of treatment blocks with the treatments systematically alternating from block to block. (Treatment A may be the reinforcement condition and B the control or nonreinforcement condition.) This is sometimes called the *ABAB design*, although it could equally well be called the BABA design. This design controls rather well for effects of practice, fatigue, and other variables that would be expected to systematically increase in magnitude over the treatment blocks but not for environmental or intra-organismic periodic fluctuations that happen to coincide with alternations in treatments. Careful determination of the length of the treatment block and control over environmental conditions may make it rather plausible that the possibility of cyclical influences with differential effects on the treatment blocks for the two treatments has been minimized. However, the way to render an experimental design invulnerable to objections regarding the preferential position that one treatment can hold in a series of treatment blocks is to use random assignment. The restriction that treatments alternate from one treatment block to the next severely limits the possibilities of random assignment. The toss of a coin can determine which treatment is given first but after that, the sequence is completely determined. Consequently, the P-value for any test statistic cannot be less than 0.50 by data permutation for the ABAB design.

11.13   Random Assignment of Treatment Blocks to Treatments

Like specific treatment times, time intervals (treatment blocks) can be assigned randomly to treatments.

Example 11.8. A child is observed for a period of 1 hour, with 12 five-minute treatment blocks used for recording the frequency of some type of undesirable behavior. Treatment A blocks are intervals of time during which there is no special intervention by the experimenter; these are the control blocks. Treatment B blocks are blocks where the experimenter introduces scolding or some other negative reinforcement expected to reduce the incidence of the undesirable behavior; consequently, a one-tailed test is appropriate. Six of the 12 treatment blocks are selected randomly for treatment A and the remaining six are assigned to treatment B. The number of possible assignments of treatment blocks to treatments is $12!/6!6! = 924$. The observed results are as follows:

Treatment Block	1	2	3	4	5	6	7	8	9	10	11	12
Measurements	15	12	8	10	7	6	8	5	10	11	7	3
Treatments	A	A	B	A	B	B	A	B	A	A	B	B

The null hypothesis is that for each treatment block, the measurement is the same as it would have been for any alternative assignment, i.e., the measurements associated with the treatment blocks are independent of the way in which the treatments are assigned to blocks. The 924 data permutations are the 924 distinctive associations of six As and six Bs with the above sequence of 12 measurements. Treatment A was predicted to provide greater frequencies (larger measurements), and so T_A can be used as the one-tailed test statistic. For the obtained data, $T_A = 15 + 12 + 10 + 8 + 10 + 11 = 66$. The proportion of the 924 data permutations with as large a value of T_A as 66 is the P-value. Determining the P-value associated with T_A or the two-tailed test statistic, $\Sigma(T^2/n)$, can be accomplished by using Program 4.4.

The test carried out for the 12 treatment blocks is the same as if there were 12 treatment times and a single measurement was made for each time. That the "treatment time" is extended over a five-minute interval and that the "measurement" for a treatment time is the total number of responses of a certain type during the five-minute interval is irrelevant to the determination of the P-value. The random assignment of treatments to treatment blocks corresponds to the random assignment of subjects to treatments in a multiple-N experiment. Consequently, the characteristics of data for which the significance tables for *t* provide close approximations to the randomization test P-value for multiple-N experiments are those for which the tables provide close approximations for N-of-1 experiments of this type. Some of those characteristics are large samples, equal sample sizes, relatively continuous data (for example, in contrast to dichotomous data), and an absence of outlier scores.

11.14 Randomization Tests for Treatment Intervention

Sometimes in behavior modification studies, an experimenter will select one of the treatment blocks as the time to introduce an experimental treatment that then remains in effect until the end of the experiment. Evidence of a treatment effect involves a comparison of measurements from blocks before treatment intervention (introduction of the experimental treatment) with those blocks following treatment intervention. This comparison is sometimes made through examining a graph or sometimes it is based on computation. Occasionally, *t* or some conventional test statistic is computed but frequently, there is no attempt to conduct a statistical test of any type. For the determination of the P-value by a randomization test procedure, the intervention must be random but random intervention is rare. More commonly, an investigator introduces an experimental treatment when the baseline or control observations are expected to be stable. But whether the determination of stability is subjective or objective, there is the possibility that the use of the baseline stability criterion leads to the introduction of the treatment at a time when the level of the responses would have shown a sharp change even without intervention. To argue on empirical grounds that such a possibility is unlikely to lead to bias in a particular experiment is like arguing in a multiple-N experiment that a certain systematic assignment of subjects is in fact as good as random assignment; it may persuade a few persons that there is no bias but it does not provide a sound basis for performing a statistical test. Therefore, it must be appreciated that in the following description of the application of a randomization test procedure, random intervention is included because it is necessary for a statistical test and not because it is in any sense typical. What is to be described is a way to conduct an intervention experiment and analyze the data that will provide a valid determination of the P-value of the treatment effect.

Example 11.9. An experimenter decides to test the effectiveness of an experimental treatment (a method of reinforcing behavior) for increasing the frequency of a certain type of desired behavior. Twenty five-minute treatment blocks are specified, for each of which a count of instances of the desired behavior is recorded. When the experimental treatment, which is a method designed to stimulate the desired behavior, is introduced it stays in effect over the remainder of the treatment blocks. Thus, the intervention of the treatment divides the treatment blocks into two classes: control blocks, which are those prior to treatment intervention, and experimental blocks, which are those after intervention. To prevent the control or experimental samples from being too small, the experimenter randomly introduces the treatment with a sample-size constraint to make sure there are at least five control and at least five experimental blocks. This sampling constraint is imposed by restricting selection to one of the blocks within the range of blocks 6 to 16. The treatment is introduced at the beginning of the selected block and remains in effect over all subsequent blocks. There are thus 11 possible assignments of the treatment intervention to the blocks.

Suppose that the experimenter selected block 8 for the intervention and obtained the following results:

Block	1	2	3	4	5	6	7	8	9	10	11	12	13	14	15	16	17	18	19	20
Data	2	3	4	3	2	3	4	*8*	*9*	*8*	*9*	*10*	*8*	*9*	*9*	*8*	*10*	*9*	*8*	*8*

The *italicized* numbers indicate the experimental treatment blocks and the measurements for those blocks. For a one-tailed test where the experimental treatment blocks are expected to provide the larger measurements, $(\bar{X}_E - \bar{X}_C)$, the mean of the experimental measurements minus the mean of the control measurements, can be used as a test statistic. The 11 data permutations for which the test statistic is computed are associated with the 11 possible times that could be selected for the intervention of the experimental treatment: the beginning of any one of block 6 to block 16. The test statistic value for the first data permutation is the mean of the last 15 measurements minus the mean of the first five, which is $120/15 - 14/5 = 5.20$. For the second data permutation, the test statistic value is $117/14 - 17/6 = 5.52$. The third data permutation is that for the obtained results, which gives an obtained test statistic value of $113/13 - 21/7 = 5.69$. (It might be more convenient to have the first data permutation associated with the earliest potential point of intervention than associated with the obtained results.) Computation of the remaining test statistic values would show that only the obtained data division provides a test statistic value as large as 5.69, and so the P-value for the results is 1/11, or 0.091.

A two-tailed test statistic for such a design is the absolute difference between means, $|\bar{X}_E - \bar{X}_C|$. For either a one- or a two-tailed test, a special computer program is required for treatment intervention designs because of the unusual method of random assignment, which dictates a special method of permuting the data. However, the data permutation procedure is simple enough to make the development of either a systematic or a random data permutation program fairly easy.

11.15 Effects of Trends

A consistent upward or downward trend over the duration of an experiment can make a statistical test less powerful (less sensitive) than it would be otherwise, even when it does not affect the validity of the test. Experiments extended over a considerable period of time are especially susceptible to the influence of long-term trends. The two test statistics $(\bar{X}_E - \bar{X}_C)$ and $|\bar{X}_E - \bar{X}_C|$ in Example 11.9 seem appropriate for detecting treatment effects when there is no general upward or downward trend in measurements over time that is independent of the treatments. However, $|\bar{X}_E - \bar{X}_C|$ lacks sensitivity in the presence of such trends.

Example 11.10. Let us consider a situation for which the two-tailed test statistic $|\bar{X}_E - \bar{X}_C|$ fails to reflect an apparently substantial treatment effect. Consider the following outcome of a treatment intervention study:

Block	1	2	3	4	5	6	7	8	9	10	11	12	13	14
Data	9	8	7	6	5	4	3	9	8	7	6	5	4	3

Treatment intervention was at the beginning of block 8, and so the measurements for the first seven blocks are control measurements and the *italicized* measurements for the last seven blocks are experimental measurements. The magnitude of the measurements was dropping consistently until treatment intervention, at which time there was a sharp rise although the decline resumed after the rise. The results suggest that fatigue, boredom, or some other measurement depressant caused the measurements to decline in size over time, but that the treatment intervention raised the magnitude of the measurements even though the decline resumed. The two-tailed test statistic $|\bar{X}_E - \bar{X}_C|$ for the obtained results has the smallest possible value, 0, and every other data permutation provides a larger value. Thus instead of reflecting a treatment effect through its smallness, the P-value has the largest possible value, 1. If the treatment effect in a treatment intervention experiment is added to a general trend effect — as in this sample — the test statistic $|\bar{X}_E - \bar{X}_C|$ can be quite insensitive to a treatment effect.

Analysis of covariance F, with the block number as a covariate, would seem to be a useful two-tailed test statistic for treatment intervention experiments; see Winer (1971) for a discussion of analysis of covariance. By using the block number as a covariate, there is statistical control over temporal trends within treatment conditions. For the data in Example 11.10 where there is a strong trend, the value of F for the analysis of covariance is larger for the obtained results than for any of the other data permutations, confirming the general impression given by the data that there was an effect of the experimental treatment. Of course, the P-value associated with an analysis of covariance F can be determined by data permutation on the basis of the data permutation procedure described in Example 11.9, with analysis of covariance F computed for each data permutation.

An alternative to analysis of covariance is the use of difference scores for either treatment intervention or other N-of-1 designs. The following example shows how different scores can be helpful for other designs.

Example 11.11. A physician wants to find out whether a certain drug will help a patient. From 10 consecutive Mondays, five are randomly drawn and are used as the dates for the administration of the drug. The remaining five Mondays are designated as days on which a placebo is given instead of the drug.

The physician expects the drug to manifest its effect about a day after its administration and so takes the physiological measurements of the treatment effect on the 10 Tuesdays following the Mondays used in the experiment. If an independent t test was used on the data with the P-value determined by data permutation, the procedure would not be basically different from that in Example 11.2, where the experiment was not extended over a very long period of time. However, the physician does not analyze only those 10 Tuesday measurements but takes the same type of measurements on the Sundays preceding the 10 Mondays used in the experiment and derives 10 "change measurements" from the 10 Sunday-to-Tuesday time spans enclosing the Mondays. The measurement on a Sunday is subtracted from the measurement on the following Tuesday to give a change measurement for the enclosed Monday. Then the 10 change measurements are permuted as if they were single measurements and the P-value for an independent t test is determined by using Program 4.4.

The reason for using change measurements instead of Tuesday measurements alone is that the physician believed that even without the experimental treatment, the subject might consistently improve (or deteriorate) over the 10-week experimental period. The effect of such a consistent trend on the data analysis can be seen by considering what would happen if the measurements for both the placebo and the drug administrations consistently increased over the 10 weeks. Even if the drug tended to produce larger measurements, the upward trend in all measurements could conceal this because the later placebo measurements might be generally higher than the early drug measurements, and such an overlap would render a test based on Tuesday measurements alone relatively insensitive. For a uniform linear trend over the experimental period, the use of Sunday-to-Tuesday change scores controls for the overlap of drug and placebo measurements resulting from the general trend and provides a more sensitive test.

Another procedure that can be used to control for general trends over a period of time is to use a time-series test and determine the P-value of the test statistic by data permutation. The general objective of time-series analysis when used for testing for treatment effects is the same as that of difference scores: to prevent concealment of treatment effects resulting from a general trend in the data that is not a function of the treatments.

11.16 Randomization Tests for Intervention and Withdrawal

When there is treatment intervention without withdrawal, there are only as many data permutations as there are possible intervention blocks, so the randomization test cannot be very sensitive unless there is a large number of blocks. For example, there must be at least 20 possible blocks for intervention for the possibility of statistical significance at the 0.05 level. However, when it is feasible to have one or more withdrawals of treatment the sensitivity of a randomization test can be increased substantially.

Example 11.12. An experimenter specifies 20 treatment blocks, for each of which the total frequency of a certain type of behavior in a child is to be recorded. Treatment intervention is to be followed by withdrawal, with the following constraints: there will be at least five control treatment blocks preceding the treatment intervention, at least five control treatment blocks following withdrawal of the treatment, and at least five experimental blocks between the two series of control blocks. The block selected for intervention restricts the blocks that can be selected for withdrawal. (Intervention or withdrawal will occur at the beginning of the selected block.) If the point of intervention was block 6, the point of withdrawal would have to be within block 11 to block 16 to satisfy the imposed constraints. But if block 7 was the point of intervention, block 11 could not be a point of withdrawal; only block 12 to block 16 would be acceptable as points of withdrawal. In an earlier discussion of this example (Edgington, 1975), there was reference to random selection of an intervention block followed by random selection of a withdrawal block. Ronald Crosier (1982) pointed out that to make all combinations of intervention and withdrawal equally probable, the assignment must involve a pair of blocks with the first block being an intervention block and the second a withdrawal block. Thus, the experimenter chooses from the following pairs of blocks, where the first block in a pair is an intervention block and the second a withdrawal block: (6, 11), (6, 12), (6, 13), (6, 14), (6, 15), (6, 16), (7, 12), (7, 13), (7, 14), (7, 15), (7, 16), (8, 13), (8, 14), (8, 15), (8, 16), (9, 14), (9, 15), (9, 16), (10, 15), (10, 16), (11, 16). Therefore, there are 21 equally probable assignments of intervention-withdrawal combinations (pairs), each with a probability of $1/21$. Note that random selection of the intervention block followed by random selection of the withdrawal block would have resulted in unequal probabilities for the 21 pairs, ranging from the probability for pair (6, 11), which is $1/6 \times 1/6 = 1/36$ to the probability for pair (11, 16), which is $1/6 \times 1 = 1/6$.

The experimenter randomly selected pair (7, 14), so block 7 was the first block receiving the experimental treatment and block 14 was the first block under the control condition after withdrawal of the experimental treatment. The following results were obtained:

Block	1	2	3	4	5	6	7	8	9	10	11	12	13	14	15	16	17	18	19	20
Data	3	4	4	3	4	3	5	5	6	6	7	5	6	4	3	4	4	3	3	3

The *italicized* numbers indicate the experimental treatment blocks and the measurements for those blocks. For a one-tailed test where the experimental treatment blocks are expected to provide the larger measurements, the test statistic $(\bar{X}_E - \bar{X}_C)$ can be used. The obtained value is the mean of the *italicized* measurements minus the mean of the 13 measurements that precede and follow them, which is $5.71 - 3.46 = 2.25$. The test statistic is computed for each of the data permutations. The first permutation would have block 6 to block 10 for the experimental blocks, the second would have block 6 to block 11 for the experimental blocks, and so on, the last permutation having block 11 to block 15 for the experimental blocks. There are 21 data permutations for which $(\bar{X}_E - \bar{X}_C)$ is computed and of the 21, three provide a value as large as 2.25. Thus, the P-value is 3/21, or about 0.143. $|\bar{X}_E - \bar{X}_C|$ can be used as a two-tailed test statistic.

Treatment intervention with withdrawal allows for more possible assignments, and therefore provides more data permutations than treatment intervention without withdrawal — but there are still not enough data permutations to permit a sensitive test. However, several treatment interventions and withdrawals can considerably increase the power of treatment intervention experiments. The rationale given above can easily be extended to ABAB designs and multiple baseline designs (Koehler and Levin, 1998; Onghena, 1992; 1994).

Regardless of the practicality or impracticality of designs where there is intervention plus withdrawal, they serve a useful function in illustrating the way randomization tests can be employed with complex random assignment schemes. The opportunity of conducting valid statistical tests of treatment effects in experiments employing a sequence of random assignments, where each assignment can substantially restrict the subsequent random assignment, should be of interest in a number of multiple-N as well as N-of-1 investigations.

11.17 Multiple Schedule Experiments

Hersen and Barlow (1976) and Kazdin (1976; 1980) pointed out limitations of N-of-1 experimental designs in which treatments are withdrawn during the experiment. One limitation is the difficulty in assessing the relative effectiveness of treatments, such as certain drugs, that might have enduring carry-over effects. Another limitation is the ethical objection to withdrawing a treatment that might be effective. Therefore, it is sometimes desirable to employ an N-of-1 design in which treatment variables are not withdrawn, especially if a statistical test still can be applied to test for treatment effects.

The treatment intervention design in Section 11.14 permits a valid determination of treatment effect without the necessity of treatment withdrawal, but that design requires a large number of treatment blocks to ensure a relatively sensitive randomization test. Multiple schedule designs (Hersen and Barlow, 1976) also involve the application of a treatment without withdrawal and require only a moderate number of treatment blocks for an adequate test. In a multiple schedule experiment, there is more than one possible source of a treatment (e.g., reinforcement) and the source alternates during the course of the experiment, with the treatment being almost continuously provided by some source. Alternating the source of treatment permits the effect of the treatment to be distinguished from the effect of its source. For instance, if two persons alternate in providing reinforcement and nonreinforcement, and the frequency of a certain type of behavior tends to be consistently higher under the reinforcing condition, there is strong evidence of an effect of reinforcement. The following example was presented by Edgington (1982).

Example 11.13. A multiple schedule design was used to assess the effect of social reinforcement on the behavior of a claustrophobic patient (Agras et al., 1969). The patient was put into a small windowless room and asked to remain in the room until she began to feel anxious, at which time she could open the door and leave. The patient was assessed four times a day to determine the time spent in the room. Two therapists (one at a time) were with the patient each day. The reinforcing therapist praised the subject for staying in the room, whereas the nonreinforcing therapist was pleasant but did not praise the patient. The reinforcing roles of the therapists were switched occasionally. The numbers of seconds spent in the room for day 2 through day 14 are given in Table 11.3, which shows results pooled within days to provide a single measurement for the reinforcing sessions and a single measurement for the nonreinforcing sessions for each day. (Day 1 did not involve reinforcement, so it was excluded.)

The data in Table 11.3 can be used to illustrate the application of a randomization test for a multiple schedule experiment. For this purpose, we will hypothesize the following random assignment. Each day, the experimenter tosses a coin to determine which of two times of day (morning or afternoon) will be the reinforcing session and then tosses the coin again to determine which time of day will be associated with Therapist 1. Two functions are served by the coin tosses: they randomly associate treatment times with treatment conditions within each day and they ensure that systematic differences

TABLE 11.3

Time (Seconds) Spent in Room

	Days												
	2	3	4	5	6	7	8	9	10	11	12	13	14
Reinforcement	250	370	460	520	180	250	290	300	320	240	300	440	500
Nonreinforcement	270	300	420	470	150	200	240	250	280	280	290	380	430

between reinforcing and nonreinforcing sessions are not a function of a particular therapist being systematically associated with a reinforcing condition.

Consider a test of this H_0: for each session (morning or afternoon), the amount of time the patient spends in the room is the same as it would have been under the alternative treatment condition. For example, this implies that on day 2, if the reinforcing conditions had been switched for the morning and afternoon sessions the woman would have spent 270 seconds in the room under the reinforcing condition and 250 seconds under the nonreinforcing condition. Therefore, under the null hypothesis the measurements in Table 11.3 for each day are the same as they would have been under any of the $2^{13} = 8192$ possible assignments of the reinforcing condition to morning or afternoon sessions for the 13 days; only the designation of a measurement as being associated with reinforcement or nonreinforcement would have been different. Program 6.3 for correlated t can be employed using as a test statistic T_L, the total of the measurements for the treatment predicted to provide the larger measurements — which in this case is the reinforcement condition. The proportion of the 8192 data permutations derived by switching measurements within days between reinforcement and nonreinforcement groups that provide as large a total for the reinforcement condition as the obtained total is found, by the use of Program 6.3, to be 0.0032, which is thus the one-tailed P-value.

The test of the effect of reinforcement described in the preceding paragraph did not require the random assignment of therapists to time of day that was postulated. One therapist could have been associated systematically with morning sessions and the other with afternoon sessions, and a single toss of a coin then would assign a therapist-plus-treatment-time experimental unit to the reinforcement condition. However, using two independent random assignments, one being the assignment of time of day to reinforcement condition and the other the assignment of therapists to time of day, allows us to carry out a second test to determine whether there was a therapist effect — that is, a differential effect of the two therapists.

The data in Table 11.3 could be permuted for this test, also. Instead of having the data in rows for "reinforcement" and "nonreinforcement," the top row could be for Therapist 1 and the bottom row for Therapist 2. For example, if Therapist 1 was the reinforcing therapist for days 2, 3, 5, and 7, the measurements for those days would be in the same rows as in Table 11.3, but the values 460 and 420 for day 4 would switch rows and the values 150 and 180 for day 6 also would switch rows. The null hypothesis would be that no matter what effect time of day or the reinforcing condition might have, the measurement for each session is the measurement that would have been obtained under any alternative assignment of therapists. Program 6.1 or Program 6.3 could be applied to perform the 8192 permutations of measurements within days and determine the P-value of the one- or two-tailed test statistics. The P-values obtained would be those associated with t or $|t|$ for a correlated t test, determined by the randomization test procedure.

11.18 Power of N-of-1 Randomization Tests

In addition to N-of-1 counterparts of standard multiple-N randomization tests, such as those in the earlier chapters, there are many randomization tests that have been developed especially for N-of-1 designs. And for any particular design, the researcher has the possibility of selecting among various test statistics. It is desirable to have a range of tests from which to select but the possibility of alternative tests raises the question of which are best. And one of the considerations in selecting a test is its sensitivity to the type of treatment effect that is expected.

The power of parametric statistical tests has received much attention, and the power of rank tests relative to their parametric counterparts also has been investigated. Alternatively, the power of N-of-1 randomization tests without multiple-N counterparts has been largely ignored. For a comprehensive discussion and extensive simulation results, the interested reader is referred to Onghena (1994), Ferron and Ware (1995), Ferron and Onghena (1996), and Ferron and Sentovich (2002).

11.19 Replicated N-of-1 Experiments

Example 11.14. A study was carried out to determine the effect of artificial food colors on aversive (undesirable) behavior in children, such as whining, running away, or breaking and throwing things (Weiss et al., 1980). The children in the study had not been diagnosed as hyperactive but the researchers expected artificial food additives to have a similar effect on their behavior to that which additives have been observed by some investigators to have on hyperactive children.

The study consisted of 22 N-of-1 experiments, each carried out on a child between 2.5 and 7 years of age. On each of 77 days, a child consumed a bottle of soft drink containing either natural or artificial food coloring. The authors stated that "the two drinks were indistinguishable by sight, smell, taste, or stain color," and neither the children nor their parents, whose record of the child's behavior was the principal source of data, knew which drink was given on any particular day. Eight days were randomly selected for giving the artificially colored drink, and the other soft drink was given on the rest of the days. (Using a large number of control days to compensate for a small number of experimental days, as in this study, can be a useful tactic for providing a relatively sensitive test of treatment effect in an N-of-1 experiment when frequent administration of the experimental treatment is undesirable.)

Ten different measures served as dependent variables, many of them consisting of the frequency of occurrence of a certain type of aversive behavior.

Using 10,000 data permutations, Program 4.5 was applied to determine the P-value for each child on each dependent variable. "Twenty of the children displayed no convincing evidence of sensitivity to the color challenge." A three-year-old boy had P-values of 0.01 and 0.02 for two types of aversive behavior, which is not very impressive considering the multiplicity of testing: 10 dependent variables for each of 22 children. However, a 34-month-old girl provided very persuasive results, even when the number of children tested is taken into account. The P-value for three of the ten dependent variables for that child was 0.0001, and P-values for five others were 0.0003, 0.0004, 0.0006, 0.001, and 0.03. The authors stressed the need for more stringent testing of additives because the food additive used in the experiment was "about 50 times less than the maximum allowable intakes (ADIs) recommended by the Food and Drug Administration." The N-of-1 experimental results of an exceptionally sensitive child showed an effect of a small dose of food coloring that might have been missed in other experimental designs.

11.20   N-of-1 Clinical Trial Facilities

A publication has summarized three years of experience of a facility established at McMaster University "to facilitate clinicians' involvement with n-of-1 studies" in the community (Guyatt, et al., 1990). Funded by a research grant to determine the value of randomized N-of-1 clinical trials, the facility performed clinical trials for physicians and taught them how to perform their own trials. During the first three years of operation of the facility ("service"), 70 N-of-1 experiments were begun and 57 of them were completed. The experiments were conducted on patients with a wide range of medical problems for which their physicians desired experimental determination of the effectiveness of certain forms of therapy. The three-year report by Guyatt and his colleagues discussed such considerations of the facility as cost-effectiveness of clinical trials on the individual as well as their procedures and experimental outcomes, and stated: "We interpret the results as supporting the feasibility and usefulness of n-of-1 trials in clinical practice." McMaster University continues to operate its N-of-1 facility (Guyatt, 1994).

An "N-of-1 clinical trial service" of a similar nature was established at the University of Washington with support of a research grant (Larson, 1990) and that service, which serves a similar function of helping physicians in the community in the use of randomized N-of-1 clinical trials, continued (Larson, 1994).

The success of the services at McMaster University and the University of Washington should be encouraging to persons wanting to set up similar facilities elsewhere. The feasibility and value of randomized N-of-1 clinical trials has been thoroughly assessed at both institutions, and the services now merit continuation for their value to the physicians they serve. Success in establishing and continuing the operation of the special facilities depends on collaboration among medical and statistical personnel as well as the support of an administrative structure, which is reflected in the following brief history of the McMaster facility.

D. Wayne Taylor, a professor of biostatistics in the Department of Epidemiology and Biostatistics, gave a "rounds" presentation in the fall of 1984 on N-of-1 designs, a topic he had included for several years in his course on clinical trials, using material from the first edition of *Randomization Tests* (Taylor, 1994). After the presentation, Robin McLeod, a physician at the Toronto General Hospital and a student in Taylor's course, discussed the possibility of doing the study on the effect of metranidazole described in Example 11.3 and the study began later in 1984. Furthermore, after the "rounds" talk, the interest of Gordon Guyatt, a physician and head of the department, led to preparations to establish a randomized N-of-1 clinical trials facility. As Guyatt, et al. (1990) indicated, the cooperation of many different professionals at the University was gained. Taylor and McLeod were early members of the "N of 1 study group" formed in the department (Guyatt, et al., 1986). Thus, the development of the facility at McMaster University depended on: a statistician with expertise in randomized N-of-1 experimental design and the randomization tests associated with them; a battery of physicians, pharmacologists, and others to prepare medications and placebos, provide medical advice, and conduct clinical trials; and administrative skills in gaining support from a research grant, overseeing the operation of the facility, and dealing with a multitude of other financial and personnel matters associated with the operation of the N-of-1 service.

Because randomization tests can be adapted to any form of randomized N-of-1 design, a statistician with such expertise can not only stimulate interest in the establishment of N-of-1 clinical trial facilities but also continue to help out, through consultation on special designs and statistical tests that are required from time to time.

11.21 Single-Cell and Other Single-Unit Neuroscience Experiments

Earlier in the discussion of the need to study individuals to investigate theories and hypotheses concerning individual behavior, the masking of individual behavior through reliance on multiple-N experiments was pointed out. Similar considerations apply in the area of neuroscience. Recordings of neural activity from a vast number of neurons in the brain does not reflect the activity within the individual neurons, which is frequently the process of interest. We will refer to studies of individual neurons as "single-cell research." The N-of-1 experiments in the preceding sections of this chapter have concerned behavior, whereas the single-cell research concerns specific physiological processes. The following presentation is based on an article by Edgington and Bland (1993).

Much of our current knowledge of how the central nervous system processes information has been provided by the extracellular recording of single-cell discharges, occurring either spontaneously or in response to a stimulus. Statistical measures of "spike trains" (single-cell discharges) that not only characterized the discharge patterns but also gave information about input-output relations and underlying mechanisms were developed in the 1990s. Important contributions were made by a number of authors, particularly Gerstein and Kiang (1960), Moore et al. (1966), Perkel et al., (1967a; 1967b), Gerstein and Perkel (1969), Gerstein (1970), and Moore et al. (1970).

Interest in single-cell and other single-unit research has continued and there have been many recent articles dealing with methodological issues associated with recording the output of single units. Those articles considered various types of single units, including single muscle units in rats (Chanaud et al., 1987) and neurons in monkeys (Montgomery, 1989) and cats (May et al., 1991). Individual interneurons of cockroaches (Mony et al., 1988) and motoneurons in flying locusts (Wolf and Pearson, 1987) were also single units that were intensively studied. However, statistical tests of causal relationships to determine whether a particular stimulus or treatment has an effect on single-unit activity have not been prominent. Time series procedures have been employed to represent spike trains but those procedures are not suitable as statistical tests of causal relationships. The appropriate use of randomization of treatment times in single-unit experiments in neuroscience permits the valid employment of randomization tests to test the effects of treatment manipulations.

Permutation tests that serve as guides for further research but that do not test hypotheses about experimental treatment effects have been used frequently in the neurosciences. The reference set of a test statistic is derived under the null hypothesis of some specified type of randomness in the spike train production. To test the null hypothesis that "each interspike interval does not depend on the values of other intervals," Dayhoff and Gerstein (1983) regarded all possible orders of a sequence of interspike intervals within an observed spike train as appropriate for providing a reference set of data permutations. The way the data are permuted for tests of randomness depends on the type of random process that is postulated in the null hypothesis. Lindsey et al. (1992) permuted data to test a hypothesis that concerned the synchrony (temporal correlation) of firing of neurons within a defined group of neurons, so their data were permuted differently from the way Dayhoff and Gerstein permuted their data.

Rejection of permutation test null hypotheses of randomness of processes can serve as a guide as to whether an observed relationship is worth investigating further but the only permutation tests for testing causal hypotheses are randomization tests, i.e., permutation tests where the data permutation is based on the randomization performed in the experiment.

A randomized N-of-1 experiment that focused on the effect of a treatment manipulation on neural activity was conducted by Grunewald et al. (1984) using a "single-finger" design. The random assignment consisted in the induction of five velocities of movement of a subject's finger repeatedly in random order to determine the effect of the velocity on certain aspects of EEGs. The correlation between movement velocities and EEG measurements was computed and the P-value was determined by the randomization test procedure.

Random assignment in a single-cell experiment can be performed in a manner analogous to random assignment in other N-of-1 experiments, permitting valid determination of the P-value by means of a randomization test. Bland et al. (1989) studied variation in neural activity when the electrical current applied to a single neuron in the hypothalamus of a rat was varied. The levels of current administered were varied randomly over time and the P-value of the relationship between current levels and the frequency of spike discharge was determined by application of a randomization test.

In single-cell experiments, a momentary introduction of a stimulus can have an effect on a spike train that persists for some time after termination of the stimulus, and such delayed effects can make it difficult to determine statistically the effect of the experimental stimulus. "Washout periods" of the type used by McLeod in her experiment described in Example 11.3 can be helpful in minimizing such carryover effects. In clinical trials of drugs, the washout periods are periods of time that allow the effects of the drug to wear off, and the incorporation of washout periods into single-cell experimental designs can serve a similar function. Data from a washout period can be excluded from the data analysis.

Whether a stimulus is momentarily introduced and withdrawn or introduced and left in effect over a span of time in a single-cell experiment, only one numerical value for each random determination can be used in a randomization test (or one set of values for multivariate tests). Suppose that prior to a single-cell experiment, we randomly assigned five blocks of time to an experimental treatment condition and five to a control (or alternative treatment) condition. Within each block of time, there will be numerous spikes in a spike train but it is necessary to summarize or condense the variable of interest, such as interspike interval, into a single numerical value for a block. There would be a total of 10 numerical values for 10 blocks of time to which treatment conditions have been randomly assigned. The numerical "measure" of a block characteristic could be the mean of interspike intervals completely contained within a block (except for those intervals in the washout portion of the block) if the experimental treatment was expected to affect interspike intervals uniformly. Alternatively, if the treatment was expected to increase variability of interspike intervals, a measure of variability such as range or variance could be computed for each block separately for interspike intervals completely contained within a block. A randomization test could then be applied to those numerical values. If several levels of stimulus intensity are used and the interspike interval is expected to be positively correlated with stimulus intensity, the numerical values could be the data for a randomization trend test. For single-unit experiments, blocks of time incorporating washout periods should be specified prior to the experiment, treatment conditions for the blocks should be randomly assigned, and after excluding the washout period, data-relevant characteristics of the block should be condensed into a single measurement for the block, that is, a single numerical value representing the block as a whole.

Randomization tests can employ special test statistics to reflect the treatment effect, making them especially sensitive to treatment effects when a specific prediction about effects is correct. For example, suppose that in a single-cell experiment a treatment is expected to increase interspike intervals and reduce interspike variability. A multivariate randomization test statistic could take into account the predicted direction of effect for both dependent variables.

11.22   Books on N-of-1 Design and Analysis

This chapter has touched on only a few of the N-of-1 designs that are amenable to randomization tests but there are numerous possibilities in books on N-of-1 design and analysis. Hersen and Barlow (1976) produced the first book that went beyond visual analysis of data to cover rigorous randomized single subject designs and analysis by means of randomization tests; a revised version of that book was subsequently released (Barlow and Hersen, 1984). Within a few years of publication of the Hersen and Barlow book, Kratochwill (1978) and Kazdin (1982) offered books with similar discussions of N-of-1 theory and methods. Those books and a later book by Kratochwill and Levin (1992) were directed at practitioners of behavior modification in the fields of psychology and education. McReynolds and Kearns (1983) wrote a book especially for researchers on communication disorders, and Tawney and Gast (1984) focused on research in special education. Behling and Merves (1984) wrote their book for clinical researchers, and Bromley (1986), Valsiner (1986), Krishef (1991), and Franklin, Allison, and Gorman (1996) discussed issues primarily of interest to psychologists and social science researchers. Todman and Dugard (2001) focused particularly on randomization tests in single-case and small-*n* experimental designs and offered convenient Microsoft Excel® routines for many common designs. Finally, Onghena and Edgington (2005) provided a tutorial on single-case randomization tests that might be of particular interest to researchers involved in pain research.

11.23   Software for N-of-1 Randomization Tests

As repeatedly illustrated in the previous sections, the general randomization test programs on the included CD (especially the ones related to Chapter 4, Chapter 5, and Chapter 6) can be brought into action to provide the P-values for N-of-1 randomization tests if there is an equivalence between the N-of-1 design and one of the common multiple-N designs. However, for many of the common N-of-1 designs (e.g., treatment intervention designs and multiple baseline designs) such an equivalent multiple-N design is lacking. For this purpose, a specific package of programs for "single-case randomization tests," SCRT, can be used (Onghena and Van Damme, 1994). This program, together with a small README file containing basic instructions, is available in the SCRT folder of the accompanying CD.

The SCRT program can be used to design and analyze most of the common N-of-1 designs. With a statistics editor, one can define one's own test statistic and one can choose between a systematic and a random procedure to generate the data divisions. An experienced user can even exploit the flexibility of SCRT to analyze data collected in multiple-N designs if there is a clear equivalence with some of the N-of-1 designs (and usually there is). SCRT also offers a facility to combine the P-values of independent N-of-1 replications but for that specific purpose, the COMBINE program on the CD (see also Chapter 6) is more general and contains more options.

Some others programs are sufficiently worthwhile to be mentioned. For example, researchers interested in regulated randomization tests for multiple-baseline designs might consider the very user-friendly RegRand program developed by Koehler and Levin (2000). If one prefers to work within a Microsoft Excel® environment, then the routines as described in Todman and Dugard (2001) are an excellent choice.

11.24 Questions and Exercises

1. "Try this new product for 10 days and come back for a test to see if it helped relieve your symptoms." What are limitations in this approach by a physician and how would you overcome them in an N-of-1 experiment?

2. An experimenter modifies Fisher's lady-tasting-tea experiment by correctly informing the lady that on each of eight taste tests it will be randomly determined whether the cup of tea that is tasted had milk added to tea or tea added to milk. What is the smallest possible P-value?

3. Instead of testing the lady's ability to identify the way the tea and milk were combined, as was Fisher's objective, suppose we conducted the experiment to find out whether the lady was able to distinguish the four cups with tea added from the four cups with milk added. What would be the effect on the smallest possible P-value? Why?

4. What is the null hypothesis in an N-of-1 experiment with only one subject, regardless of the treatments or the test statistic?

5. The effectiveness of a treatment shortly after administration can be quite different from its effectiveness later. What can an experimenter do, if anything, to take such a possibility into consideration? If nothing, why not?

6. Describe the random assignment procedure Weiss used in his replicated N-of-1 study of the effects of food additives.

7. In Weiss's experiment referred to in Question 6, which P-values would be independent, so that combining of P-values could be employed, and which would be dependent?

8. What random assignment procedure did Crosier recommend for treatment intervention and withdrawal experiments to make all assignments equally probable?

9. Smith used a randomized block design with three-day blocks to determine the relative effects of drugs in treating narcolepsy. What were the two places where he introduced randomization into the experiment?

10. Describe an N-of-1 experiment to find out how to improve the swimming of an outstanding Olympic swimmer.

REFERENCES

Agras, W.S., Leitenberg, H., Barlow, D.H., and Thomson, L.E., Instructions and reinforcement in the modification of neurotic behavior, *Am. J. Psychiatr.*, 125, 1435, 1969.

Barlow, D.H. and Hersen, M., *Single Case Experimental Designs: Strategies for Studying Behavior*, Pergamon Press, New York, 1984.

Behling, J.H. and Merves, E.S., *The Practice of Clinical Research: The Single Case Method*, University Press of America, Lanham, MD, 1984.

Bland, B.H., Colom, L.V., and Ford, R.D., Responses of septal theta-on and theta-off cells to activation of the dorsomedial-posterior hypothalamic region, *Brain Res. Bull.*, 24, 71, 1989.

Bromley, D.B., *The Case-Study Method in Psychology and Related Disciplines*, John Wiley & Sons, New York, 1986.

Chanaud, C.M., Pratt, C.A. and Loeb, G.E., A multiple-contact EMG recording array for mapping single muscle unit territories, *J. Neurosci. Meth.*, 21, 105, 1987.

Crosier, R., Personal correspondence, 1982.

Dayhoff, J.E. and Gerstein, G.L., Favored patterns in spike trains, *I. Detection. J. Neurophys.*, 49, 1334, 1983.

Dukes, W.F., N = 1, *Psych. Bull.*, 64, 74, 1965.

Edgington, E.S., Statistical inference from N = 1 experiments, *J. Psychol.*, 65, 195, 1967.

Edgington, E.S., Randomization tests for one-subject operant experiments, *J. Psychol.*, 90, 57, 1975.

Edgington, E.S., Nonparametric tests for single-subject multiple schedule experiments, *Behav. Assess.*, 4, 83, 1982.

Edgington, E.S., Statistics and single case analysis, in: *Progress in Behavior Modification*, Vol. 16, Miltersen, R., Eisler, M., and Miller, P.M. (Eds.), Academic Press, New York, 1984.

Edgington, E.S., Randomized single-subject experimental designs, *Behav. Res. Ther.*, 34, 567, 1996.

Edgington, E.S. and Bland, B.H., Randomization tests: Application to single-cell and other single-unit neuroscience experiments, *J. Neurosci. Meth.*, 47, 169, 1993.

Ferron, J. and Onghena, P., The power of randomization tests for single-case phase designs, *J. Exp. Educ.*, 64, 231, 1996.

Ferron, J.M. and Ware, W.B., Analyzing single-case data: The power of randomization tests, *J. Exp. Educ.*, 63, 167, 1995.

Ferron, J. and Sentovich, C., Statistical power of randomization tests used with multiple-baseline designs, *J. Exp. Educ.*, 70, 165, 2002.

Fisher, R.A., *The Design of Experiments* (6th ed.), Hafner, London, 1951.

Franklin, R.D., Allison, D.B., and Gorman, B.S. (Eds.), *Design and Analysis of Single-Case Research*, Erlbaum, Mahwah, NJ, 1996.

Gerstein, G.L., Functional association of neurons: detection and interpretation, in: *The Neurosciences: Second Study Program*, Schmitt, F.O. (Ed.), Rockefeller University Press, New York, 1970, 648.

Gerstein, G.L. and Kiang, N.Y.-S., An approach to the quantitative analysis of electrophysiological data from single neurons, *Biophys. J.*, 2, 15, 1960.

Gerstein, G.L. and Perkel, D.H., Simultaneously recorded trains of action potentials: Analysis and functional interpretation, *Science*, 164, 828, 1969.

Gridgeman, N.T., The lady tasting tea, and allied topics, *J. Am. Statist. Assn.*, 777, 1959.

Grunewald, G., Grunewald-Zuberbier, E., Schuhmacher, H., Mewald, J., and Noth, J., Somatosensory evoked potentials to mechanical disturbances of positioning movements in man: Gating of middle-range components, *Electroenceph. Clin. Neurophysiol.*, 5, 525, 1984.

Guyatt, G., Personal correspondence, 1994.

Guyatt, G., Keller, J.L., Jaeschke, R., Rosenbloom, D., Adachi, J.D., and Newhouse, M.T., The *n*-of-1 randomized controlled trial: clinical effectiveness, *Ann. Intern. Med.*, 112, 293, 1990.

Guyatt, G., Sackett, D., Taylor, D.W., Chong, J., Roberts, R., and Pugsley, S., Determining optimal therapy — randomized trials in individual patients, *N. Eng. J. Med.*, 314, 14, 889, 1986.

Hersen, M. and Barlow, D.H. (Eds.), *Single-Case Experimental Designs: Strategies for Studying Behavior Change*, Pergamon, Oxford, 1976.

Hogben, L.T., *Statistical Theory*, Allen and Unwin, London, 1957.

Kazdin, A.E., Statistical analysis for single-case experimental designs, in: *Single-Case Experimental Designs: Strategies for Studying Behavior Change*, Hersen, M. and Barlow, D.H. (Eds), Pergamon, Oxford, 1976.

Kazdin, A.E., Obstacles in using randomization tests in single-case experimentation, *J. Educ. Statist.*, 5, 253, 1980.

Kazdin, A.E., *Single Case Research Designs: Methods for Clinical and Applied Settings*, Oxford University Press, New York, 1982.

Koehler, M.J. and Levin, J.R., Regulated randomization: A potentially sharper analytical tool for the multiple-baseline design, *Psychol. Meth.*, 3, 206, 1998.

Koehler, M.J. and Levin, J.R., RegRand: Statistical software for the multiple-baseline design, *Behav. Res. Meth. Ins. C.*, 32, 367, 2000.

Kratochwill, T.R., *Single Subject Research: Strategies for Evaluating Change*, Academic Press, New York, 1978.

Kratochwill, T.R. and Levin, J.L., *Single-Case Research Design and Analysis: New Directions for Psychology and Education*, Lawrence Erlbaum Associates, Hillsdale, NJ, 1992.

Krishef, C.H., *Fundamental Approaches to Single Subject Design and Analysis*, Krieger, Malabar, FL, 1991.

Larson, E.B., N-of-1 clinical trials: A technique for improving medical therapeutics, *West. J. Med.*, 152, 52, 1990.

Larson, E.B., Personal correspondence, 1994.

Levin, J.R., Marascuilo, L.A., and Hubert, L.J., N = nonparametric randomization tests, in: *Single-Subject Research: Strategies for Evaluating Change*, Kratochwill, T.R. (Ed.), Academic Press, New York, 1978.

Lindsey, B.G., Hernandez, Y.M., Morris, K.F., Shannon, R., and Gerstein, G.L., Respiratory-related neural assemblies in the brain stem midline, *J. Neurophys.*, 67, 905, 1992.

May, B.J., Aleszczyk, C.M., and Sachs, M.B., Single-unit recording in the ventral cochlear nucleus of behaving cats, *J. Neurosci. Meth.*, 40, 155, 1991.

McLeod, R.S., Cohen, Z., Taylor, D.W., and Cullen, J.B., Single-patient randomised clinical trial: Use in determining optimum treatment for patient with inflammation of Kock continent ileostomy reservoir, *Lancet*, 1(8483), 726, March 29, 1986.

McReynolds, L.V. and Kearns, K.P., *Single-Subject Experimental Designs in Communicative Disorders*, University Park Press, Baltimore, MD, 1983.

Montgomery, E.B., A new method for relating behavior to neuronal activity in performing monkeys, *J. Neurosci. Meth.*, 28: 197, 1989.

Mony, L., Hue, B., and Tessier, J.-C., Synaptic currents recorded from the dendritic field of an insect giant interneurone, *J. Neurosci. Meth.*, 25, 103, 1988.

Moore, G.P., Perkel, D.H., and Segundo, J.P., Statistical analysis and functional interpretation of neuronal spike data, *Ann. Rev. Physiol.*, 28, 493, 1966.

Moore, G.P., Segundo, J.P., Perkel, D.H., and Levitan, H., Statistical signs of synaptic interaction in neurons, *Biophys. J.*, 10, 876, 1970.

Neyman, J., *First Course in Probability and Statistics*, Henry Holt, New York, 1950.

Onghena, P., Randomization tests for extensions and variations of ABAB single-case experimental designs: a rejoinder, *Behav. Assess.*, 14, 153, 1992.

Onghena, P., The power of randomization tests for single-case designs, Ph.D. thesis, University of Leuven, Belgium, 1994.

Onghena, P., Single-case designs, in: *Encyclopedia of Statistics in Behavioral Science*, Vol. 4, Everitt, B. and Howell, D. (Eds.), John Wiley & Sons, New York, 2005, 1850–1854.

Onghena, P. and Edgington, E.S., Customization of pain treatments: Single-case design and analysis, *Clin. J. Pain*, 21, 56, 2005.

Onghena, P. and Van Damme, G., SCRT 1.1: Single Case Randomization Tests, *Behav. Res. Meth. Ins. C.*, 26, 369, 1994.

Perkel, D.H., Gerstein, G.L., and Moore, G.P., Neuronal spike trains and stochastic point processes. I. The single spike train, *Biophys. J.*, 7, 391, 1967a.

Perkel, D.H., Gerstein, G.L., and Moore, G.P., Neuronal spike trains and stochastic processes. II. Simultaneous spike trains, *Biophys. J.*, 7, 419, 1967b.

Pitman, E.J.G., Significance tests which may be applied to samples from any populations, *J.R. Statist. Soc. B.*, 4, 119, 1937.

Smith, C.M., Controlled observations on the single subject, *Can. Med. Assn. J.*, 88, 410, 1963.

Tawney, J.W. and Gast, D.L., *Single Subject Research in Special Education*, C.E. Merrill, Columbus, OH, 1984.

Taylor, D.W., Personal correspondence, 1994.

Todman, J.B. and Dugard, P., Single-Case and Small-N Experimental Designs: A Practical Guide to Randomization Tests, Erlbaum, Mahwah, NJ, 2001.

Valsiner, J., *The Individual Subject and Scientific Psychology*, Plenum Press, New York, 1986.

Weiss, B., Williams, J.H., Margen, S., Abrams, B., Caan, B., Citron, L.J., Cox, C., McKibben, J., Ogar, D., and Schultz, S., Behaviorial responses to artificial food colors, *Science*, 297, 1487, 1980.

Winer, B.J., *Statistical Principles in Experimental Design* (2nd ed.), McGraw-Hill, New York, 1971.

Wolf, H. and Pearson, K.G., Intracellular recordings from interneurons and motoneurons in intact flying locusts, *J. Neurosci. Meth.*, 21, 345, 1987.

Wrighton, R.F., The theoretical basis of the therapeutic trial, *Acta Geneti. Statisti. Med.*, 4, 312, 1953.

12

Tests of Quantitative Laws

In Chapter 11, the N-of-1 designs and tests were presented as means of detecting causal relationships between certain treatment manipulations and certain types of responses. However, N-of-1 experimentation for assessing theories should go beyond simply testing the null hypothesis of no differential treatment effect and the same is true of multiple-subject experiments. An investigator can be certain of the existence of a causal relationship between a treatment and a response but may question a law (or model) of that causal relationship. Such laws may be quantitative and this chapter concerns randomization tests of those laws. (Although some relationships that formerly were called laws are now called models, that distinction will be ignored and all tests of quantitative relationships between treatments and effects will be called tests of quantitative laws.)

Quantitative laws in the physical sciences, such as the law of gravity or Boyle's law, have an acceptance that is not characteristic of quantitative laws for biological organisms. This chapter concerns the testing of quantitative laws regarding the relationship between treatment magnitudes and quantitative responses of subjects. The randomization tests to be discussed test the null hypothesis that the relationship between treatment and response measurements is that specified by a particular law, and rejection of the null hypothesis implies acceptance of the complementary hypothesis that the law did not hold in the experiment that was conducted. Because the null hypothesis concerns the lawful relationship applying to all experimental units, the complementary hypothesis is that the law does not hold for one or more experimental units.

Thus, randomization tests of quantitative laws employ the quantitative relationship specified by the law as the null hypothesis that is the basis for permuting experimental data to provide the reference set. This type of null hypothesis is fundamentally different from the traditional randomization test null hypothesis of no treatment effect and consequently requires a discussion of its rationale prior to application of these special randomization tests.

12.1 Generic and Specific Null Hypotheses

Before examining the multiplicity of null hypotheses that can be tested by randomization tests, it will be useful to reconsider the randomization test null hypothesis of no treatment effect. Rejection of that null hypothesis implies acceptance of the complementary hypothesis of a differential treatment effect on the measurements for at least some of the experimental units, regardless of the test statistic. For example, even if the test statistic is sensitive to interaction or a linear trend, the complementary hypothesis is simply the existence of a treatment effect, not the existence of an interaction or a linear trend.

Traditionally, the null hypothesis for a randomization test has been regarded as that of no treatment effect or, equivalently, no differential treatment effect. In its literal sense, it would be a very broad, generic hypothesis, implying that the experimental unit is not affected in any way whatsoever by the manipulation of treatments; the effect on the experimental unit is as if the same treatment was given different treatment designations. Clearly, such a null hypothesis was not intended by Fisher or Pitman. Their agricultural examples of the application of two different types of soil treatment would always provide chemical or other differences in the plots that precise measurement could distinguish as associated with the two different substances applied to the plots. Fisher was not interested in the generic null hypothesis of no difference whatsoever in effects but in a more specific null hypothesis, such as no difference in yield measured in bushels per acre. Similarly, in external application of ointment to animals, it is not the null hypothesis of absolutely identical effects that is likely to be of interest. There might be a number of effects that are not relevant. For example, whether there is a slight difference from one ointment to another in the color or odor of the skin after the ointment is applied is likely to be irrelevant. The more specific null hypothesis of interest may be that there is no effect of treatment differences on a certain characteristic, such as the healing of a wound.

The specificity of the null hypothesis when stated in terms of invariance of specific measurements over various possible random assignments provides insight into the wide scope of null hypotheses that can be tested by means of randomization tests. The null hypothesis that the measurements are invariant under alternative assignments can be tested by a randomization test, no matter what is measured. The "measurements" need not be directly obtained data but can be quantitative expressions of the relationship between the obtained data and other variables. In Example 8.4, the quantities that were permuted in the randomization test were proportions of playing time spent with certain toys. The null hypothesis did not concern the amount of time spent playing with certain toys but the proportion of time that was spent with those toys. The proportions of total playing time were the "measurements" postulated by the null hypothesis to be invariant over all possible assignments to treatments. In Section 6.18 is another example of a null hypothesis specifying invariance of a derived measurement: the "measurement" is the difference between two measurements for a subject, one under each of two levels of a factor, the null hypothesis being one of no interaction, not no effect of either factor.

12.2 The Referent of a Law or Model

When a certain relationship between quantitative levels of a treatment and response magnitudes is widely accepted, the relationship is sometimes called a law. Suppose that we believe a law not to be true and would like a test sensitive to some alternative to the law. The null hypothesis would be that the law holds and a randomization test could be conducted. Here we will deal with the testing of quantitative laws, the desire being to find strong evidence against a particular lawful relationship.

With quantitatively defined treatment levels and quantitative data, the measurements hypothesized to be invariant over treatment assignments can be measurements derived from some mathematical relationship between the treatment magnitude and the fundamental measurements. One relationship between a treatment magnitude and the magnitude of response that is characteristic of many laws assumed to apply to biological organisms is that of a response being proportional to a stimulus: $R = kS$, where R is the response magnitude, k is an unknown constant, and S is the stimulus magnitude. The same relationship may be expressed as $R/S = k$. Several examples of this and related laws will be discussed in this chapter.

Some laws involve a stochastic or random element, like laws of genetics. The randomization tests to be presented are not for testing such laws but for testing deterministic laws. Of course, a lawful relationship between a stimulus and response must refer to the relationship between the variation in response associated with variation in the stimulus. In the following discussion, the variation to which the null hypothesis refers is hypothetical. Specifically, the null hypothesis is that the relationship between the alternative stimulus magnitudes that could have been presented to any subject-plus-treatment-time experimental unit and the responses that would have been given by that subject at that time is the lawful relationship under test.

12.3 Test of Incremental Effects

To illustrate the use of randomization tests for testing quantitative relationships between treatment and response magnitudes, a simple relationship to be tested will be hypothesized. The relationship is similar to what is called unit additivity, which states that the difference in response to treatments is the same for all subjects and, presumably, from time to time within subjects. But the hypothesis of unit additivity cannot be tested by means of a randomization test, whereas the hypothesis of incremental effects can be. What we will call a test of incremental effects allows for variation from subject to subject and from time to time within subjects in the difference in response to treatments. The hypothesis of incremental effects is this: $S - R = k$ for each subject at the time the subject is administered S, the stimulus. In other words, it is hypothesized that for any subject at the time that subject is administered the stimulus magnitude S, the difference between the stimulus and the response magnitude R is the same as it would have been for any other stimulus magnitude. Such a hypothesis might be reasonable when the stimulus and response are expressed in the same units, such as the amount of a substance ingested and the amount of that substance that is excreted. It might be thought that for a particular subject at a given time, there is a fixed amount of the substance that would be assimilated regardless of the amount ingested, and that therefore the amount excreted would increase by the same amount as the amount ingested increased if all dosages were above the amount that could be assimilated.

Example 12.1. An experimenter randomly assigns five subjects to 40, 50, 60, 70, and 80 units of a substance, with one subject per dosage, to test the null hypothesis that the amount assimilated within the subject would be constant over those treatment levels. As the amount assimilated is not readily ascertainable, it is determined as the amount administered minus the amount excreted. It is expected that the amount assimilated will increase somewhat with increased quantities administered. This table shows the results:

Amount ingested	40	50	60	70	80
Amount excreted	18	26	31	40	39
S – R	22	24	29	30	41

The amount excreted increased as the amount ingested increased, which is consistent with the null hypothesis, but the arithmetic difference (S – R) which was to reflect the amount assimilated was supposed to be constant whereas it tends to increase with the amount ingested. As it was predicted that (S – R) would increase as the amount ingested increased, the test statistic could be the correlation between the amount ingested and (S – R), and the P-value for a randomization test could be determined by Program 8.1. No program would be necessary in this instance as the correlation is maximal for the 5! = 120 data permutations. The P-value is 1/120, or about 0.008.

In Example 12.1, (S – R) can be called a measurement but it is not a direct measurement of assimilation but a way of calculating it. Thus, the test is conceptually different from testing the null hypothesis of no treatment effect on amount assimilated because the amount assimilated cannot be directly derived. However, there are situations where (S – R) is simply a relationship between stimulus and response rather than a means of inferring some characteristic that, in principle, is directly measurable. For example, suppose homeostatic temperature control in experimental animals is rendered inoperable and temperature in one part of the body is varied to test the hypothesis that the temperature in an adjacent organ is affected incrementally. (S − R) would be the induced temperature minus the temperature measured in a nearby organ of the body, and the null hypothesis would be invariance of the difference in temperature over the range of induced temperatures. The test of this hypothesis would provide the same P-value independently of whether the measurements were in Fahrenheit or Celsius units, unlike a test that hypothesized invariance of the ratio of S to R, as in Weber's and other laws that will be discussed in the following sections.

12.4   Weber's Law

Weber's law of perceptual discrimination states that $\Delta I/I = k$, where ΔI is the incremental change in an intensity I that is required for the change to be detected and k is a constant. Thus, according to Weber's law the change in a sensory stimulus that is barely detectable is proportional to the stimulus magnitude. According to the law, if the intensity of a stimulus of 20 units had to be increased to 22 units for the change in intensity to be detected, 200 units of the same type of stimulus would have to be increased to 220 units to be detected.

Although Weber's law is rarely discussed in modern introductory psychology texts, for many years it was an essential topic for introductory psychology. Debates raged for some time over the tenability of Weber's law for various sense modalities and for various ranges of intensity within a sense modality before the law was subjected to a statistical test. Quantitative analyses were conducted but statistical tests with the law as a null hypothesis were missing. The following example illustrates an early statistical test intended to fill the void.

Example 12.2. Weber's law was presented for many years as valid for certain sense modalities, especially vision. However, restrictions on the range over which Weber's law was assumed to hold were stated because at the extremely low and extremely high intensities, $\Delta I/I$ was known to be high. Nevertheless, within the "middle range of intensity" Weber's law was said to hold for Konig's visual brightness discrimination data for white light (Blanchard, 1918). Table 12.1, in which the observed value of $\Delta I/I$ is called the Weber fraction, shows Konig's data for all but the extremely high and low intensity levels.

For a N-of-1 randomization test, where intensity is randomly distributed over treatment times, we can test Weber's law with the above data from a single subject without being concerned about whether other subjects would provide the same Weber fraction. The following testing procedure is that which was applied to the data in Table 12.1 first as a test of randomness of

TABLE 12.1

Visual-Brightness Discrimination for White Light (Data in Millilamberts)

Log intensity	−1.70	−1.40	−1.10	−0.70	−0.40	−0.10	0.30	0.60	
Weber fraction	0.0560	0.0455	0.0380	0.0314	0.0290	0.0217	0.0188	0.0175	
Sign of difference		−	−	−	−	−	−	−	+

Log intensity	0.90	1.30	1.60	1.90	2.30	2.60	2.90	3.30	3.60
Weber fraction	0.0178	0.0176	0.0173	0.0172	0.0170	0.0191	0.0260	0.0266	0.0346
Sign of difference		−	−	−	−	+	+	+	+

fluctuation of the Weber fraction over intensity levels (Edgington, 1961a) and later as a randomization test of Weber's law (Edgington, 1969). Of course, the randomization test was based on random assignment but the actual experimental random assignment information was not available; consequently, it was necessary to postulate a random assignment procedure to demonstrate the application of a randomization test. The statistical test applied to the data uses aspects of the data that investigators have used qualitatively in criticizing Weber's law: the gradual fall and gradual rise of the Weber fraction values with increasing intensity (Boring, 1942; Holway and Pratt, 1936).

Suppose the experimenter randomly assigned 17 treatment times to the 17 intensity levels in Table 12.1 prior to presenting the light intensities to the subject. The null hypothesis states that the Weber fraction is a constant for any treatment time over the light intensities that could be administered. If that null hypothesis is true, rises and falls in the Weber fractions over intensities (indicated by "+" or "−") were randomly determined by the random assignment of treatment times to intensities. As gradual rises and falls are expected, the test statistic is the number of runs of signs of differences between adjacent Weber fractions, with the P-value being the proportion of the 17! pairings of the intensity levels and Weber fractions that provide as few runs of plus or minus signs as four, the number for the obtained results. (The first run is seven "−" signs, the next run is a single "+" sign, the third run is four "−" signs, and the fourth run is four "+" signs.)

A random data permutation procedure that could be employed for the randomization test would be the one in Program 8.2 for correlation but the test statistic would be the number of runs, that is, the number of successive groups of "+"s and "−"s. Lacking the availability of computer programs for randomization tests, Edgington (1961b) determined by using a recursion formula the proportion of the 17! permutations of the Weber fractions in Table 12.1 that would provide as few as four runs, thus getting the P-value that a systematic randomization test would provide. However, the direct computation by means of the recursion formula assumed no tied Weber fractions (which is true of the results in Table 12.1), thus ensuring that for all 17! permutations, successive observations would have either a positive or a negative difference. Following the "no-ties" assumption that is characteristic of tables for early rank tests, Table 12.2 was derived by use of the

TABLE 12.2

Probability of *r* of Fewer Runs of Signs of First Differences for *o* Observations

Number of Runs (r)	Number of Observations (o)																			
	1	2	3	4	5	6	7	8	9	10	11	12	13	14	15	16	17	18	19	20
1	—	1.0000	0.3333	0.0833	0.0167	0.0028	0.0004	0.0000	0.0000	0.0000	0.0000	0.0000	0.0000	0.0000	0.0000	0.0000	0.0000	0.0000	0.0000	0.0000
2	—	—	1.0000	0.5833	0.2500	0.0861	0.0250	0.0063	0.0014	0.0003	0.0001	0.0000	0.0000	0.0000	0.0000	0.0000	0.0000	0.0000	0.0000	0.0000
3	—	—	—	1.0000	0.7333	0.4139	0.1909	0.0749	0.0257	0.0079	0.0022	0.0005	0.0001	0.0000	0.0000	0.0000	0.0000	0.0000	0.0000	0.0000
4	—	—	—	—	1.0000	0.8306	0.5583	0.3124	0.1500	0.0633	0.0239	0.0082	0.0026	0.0007	0.0002	0.0001	0.0000	0.0000	0.0000	0.0000
5	—	—	—	—	—	1.0000	0.8921	0.6750	0.4347	0.2427	0.1196	0.0529	0.0213	0.0079	0.0027	0.0009	0.0003	0.0001	0.0000	0.0000
6	—	—	—	—	—	—	1.0000	0.9313	0.7653	0.5476	0.3438	0.1918	0.0964	0.0441	0.0186	0.0072	0.0026	0.0009	0.0003	0.0001
7	—	—	—	—	—	—	—	1.0000	0.9563	0.8329	0.6460	0.4453	0.2749	0.1534	0.0782	0.0367	0.0160	0.0065	0.0025	0.0009
8	—	—	—	—	—	—	—	—	1.0000	0.9722	0.8823	0.7280	0.5413	0.3633	0.2216	0.1238	0.0638	0.0306	0.0137	0.0058
9	—	—	—	—	—	—	—	—	—	1.0000	0.9823	0.9179	0.7942	0.6278	0.4520	0.2975	0.1799	0.1006	0.0523	0.0255
10	—	—	—	—	—	—	—	—	—	—	1.0000	0.9887	0.9432	0.8464	0.7030	0.5369	0.3770	0.2443	0.1467	0.0821
11	—	—	—	—	—	—	—	—	—	—	—	1.0000	0.9928	0.9609	0.8866	0.7665	0.6150	0.4568	0.3144	0.2012
12	—	—	—	—	—	—	—	—	—	—	—	—	1.0000	0.9954	0.9733	0.9172	0.8188	0.6848	0.5337	0.3873
13	—	—	—	—	—	—	—	—	—	—	—	—	—	1.0000	0.9971	0.9818	0.9400	0.8611	0.7454	0.6055
14	—	—	—	—	—	—	—	—	—	—	—	—	—	—	1.0000	0.9981	0.9877	0.9569	0.8945	0.7969
15	—	—	—	—	—	—	—	—	—	—	—	—	—	—	—	1.0000	0.9988	0.9917	0.9692	0.9207
16	—	—	—	—	—	—	—	—	—	—	—	—	—	—	—	—	1.0000	0.9992	0.9944	0.9782
17	—	—	—	—	—	—	—	—	—	—	—	—	—	—	—	—	—	1.0000	0.9995	0.9962
18	—	—	—	—	—	—	—	—	—	—	—	—	—	—	—	—	—	—	1.0000	0.9997
19	—	—	—	—	—	—	—	—	—	—	—	—	—	—	—	—	—	—	—	1.0000

recursion formula to determine the probability of getting as few runs as any obtained number for various numbers of successive observations. (Like tables for rank tests, Table 12.2 can give invalid results when there are ties.) The table is organized in terms of the least number of runs because treatment effects are expected to result in gradual changes, resulting in fewer runs than would be expected by chance rather than in a large number of runs. The P-value for getting as few as four runs in 17 successive observations is obtained from the intersection of row 4 with column 17, which is 0.0000, meaning that the value is less than 0.0001. Such a small P-value certainly would cast doubt on Weber's law being true.

That the above test of Weber's law is fundamentally different from testing the traditional null hypothesis of no treatment effect is easy to show. Suppose the fundamental measurement of response, ΔI, was not affected at all by the intensity variation. If that were the case, $\Delta I/I$ would tend to systematically decrease, almost certainly providing a highly significant downward trend because of no effect on ΔI. To accommodate this extension of the randomization test methodology (without in any way adding to the assumptions), it is advisable to rephrase the common null hypothesis for randomization tests: the null hypothesis is that a specified relationship between a treatment variable and observations of the experimental units is invariant under all random assignments. The traditional null hypothesis is a special case associated with the situation where the "specified relationship" is independence or unrelatedness.

The runs test was used instead of correlation to test Weber's law because it was not expected that the Weber fractions would show a monotonic (consistently upward or downward) trend. If a monotonic trend in the Weber fraction was expected over the entire range of intensities, the rank correlation between I and ΔI would have been a suitable randomization test statistic. If a linear correlation was expected, randomization test Program 8.2 could be used with either a one-tailed or two-tailed test statistic. However, although $\Delta I/I$ and $I/\Delta I$ are equivalent test statistics for the runs test, they are not equivalent for product-moment correlation and some researchers may prefer to correlate the logarithm of the Weber fraction with I rather than arbitrarily choose between $\Delta I/I$ and $I/\Delta I$. (Of course, the logarithms of the two ratios are equivalent test statistics for product-moment correlation.)

Table 12.2, the probability table for the runs test, is based on the assumption of no ties among any — not just adjacent — values in the ordered series. If ties occur, the data can be reduced by discarding tied values until all values are different; then the number of runs for that subset is the appropriate test statistic, along with the size of the subset, to determine the P-value from Table 12.2.

To test the null hypothesis of invariance of the Weber fraction within individual subjects, it is possible to replicate the test over several subjects and combine the P-values. As the number of data permutations (which is related to sensitivity) would tend to be large for each subject, use of the formula in Section 6.28 for combining continuous P-values would suffice.

Example 12.3. For a single series of treatment values, it may be desirable to control for carryover effects by spacing the treatment sessions, but an alternative approach would be to use several subjects for a single series. In such a case, the null hypothesis would be this: the measure of relationship R (such as the Weber fraction or some related ratio) for each subject at each of the times that subject was administered a treatment was the same as it would have been at that time for that subject under any of the alternative treatments. If we have four subjects, with each to receive five treatment magnitudes, and there are 20 different treatment magnitudes, we could first randomly select five treatment times from 20 different treatment times without replacement for each of the four subjects, providing 20 experimental units, each consisting of a subject plus a treatment time. Those 20 experimental units then would be randomly assigned to the 20 treatment magnitudes. This procedure ensures that there are 20! equally probable assignments of experimental units to the treatments, thus permitting the same test to be used as if the 20 subject-plus-treatment times to be randomly assigned to treatments were all for one subject. As with other multiple-subject experiments, the subjects should be as homogenous as possible with respect to characteristics likely to affect their responses, so that between-subject variability will not greatly limit the power of the test.

Of course, it also would be possible to have subjects randomly assigned to treatment levels with each subject being exposed to only the one intensity of treatment and still perform a randomization test. The test would be conducted in the same manner as if there were treatment times for a single subject randomly assigned to the treatment levels. With assignment of several subjects between only two stimulus intensities, a randomization t test could be conducted on the Weber fractions or their logarithms.

12.5 Other Psychophysical Laws

An alternative to Weber's law that has at times been proposed is $\Delta I/(I + r) = k$, where r is a "baseline intensity," a stimulus intensity inherent in the organism. An example of such an intensity would be the weight of a person's arm during a comparison of weights of lifted objects. Now if the fixed value of r was known, it could be added to the value of I in the denominator and the value of the modified Weber fraction would be computed for all treatment (intensity) levels. But if r is unknown, there is no direct test of the law by means of a randomization test because there would be no numerical values of $\Delta I/(I + r)$ to permute.

Another alternative to Weber's law is what Ganz (1975) calls Barlow's quantum fluctuation model: $\Delta I/\sqrt{I} = k$. Also called the square-root law, this law can be tested in the same manner as Weber's law, employing the same experimental design and essentially the same randomization test. The only difference in the randomization test would be the employment of $\Delta I / \sqrt{I}$ as a test statistic instead of $\Delta I / I$.

In addition to Weber's law and its analogs, there are numerous other laws related to sensory thresholds. One of these is Bloch's law of visual detection (Ganz, 1975) which specifies $It = R$, where R is the visual response, I is the intensity, and t the duration of a flash of light. (Also known as the Bunsen-Roscoe law, the same relationship applies to the responsiveness of various photochemical processes, in photography as well as in vision, to light flashes.) Stated differently, Bloch's law is that the response is proportional to the product of the intensity and duration of a light flash.

Example 12.4. A two-factor experiment, where both intensity and flash duration are varied, can be employed to test the law. Consider a factorial design with five levels of intensity and five levels of flash duration in which the 25 combinations of intensity and flash duration are randomly assigned to treatment times for a subject. For each of the 25 treatment conditions, the subject provides a fraction analogous to the Weber fraction (which could be called a Bloch fraction) defined as R/It, the response for that treatment condition divided by the product of the intensity and duration for that condition. The null hypothesis is the identity of the Bloch fraction over all 25 treatment combinations at any given treatment time. Variation in the fractions actually obtained can be consistent with the null hypothesis because that variation could be due to differences in treatment times. According to the null hypothesis, at each of the 25 treatment times the Bloch fraction is invariant over all treatment combinations that could have been given at that time. A systematic randomization test could be based on the 25! permutations of the Bloch fractions over the 25 treatment combinations. The runs test applied to the test of Weber's law or some correlational procedure could be used, the choice being determined by the type of departure from Bloch's law that is expected. For the runs test, the Bloch fractions would be ordered according to the magnitude of their associated It products.

Tests of proportionality of the relationship between each factor and the response could be carried out as in a test of main effects in a factorial design. For example, to test the proportionality of the I effect without any assumptions about the effect of t, the reference set could be derived by permuting the values of R/I for each of the 25 measures over the five levels of I within each of the five levels of t, thus providing $(5!)^5$ data permutations. If it was expected that R/I would tend to increase over increasing magnitudes of I, a correlation trend test for repeated measures would be appropriate, the test statistic being the product of the trend coefficients (e.g., 1, 2, 3, 4, and 5) and the intensity level totals of R/I or the logarithms of R/I.

Ricco's law (Ganz, 1975) is an analog of Bloch's law related to spatial summation that specifies $AI = R$, where again, R is the visual effect, I is intensity, and A is the area of the visual stimulus. The experimental design and statistical tests could be organized in the same manner as for Bloch's law.

12.6 Foraging Behavior of Hawks

Ecologists have contributed many quantitative models or laws for describing relationships between animal behavior and environmental factors in natural settings. The following example illustrates at one and the same time the utility of randomization tests for the assessment of those models and the difficulty in providing experimental designs that are appropriate.

Aronson and Givinish (1983) referred to data obtained by Andersson (1978; 1981) on the foraging (prey-seeking) behavior of hawks, some from unobtrusive observation of natural behavior and some from experiments. Andersson had provided several explanations for an inverse relationship between search time (per unit of search area) and distance from the nest. Aronson and Givinish noted that Andersson had not tested a null hypothesis and commented:

> Recently, several ecologists have emphasized the need for having such null hypotheses against which to test the predictions of alternative hypotheses (Connor and Simberloff, 1978; Poole and Rathcke, 1979; Strong et al., 1979; Cole, 1981; DeVita et al., 1982). Here we use Andersson's (1981) data to test his (alternative) hypothesis against three null hypotheses regarding central-place foraging behavior.

Andersson had proposed a model for maximizing prey capture for a given effort, which was that the time spent foraging (per unit of area) should decrease linearly with distance from the "central place" or nest. Aronson and Givinish graphed Andersson's data and noted that the downward trend in search time over distance from the nest was not linear but curvilinear.

Although proportionality can easily be tested by means of a randomization test, that is not so for a test of linearity. The null hypothesis of a downward linear trend with no assumptions about the intercept or slope could be that at any given time, the graph of the search times plotted against all distances to which that time could be assigned would be linear. Unfortunately, that hypothesis in conjunction with the data does not provide the basis for transforming the data into some form that would be invariant over all possible randomizations, and so there is no apparent way to test the null hypothesis of linearity of a downward trend by means of a randomization test in the absence of special assumptions.

Example 12.5. Aronson and Givinish formulated a number of models that seemed to fit Andersson's data better than the linear model. One of these was the "equal-time hypothesis," which was that the time a hawk spent searching a plot of a given size should be inversely proportional to distance from the nest. (In the following discussion, we will refer to foraging time per unit of area simply as "foraging time.") We will define our null hypothesis for the hawk's behavior as referring to a relationship existing at a particular time. H_0 then is that *distance × foraging time* is constant over distance. Suppose we expect the time to drop less sharply over distance and we use experimental conditions in which we randomly determined when to release the hawk at 30, 50, 70, 90, or 110 meters from the nest and obtained these results:

Distance	30	50	70	90	110
Foraging time	100	85	75	70	60
dt	3000	4250	5250	6300	6600

where dt is the product of the distance and foraging time. Under the null hypothesis, the dt product at any of the release times is the same as it would have been for that release time at any alternative distance. Thus, the null hypothesis permits variation over time in the dt product but not over distances at a particular time. With our expectation that dt would increase rather than remain constant over the distances, a reasonable test statistic to employ would be the correlation coefficient (r or a rank correlation coefficient) for the correlation between distance and dt, and we would compute the correlation for all 120 ways of pairing the dt values with the five distance values. Having predicted that the time would drop less sharply than a drop inversely proportional to the distance, we expect a positive correlation between dt and distance. Because there is a consistent rise in dt with increasing distance, the P-value would be 1/20, or about 0.008. Thus, we could reject at the 0.01 level the null hypothesis of foraging time being inversely proportional to distance.

12.7 Complications

The model of an inversely proportional relationship between time and distance from the nest in the preceding example was readily testable, but the place at which the hawk is released is "confounded" with the distance from the hawk's nest. It was not distance alone that was varied; other properties also change when we vary the distance at which a hawk is released. There is also variation in the trees, shrubs, grass, and other cover for prey when distance from the nest is varied, so the characteristics of the release sites must be regarded as part of the treatment. Whether we are testing the traditional randomization test null hypothesis of no differential effect of a variable or the null hypothesis of a particular quantitative law, randomization cannot control for the effects of extraneous variables systematically associated with a treatment condition. Random assignment of experimental units, even complex units incorporating many important variables, will not control for the influence of extraneous variables associated with a treatment condition; other types of experimental control are required.

However, rigorous experiments can be conducted in naturalistic settings and behavioral laws can be tested there. For example, if the law to be tested with hawks was that the foraging time at a location would be proportional to the dosage of a given stimulant, the random determination of what dosage would be administered at given times could provide a sound basis for a randomization test of that proportionality. In the settings under consideration, drug dosage is easier to manipulate independently of alternative important influences on results than is distance from the nest.

Randomization tests can be devised to test a variety of forms of lawful relationships between treatments and effects but their utility will be limited by the experimental design considerations just discussed. Whether the relationship is one of simple proportionality, as in the foraging behavior of hawks, or a more complex one involving exponential or logarithmic components, there is likely to be greater success in testing the laws that concern physiology or processes closely related to fundamental units such as photons than laws that concern more molar behavior.

12.8   Questions and Exercises

1. For the test of Weber's law, explain why the smallness of the number of runs rather than the largeness of the number of runs is likely to indicate deviation from the law.

2. In Table 12.1, it is obvious that the Weber fraction is not constant over the levels of intensity, so why is it necessary to conduct a statistical test?

3. Describe a randomization test of Weber's law when it expected that there will be larger Weber fractions for the low and the high intensities.

4. Rearrange the symbols to show $R = kS$ as a type of proportional relationship, where R is the response, k is a constant, and S is the stimulus.

5. Evidence against the null hypothesis of no treatment effect can readily lead to further research on the nature of the effect. But how can evidence against a scientific law that specifies a particular type of effect have any constructive value? Give examples.

6. Suppose that before determining the formula $A = \pi r^2$ for the area of a circle, it was proposed that the area of a circle would be proportional to the radius, that lawful relationship being symbolized as $A = kr$. A randomized experiment was conducted to control for differences in the measuring instruments and techniques over time. How could a randomized experiment be employed to test the $A = kr$ relationship if it was expected that $A = kr$ underestimated the area?

7. For the situation described in Question 6, how could a randomized experiment be employed if it was expected that $A = kr$ overestimated the area?

8. In the example of a test of incremental effects, the experimenter expected that the amount of a substance that was consumed would increase the amount excreted. How would the randomization test have differed if a negative relationship between the amount consumed and the amount excreted was expected?

9. The null hypothesis for the test of incremental effects may appear to be simply the hypothesis of no treatment effect, and it is true that the randomization test described tests the null hypothesis of no effect of the amount consumed on the amount secreted. But the null hypothesis of the relationship between the amount consumed and the amount digested (amount not secreted) is of a quite different form. What is that null hypothesis?

10. In the test of Weber's law that used the number of runs of signs of differences between adjacent measurements as the test statistic, the test statistic was categorical and unaffected by how much difference there was between adjacent measurements. Give a justification for regarding it as a test of a quantitative law when the test statistic was not quantitative.

REFERENCES

Andersson, M., Optimal foraging area: size and allocation of search effort, *Theoret. Population Biol.*, 13, 397, 1978.

Andersson, M., Central place foraging in the Whinchat. *Saxicola rubetra*, *Ecology*, 62, 538, 1981.

Aronson, R.B. and Givinish, T.J., Optimal central-place foragers: a comparison with null hypotheses, *Ecology*, 64, 395, 1983.

Blanchard, J., The brightness sensibility of the retina, *Phys. Rev.*, 11, 81, 1918.

Boring, E.G., *Sensation and Perception in the History of Experimental Psychology*, Appleton-Century, New York, 1942.

Cole, B.J., Overlap, regularity, and flowering phenomenologies, *Am. Naturalist*, 117, 993, 1981.

Connor, E.F. and Simberloff, D.S., Species number and compositional similarity of the Galapagos flora and fauna, *Ecol. Monogr.*, 48, 219, 1978.

DeVita, J., Kelly, D., and Payne, S., Arthropod encounter rate: a null model based on random motion, *Am. Naturalist*, 119, 499, 1982.

Edgington, E.S., A statistical test of cyclical trends, with application to Weber's law, *Am. J. Psychol.*, 74, 630, 1961a.

Edgington, E.S., Probability table for number of runs of signs of first differences in ordered series, *J. Am. Statist. Assn.*, 56, 156, 1961b.

Edgington, E.S., *Statistical Inference: The Distribution-Free Approach*, McGraw-Hill, New York, 1969.

Ganz, L., Vision, in *Experimental Sensory Psychology*, Scharf, B. (Ed.), Scott, Foresman & Company, Glenview, IL, 1975.

Holway, A.H. and Pratt, C.C., The Weber-ratio for intensity discrimination, *Psychol. Rev.*, 43, 322, 1936.

Poole, R.W. and Rathcke, B.J., Regularity, randomness, and aggregation in flowering phenomenologies, *Science*, 203, 470, 1979.

Strong, D.R., Jr., Szyska, L.A., and Simberloff, D.S., Tests of community-wide character displacement against null hypotheses, *Evolution*, 33, 897, 1979.

13

Tests of Direction and Magnitude of Effect

In Section 4.9, we indicated the importance of testing one-tailed null hypotheses and made the crucial distinction between "tests of one-tailed null hypotheses" and "tests of two-tailed null hypotheses that use a one-tailed test statistic." The latter are frequently called "one-tailed tests" but in fact this terminology might be misleading because these tests really concern two-tailed null hypotheses. In this chapter, we elaborate on the idea already introduced in Section 4.9 of testing genuine one-tailed null hypotheses and extend this idea to testing null hypotheses about effect magnitude.

13.1 Tests of One-Tailed Null Hypotheses for Correlated *t* Tests

In Section 4.9, it was shown that precise reference sets for one-tailed null hypotheses with a between-subjects *t* test are not possible because the null hypothesis A ≤ B contains an inequality. However, Example 4.5 demonstrated that using P-values from precise reference sets for H_0 of no treatment effect for drawing inferences about direction of effect is a valid procedure when T_A or $(\bar{A} - \bar{B})$ are used as test statistics for between-subjects *t* tests. The same reasoning applies to the correlated *t* test, as will be shown in the following example.

Example 13.1. The two-tailed, nondirectional, null hypothesis of no differential treatment effect for a correlated *t* test (A = B) is that for each subject, the measurements associated with the two treatment times are the same as they would have been if the treatment times for the A and B treatments were reversed. When it is predicted that the A treatment will tend to provide larger measurements, the one-tailed null hypothesis (A ≤ B) is that for each subject the measurement associated with each of the two treatment times would never be larger for an A treatment than for a B treatment at that time. It will be shown that the P-value for a "one-tailed test" based on the precise reference set for the two-tailed null hypothesis, using T_A or $(\bar{A} - \bar{B})$ as the one-tailed test statistic, is a conservative value for a test of the one-tailed null hypothesis.

Suppose we have five subjects, each with two treatment times, and that there is random assignment of treatment times to treatments A and B independently for each subject. The experiment is conducted and the following results are obtained:

	A	B
S_1	10	8
S_2	12	7
S_3	9	6
S_4	16	13
S_5	11	12
Totals	58	46
$\bar{A} - \bar{B}$	2.4	

A randomization test for correlated t gives a P-value of about 0.063, as there are only two of the 32 data permutations providing as large a value of T_A as 58. Of course, the use of $(\bar{A} - \bar{B})$ as an equivalent test statistic would give the same P-value. If the one-tailed null hypothesis is true, the following modified data permutation represents the randomization for which the times assigned to treatments A and B are reversed for S_1 and for S_4:

	A	B
S_1	8 or less	10 or more
S_2	12	7
S_3	9	6
S_4	13 or less	16 or more
S_5	11	12
Totals	53 or less	51 or more
$\bar{A} - \bar{B}$	0.4 or less	

The above data permutation for a particular randomization illustrates what is common to all of the randomizations: the test statistic T_A or $(\bar{A} - \bar{B})$ will never be larger under any of the hypotheses represented in the modified reference set than under the null hypothesis providing the precise reference set. Thus, as shown in Example 4.5 for the between-subjects t test, the proportion of data permutations with as large a test statistic value as the obtained value under the one-tailed correlated t null hypothesis will be no greater than the proportion based on the correlated t precise reference set. Assigning the P-value for the precise reference set, based on T_A or $(\bar{A} - \bar{B})$ as the test statistic, to the test of the one-tailed null hypothesis thus is conservative for the correlated t test as well as the between-subjects t test.

Example 4.5 and Example 13.1 concern tests of one-tailed H_0s for between-subjects and correlated t randomization tests. Any of the tests in the book, whether for multiple-N or N-of-1 experiments, for which a between-subjects or correlated t program is applicable therefore can be used to test a one-tailed null hypothesis. Of course, that includes some of the rank tests and tests for dichotomous data. For example, consider a situation where the dependent variable had two levels, "lived" and "died," with random assignment of experimental animals consistent with either between-subjects t or correlated t designs. By representing the results by 0s and 1s, it is clear that one not only can conduct a randomization test of the null hypothesis of no differential treatment effect on mortality but also can test the one-tailed null hypothesis that no animal that would have died if assigned to B would have lived if assigned to A.

The appending of "or less" and "or more" to measurements in the preceding examples illustrates the logic in inferring that one-tailed test statistic values under the one-tailed null hypothesis will not be greater than under the precise null hypothesis of no differential treatment effect. Reducing the size of A measurements or increasing the size of B measurements for any data permutation cannot increase the value of the $(\bar{A} - \bar{B})$ test statistic. The same is true of the T_A test statistic.

However, it is not just test statistics related to means that permit randomization tests of one-tailed null hypotheses. For example, the A median minus the B median would be a valid test statistic, as would the A median itself, because for those test statistics reducing the size of A measurements or increasing the size of B measurements cannot increase the value of the test statistic.

13.2 Other Tests of One-Tailed Null Hypotheses Using T_A or $(\bar{A} - \bar{B})$ as Test Statistics

In the following explanation, appending of "or less" and "or more" to individual measurements will not be employed, although the reasoning regarding the effect of increasing or decreasing measurements to accommodate one-tailed null hypothesis possibilities is based on the effect of such modification of measurements on the test statistic value. There are randomization tests that use a T_A or $(\bar{A} - \bar{B})$ one-tailed test statistic but which do not permit use of the programs for between-subjects or correlated t because of differences in the way the randomization of experimental units must be carried out. In this section, we will consider some tests of that type for which valid one-tailed statistical inferences nevertheless can be drawn.

For completely randomized factorial designs and randomized block designs with only two levels of one of the factors, tests of a one-tailed hypothesis are applicable for that factor. Of course, the t test programs could not be applied because of the difference in the way the data must be permuted for a test of main effects. Nevertheless, tests of one-tailed null hypotheses can be justified. Where level A_1 is expected generally to provide larger measurements than A_2, then using the test statistic $(\bar{A}_1 - \bar{A}_2)$ for each data permutation in the precise reference set for testing the main effect of A would serve as a test of the one-tailed null hypothesis that for no level of B would $\bar{A}_1$ provide a larger measurement than $\bar{A}_2$ for any subject. Decreasing A_1 measurements or increasing A_2 measurements for any level of B could not provide a larger overall value of the test statistic $(\bar{A}_1 - \bar{A}_2)$ or of the equivalent test statistic, T_{A1}, so the P-value provided by the reference set based on permuting the data for a nondirectional test is a conservative, valid, P-value for a one-tailed test.

Other examples of the valid use of one-tailed test statistics to test one-tailed null hypotheses when t test programs are not applicable are tests in Chapter 11. Several of the N-of-1 experimental designs in that chapter employ a difference between means as a test statistic but require different methods of permuting the data from that associated with the between-subjects or correlated t tests. Nevertheless, randomization tests of one-tailed null hypotheses are possible based on the type of argument employed above with randomization t tests. Treatment intervention without withdrawal in Section 11.14 is such a procedure. It involves random determination of the time at which a treatment is introduced and the one-tailed test statistic is $(\bar{A} - \bar{B})$, where A is the condition (either pre- or post-treatment intervention) predicted to provide the larger measurements. To test the H_0 of no treatment effect, the reference set of data permutations is generated by dividing a sequence of measurements at all possible points at which the treatment could have been introduced. The reference set of data permutations for the one-tailed null hypothesis would be those for the precise, two-tailed null hypothesis (using the one-tailed test statistic) with "or less" appended to the A measurements and "or more" appended to the B measurements that were not associated with the same treatments for the actual assignment. Clearly, such a modification of a data permutation could not increase the value of the test statistic, so any data permutation that in the precise reference set had a test statistic value less than the obtained value could not have a value as large as the obtained value in the modified reference set. Treatment intervention and withdrawal designs also permit the use of the one-tailed test statistic to test a one-tailed null hypothesis even though, as with the treatment intervention design, the random assignment procedure is quite unconventional.

In treatment intervention designs and treatment intervention and withdrawal designs, the number of experimental units in the A or B treatment condition varies from one possible assignment to another. Consequently, T_A as a one-tailed test statistic is insensitive to treatment effects because its value is greatly dependent on the number of A measurements as well as their size. Nevertheless, it would be perfectly valid although inappropriate to employ T_A to test the one-tailed null hypothesis for those experimental designs. Alternatively, $\bar{A}$ would be a test statistic that is both appropriate and valid for testing the one-tailed null hypothesis.

13.3 Tests of One-Tailed Null Hypotheses about Differences in Variability

If an experimenter wanted a test statistic sensitive to greater variability for the A treatment, σ_A^2/σ_B^2 could be used as a test statistic. However, in the absence of random sampling the randomization test tests hypotheses about individual experimental units, not relationships among them, and so no null hypothesis about variability among experimental units could be tested by that test statistic. Instead of testing a null hypothesis of no treatment effect on variability, the test would test the same "two-tailed" null hypothesis as if a difference between means was the test statistic: the hypothesis of no differential treatment effect on any measurement with a one-tailed test statistic sensitive to greater variability of the A meaurements. A similar limitation exists for other test statistics employed in parametric tests to draw directional inferences about differences between populations whenever those test statistics refer to changes in within-treatment relationships among measurements. For instance, a test statistic for a difference in skewness could be sensitive to a specified direction of difference in skewness but it would not permit a test of a one-tailed null hypothesis.

13.4 Tests of One-Tailed Null Hypotheses for Correlation

Another type of statistical test for which there are commonly one-tailed and two-tailed test statistics is correlation. A randomization test cannot test hypotheses about correlation or direction of correlation but only about treatment effects on individual experimental units. The two-tailed null hypothesis for a randomization test is that the measurement at each treatment time is the same as it would have been under any alternative treatment level. Formulation of a one-tailed randomization test null hypothesis is more arbitrary because it is necessary to translate the concepts of positive and negative correlation into analogous directional relationships referring to individual experimental units. A one-tailed null hypothesis for a randomization test where a positive correlation is expected could be this: for none of the treatment times would a larger value of a treatment level lead to a larger response value or a smaller value of a treatment level lead to a smaller response value. Thus, any data permutation for that one-tailed null hypothesis can be derived from data permutations under the precise null hypothesis of no differential treatment effect by appending "or less" to a measurement value that is associated with a larger treatment magnitude than for the obtained results, and "or more" to a value that is associated with a smaller treatment magnitude than for the obtained results.

Appending "or less" and "or more" in accordance with the one-tailed null hypothesis does not necessarily ensure that the one-tailed test statistic ΣXY cannot have a larger value. The following example shows that under the one-tailed null hypothesis, the P-value from the precise two-tailed null hypothesis distribution (data permutations with no "or less" or "or more" qualifiers) is not necessarily a conservative value for the one-tailed null hypothesis.

Example 13.2. Three subjects with predetermined treatment times are randomly assigned to treatment levels 10, 20, and 30, with the prediction that in general the larger treatment levels will lead to larger response measurements. Below are the six data permutations under the precise null hypothesis of no treatment effect, using ΣXY as the test statistic, where the first row shows the obtained results:

Treatment Levels			
10	**20**	**30**	**ΣXY**
7	9	11	580
7	11	9	560
9	7	11	560
9	11	7	520
11	7	9	520
11	9	7	500

No other data permutation has a value of ΣXY as large as 580, the value for the obtained results. The P-value thus is $1/6$, or about 0.167. Consider now the appending of "or less" and "or more" to the above data for row 2 to row 6 to numerical values different from those in the same column for row 1:

Treatment Levels		
10	**20**	**30**
7	9	11
7	11 or more	9 or less
9 or more	7 or less	11
9 or more	11 or more	7 or less
11 or more	7 or less	9 or less
11 or more	9	7 or less

The test statistic ΣXY for the obtained results of course is still 580 but the values for the other data permutations are indefinite; for any row, the value of ΣXY could be greater than, less than, or equal to the value under the precise null hypothesis. (If the value was necessarily no greater than under the precise, two-tailed, null hypothesis, the P-value under the one-tailed null hypothesis could not be larger than the P-value under the precise null hypothesis.) Suppose the subject actually assigned to treatment level 30, who provided a measurement of 11, was the only subject for which there was a differential treatment effect and that the effect would have led to an increase from 11 to

18 if that subject had been assigned to level 20. In that case, the precise data permutation in the second row would be 7, 18, and 9, providing $\Sigma XY = 700$; the precise data permutation in the fourth row would be 9, 18, and 7, providing $\Sigma XY = 660$; and the precise data permutation in the fifth row would be 18, 7, and 9, providing $\Sigma XY = 590$. For that possibility, there would be four data permutations with a ΣXY as large as the obtained value, making the P-value 4/6, or 0.667, instead of 0.167. This counter-example shows that it is invalid to use the P-value based on a reference set of ΣXY values under the two-tailed null hypothesis as the P-value for a test of the one-tailed null hypothesis.

Bear in mind that there may be alternative one-tailed null hypotheses for randomization tests of correlation for which the P-value for ΣXY would be conservative for a test of the one-tailed null hypothesis. Obviously, under the proposed one-tailed null hypothesis there is no limit to how large ΣXY could be for any alternative data permutation. It may seem that using r as a test statistic would be more effective because there is an upper and lower limit on the value the test statistic could attain under the one-tailed null hypothesis, but the following example shows that r also is inappropriate in a test of the one-tailed null hypothesis.

Example 13.3. Consider the following data from an experiment in which there was random assignment of eight experimental units, each consisting of a subject plus a treatment time, to eight treatment levels:

	Treatment Levels							
	10	20	30	40	50	60	70	80
Data	7	9	13	12	15	4	5	20

The product-moment correlation for the obtained results is 0.253. The data permutation like the above but with the 4 and 5 for treatment magnitudes 60 and 70 reversed gives $r = 0.242$, so it would not be counted as a data permutation with r greater than or equal to the obtained value using the randomization test correlation programs. However, consider the data permutation for the one-tailed null hypothesis, where we would have "5 or more" under 60 and "4 or less" under 70. That data permutation is consistent with a measurement of 18 being associated with 60 and a measurement of 4 being associated with 70, which would provide $r = 0.450$. Thus, a randomization that would not have a test statistic value as large as the obtained value in the precise reference set could have a test statistic as large as the obtained value in the modified reference set. Therefore, there would be no justification in using the P-value based on the precise reference set under the two-tailed null hypothesis, employing r as a test statistic, as a conservative value for a test of the one-tailed null hypothesis.

Spearman's ρ and Kendall's τ, two common rank correlation coefficients described in Section 8.7 and Section 8.8, are also inappropriate for testing the one-tailed null hypothesis. The following table shows the values of r, ρ, and τ for three data permutations, the first of which represents the obtained results:

Data Permutation								r	ρ	τ
7	9	13	12	15	4	5	20	0.253	0.143	0.214
7	9	13	12	15	5	4	20	0.242	0.119	0.143
7	9	13	12	15	18	4	20	0.450	0.476	0.500

The limitations just described on determining P-values for testing the one-tailed null hypothesis are for the test statistics ΣXY, r, ρ, and τ, the test statistics sensitive to directional correlation considered in Chapter 8. Alternative test statistics may not have these limitations.

The correlation trend test program provided P-values associated with one-tailed and two-tailed test statistics, but both P-values were based on the precise reference set of data permutations; only the test statistic was different. Whether a one-tailed test statistic or a two-tailed test statistic should be employed depends on the specificity of the prediction. For example, a person may predict a linear trend, an upward linear trend, a quadratic trend, or a U-shaped trend. When the data tend to follow the more specific prediction, the use of the one-tailed test statistic is more sensitive but the data permutations in the precise reference set are generated on the basis of the null hypothesis of no treatment effect, a nondirectional null hypothesis. The one-tailed test statistic for the correlation trend test is the same as for product-moment correlation and, as is the case for correlation, the P-value based on the precise reference set is not necessarily a conservative value for a test of the one-tailed null hypothesis.

13.5 Testing Null Hypotheses about Magnitude of Effect

It can be helpful at times to statistically infer that treatment A is more effective than treatment B by more than some specified amount Δ. For example, in brain research there may be interest in the difference in time for stimulation of location A or location B to have an effect at a remote location C. (Location A or B could be the right or left ear or the right or left eye, or perhaps two locations in the cerebral cortex.) As the A-C neural connections would be different from the B-C neural connections, the null hypothesis of identical times to have an effect on C might not be considered worth testing. Nevertheless, if it is expected that there is a B-C pathway that is more direct than current theory implies, a brain researcher may be interested in testing the one-tailed null hypothesis that over a number of subjects (or a number of trials on one subject) at no particular time of stimulation would the elapsed time between stimulation of A and a response at C be more than, say, 6 milliseconds longer than for stimulation of B. The complementary hypothesis would be that on at least one occasion, the time lapse between stimulation at A and a response at C would be more than 6 milliseconds longer than for stimulation at location B.

The primary interest here is in using P-values for one-tailed test statistics in precise reference sets to test one-tailed null hypotheses of the form A − B ≤ Δ. The tests are based on precise reference sets under a null hypothesis of a specified amount of effect Δ, which is A − B = Δ. The precise sets could be employed simply to test that hypothesis as a law somewhat like the laws or models considered in Chapter 12, but as stated above the primary interest here is in using the precise tests simply to provide a P-value for a test of a one-tailed null hypothesis.

What was stated about the assumptions underlying the randomization tests in Chapter 12 is true here as well: the assumptions are the same as for the simple "bare-bones" traditional randomization tests in earlier chapters, namely the appropriate random assignment and experimental independence, the minimum assumptions that any test of experimental treatment effects requires.

13.6 Testing Null Hypotheses about Specific Additive Effects

Example 13.4. Suppose there is interest in testing the null hypothesis, A − B ≤ 5, which is the hypothesis that for none of the subjects would the measurement under treatment A be larger than under B by more than 5 points. As in Section 12.1, this is done by first testing the precise null hypothesis (A − B = 5, in this case), using a one-tailed test statistic, then using that P-value as a conservative P-value for a test of the one-tailed null hypothesis.

Consider the first step, the testing of the precise (nondirectional) null hypothesis. Data permutations in the precise reference set could be generated in this fashion from the obtained results: for each data permutation, add 5 to each A measurement that was a B measurement for the obtained results and subtract 5 from each B measurement that was an A measurement for the obtained results. To illustrate the procedure, hypothetical obtained results are shown below along with another data permutation that was derived from the obtained results to represent the randomization in which the assignments of subjects with the 7 and 6 for the obtained results were reversed, providing measurements shown in bold face type:

	Obtained Results		Another Data Permutation	
	A	B	A	B
	12	6	12	2
	7	9	11	9
	15	4	15	4
Totals	34	19	38	15
$\bar{A} - \bar{B}$		5		7.667

The P-value for a test of the null hypothesis A − B = 5 would be the proportion of the 20 data permutations with T_A as large as 34 or, equivalently, the proportion with $(\bar{A} - \bar{B})$ as large as 5.

Before considering the transformation of this precise reference set into a modified reference set for testing a one-tailed null hypothesis of magnitude of effect, let us consider a simpler procedure (Bradley, 1968; Lehmann, 1963) of getting the P-value from a precise reference set for a specified magnitude of effect. The simpler procedure does not require performing addition or subtraction on each data permutation but only a single adjustment made on the obtained results. The procedure is to subtract 5 from each of the obtained A measurements, then carry out the randomization test on the revised obtained data permutation as if testing the null hypothesis of identity of A and B effects. For example, the above obtained results would become A: 7, 2, 10; B: 6, 9, 4, and the reference set of 20 data permutations could be derived by permuting those values by use of the programs in Chapter 4. For a randomization test of the one-tailed null hypothesis, this simpler procedure provides a different distribution of test statistic values, but the P-value will be equal to that of the procedure described above for the use of either T_A or $(\bar{A} - \bar{B})$ as a test statistic.

We will now demonstrate that the simpler procedure will give the same P-value as the more complex procedure for the one-tailed test statistic T_A. Imagine two rows of data permutations, the top row for the complex procedure and the bottom row for the simple procedure, with each data permutation for the simple procedure immediately below the data permutation for the complex procedure for the same randomization. For the two data permutations associated with each randomization, each A measurement for the simpler procedure will be 5 less than for the more complex procedure because those A measurements will consist either of measurements adjusted downward by 5 or of measurements which for the more complex procedure would have been increased by 5. Consequently, $\bar{A}$ for the adjusted measurements for each data permutation will be 5 less than for the other procedure, and T_A will be 15 less than for the other procedure for each data permutation. Because the revised distribution of T_A is the same except for a shift of 15 points, the P-value — which is the proportion of data permutations with T_A as large as the value for the obtained results — is the same. The P-value for $(\bar{A} - \bar{B})$ is also the same for the complex and simple procedures.

The act of subtracting 5 from the obtained A scores prior to generating the reference set provides a P-value for the test of the null hypothesis A − B = 5, but it is the one-tailed null hypothesis, A − B ≤ 5, in which we are interested. That the P-value for the test of the one-tailed null hypothesis could be no larger than the P-value for the nondirectional null hypothesis can be seen by considering the effect of adding "or less" and "or more" to measurements in the reference set of data permutations for the simple procedure. Thus, for any Δ (i.e., magnitude of effect under test) if Δ is first subtracted from each A measurement for the obtained results, application of the programs for between-subjects t will give a valid P-value for the test of the one-tailed null hypothesis, using T_A or $(\bar{A} - \bar{B})$ as the test statistic.

Example 13.5. Of course, although the reference set for testing the additive effect A − B = Δ is different for correlated t randomization tests than for between-subjects t tests, the considerations are similar. Suppose we have a number of subjects, each taking treatment A and treatment B. Suppose the two-tailed null hypothesis is that for each subject, for each of the treatment times $A - B = 3$, where A and B for a particular treatment time represent the treatment administered and the alternative treatment that could have been administered at that time. For the direct, complex procedure, the data would be permuted within the n subjects in all 2^n possible ways and, as for between-subjects t, 3 is subtracted from each obtained A measurement that is switched to B and added to each obtained B measurement switched to A. A simpler procedure would be to adjust the obtained A measurements, as in the example for between-subjects t tests: subtract Δ, which in this case is 3, from each A measurement and use that adjusted data permutation and others based on permutations of it as the reference set. The P-value for T_A or $(\bar{A} - \bar{B})$ would be the same as for the complex procedure. And for the reasons given in the previous example, the P-value under the one-tailed null hypothesis based on either T_A or $(\bar{A} - \bar{B})$ would be no larger than under the precise, two-tailed null hypothesis, and thus would be a conservative value. Programs for correlated t tests in Chapter 6 can be applied to the adjusted obtained data permutation to provide the P-value based on the one-tailed test statistic.

To sum, to test the one-tailed null hypothesis, $A - B \leq \Delta$, subtract Δ from each A measurement and perform a randomization t test with a one-tailed test statistic. A two-tailed test statistic for testing the $A - B = \Delta$ null hypothesis sensitive to the $A - B$ difference being either greater than or less than Δ could be $|(\bar{A} - \bar{B}) - \Delta|$ for the complex data-permuting procedure or $|\bar{A} - \bar{B}|$ for the simpler procedure, where the A measurements have A subtracted from them before permuting the data. For every randomization, $|(\bar{A} - \bar{B}) - \Delta|$ for the complex procedure gives the same numerical value as $|\bar{A} - \bar{B}|$ for the simple procedure, so the P-values would be the same.

13.7 Questions and Exercises

1. What is the one-tailed null hypothesis for a randomization test in which it is expected that Treatment A will provide the larger mean?

2. What is the one-tailed null hypothesis when it is expected that A will provide the larger median?

3. If you expect a difference between the effects of two treatments to be at least 10, what is the null hypothesis?

4. What is the null hypothesis if you expect the difference in Question 3 to be exactly 10?

5. Why are the tests in this chapter not regarded as tests of quantitative laws?

6. We expect treatment A to provide hits closer to the target than treatment B. What is the null hypothesis?

7. For Question 6, what would be a two-tailed null hypothesis? What would be the one-tailed alternative hypothesis?

8. In Chapter 10, in the experiment involving animals raised in cages under colored light, what was the one-tailed null hypothesis?

9. For Question 1, there is a two-tailed null hypothesis and two different one-tailed null hypotheses that could be tested. Is this also true for the experiment in Question 6? Explain why or why not.

10. When there has been random assignment of subjects to more than two magnitudes of treatment, what is a one-tailed null hypothesis that can be tested by a randomization test?

REFERENCES

Bradley, J.V., *Distribution-Free Statistical Tests*, Prentice Hall, New York, 1968.

Edgington, E.S., *Statistical Inference: The Distribution-Free Approach*, McGraw-Hill, New York, 1969.

Lehmann, E.L., Nonparametric confidence intervals for a shift parameter, *Ann. Math. Statist.*, 34, 1963, 1507.

14

Fundamentals of Validity

Other than peripheral aspects such as asymptotic relationships to normal curve procedures, the theory of randomization tests is not extensively developed. The task therefore is not a matter of presenting theory that has been developed but of proposing theory. The function of this chapter is not to explain what randomization test theory should be because that would be premature. The aim is to provide a technique of viewing randomization tests consistent with the book as a whole and within a framework that will be helpful in understanding the basic aspects of randomization tests.

The following are fundamental propositions affecting the approach to randomization tests taken in this book:

1. In experimental research, the random selection of experimental units is so uncommon that it is useful to base randomization tests on random assignment alone, making them applicable to experimental results from nonrandomly selected experimental units.

2. Randomization tests offer a unique opportunity to see what statistical inferences about treatment effects can be made with a bare minimum of statistical assumptions.

3. Understanding the conduct of statistical tests with minimal assumptions facilitates the understanding of similar tests requiring additional assumptions.

4. A test is valid if the probability of getting a P-value as small as p under the null hypothesis is no greater than p.

5. It is the complement (negation) of the null hypothesis, not just any alternative to the null hypothesis, that is provided statistical support by the smallness of a P-value.

6. Random assignment and randomization tests can ensure validity of statistical inferences about treatment effects only to the extent that an experimenter has provided good control over variables that are not randomized.

14.1   Randomization Tests as Distribution-Free Tests

Statistical assumptions most familiar to researchers probably are distributional assumptions associated with various normal curve procedures, which
concern the nature of distributions that presumably are randomly sampled,
the most common assumptions being normality and homogeneity of variance. Although normality frequently is presented as a single assumption, it
of course includes more than a general shape of a population distribution;
the infinite largeness of the population and continuity are distributional
assumptions included within the assumption of a normal distribution.

Despite the application of the term "distribution-free" to rank tests and
permutation tests, those tests over many years were discussed in terms of
drawing inferences about populations assumed to have been randomly
sampled. The assumption of continuity of population distributions was
invoked to make the probability of tied ranks negligible, and for tests of
differences between populations it was common to use (although not
always explicitly) the assumption that a difference between populations is
a uniform shift, not a difference in shape. The second assumption, which
serves essentially the same function as the assumption of homogeneity of
variance for *t* tests, is introduced to permit inferences to be drawn about
means. It is sometimes called the *unit additivity* assumption when it refers
to treatment effects.

Because of the common practice of adding the assumption of unit additivity or other assumptions regarding the distribution of treatment effects to
the basic assumptions of randomization tests, it is important to stress that
the theory in this chapter concerns basic randomization tests without the
unit additivity assumption. No distributional assumptions are made; these
tests truly are distribution-free. Besides the great practical value of such
randomization tests, a sound understanding of the theoretical foundations
and implications of these tests allows the researcher to develop valid new
applications and evaluate existing ones.

14.2   Differences between Randomization Test Theory
and Permutation Test Theory

Throughout this book, we have discussed randomization tests as special
types of permutation tests and the production of randomization test reference sets by permuting data, but the rationale for permuting data for nonexperimental permutation tests is quite different from the rationale for
randomization tests. For randomization tests, the permuting of data generates
hypothetical outcomes for the same subjects (or other experimental units)
under alternative random assignments. Alternatively, for nonexperimental

permutation tests the permuting of data does not provide hypothetical outcomes for the same subjects but outcomes for other subjects randomly drawn from identical infinite populations.

Pitman's (1937a; 1937b; 1938) recognition that random sampling was unnecessary for a valid test of the difference between treatments in a randomized experiment was of considerable practical importance in light of the multitude of experiments in which experimental units are not randomly selected. But the theoretical importance is equally great because it shows the falseness of the almost universal assertion in textbooks that statistical inferences concern populations that have been randomly sampled. As mentioned above, until recently even rank tests were almost always discussed only in terms of random sampling of populations and inferences about those populations.

Instead of regarding randomization tests as special types of permutation tests, Kempthorne (1986, p. 524) stressed the difference in this way:

> The distinction between randomization tests and permutation tests is important. The latter are based on the assumption of random sampling, an assumption that is often patently false or unverifiable, even though necessary to make an attack on the substantive problem being addressed.

Of course, in this quotation Kempthorne is referring to the rationale underlying the tests, but in other places (e.g., Kempthorne and Doerfler, 1969) he widens the conceptual gap between permutation tests and randomization tests by discussing the generation of the randomization test reference set in terms of superimposing treatment designations on the data, rather than in terms of permuting the data.

It should be noted that a number of considerations in this chapter are applicable also to permutation tests when they are based on random sampling. Some aspects of randomization test theory are applicable to other types of permutation tests with little or no modification, whereas other aspects are not so obviously generalizable. However, detailed consideration of such generalizability is not provided here because the function of the chapter is to explore ideas relevant to the theory of randomization tests, not permutation tests in general.

14.3 Parametric Tests as Approximations to Randomization Tests

As pointed out in Chapter 1, long before randomization tests were of practical value it was widely accepted that they were the only truly valid statistical tests of treatment effects and that a parametric test (i.e., the use of parametric significance tables) was valid only to the extent that it provided similar P-values to those that would be given by a randomization test.

Studies showed that in some situations, parametric tests need not meet the parametric assumptions to provide very similar P-values. Those parametric tests were said to be "robust" with respect to the effect of violation of certain assumptions on the validity of the tests.

Comparisons of power and validity of randomization tests and parametric tests were common, but little attention was paid to the implications of lack of random sampling for interpreting results of the statistical tests. When a parametric test is valid only by virtue of approximating the P-value a randomization test would give, the interpretation of results of the parametric tests should be that provided for the same P-value from a randomization test. Use of a t table for a t test applied to a set of relatively continuous experimental data from large samples of nonrandomly selected subjects might provide a close approximation to the P-value a randomization test would give, but that fact would not justify interpreting the P-value in parametric terms. Because its validity is "borrowed" from the randomization test it approximates, determination of the P-value by use of a t table would not allow statistical inferences that the randomization test would not justify. For example, a statistical inference about a difference between means would not be justified because the null hypothesis tested and the complementary hypothesis anticipated to be true would be the same when the parametric significance table was used as when the P-value was based on a randomization test. (It would only be through making the tenuous assumption of unit additivity, which would vitiate the distribution-free nature of randomization tests, that inferences from randomization tests could relate to mean differences.) Statistical inferences about parameters require random sampling, and the closeness of P-values of parametric tests and randomization tests in the absence of random sampling only justifies use of parametric tables for inferences about the existence of treatment effects, not statistical inferences about parameters. Whether a parametric test is a test of a difference between means or of a difference between variances affects the null hypothesis that is tested when parametric assumptions are met. However, when parametric assumptions are untenable and the test is used as an approximation to a randomization test, the null hypothesis is that of the randomization test: identity of treatment effects, not identity of means or homogeneity of variance. And statistical inferences about treatment effects would have the same lack of generality as the randomization test being approximated. Similarly, the limitations on the possibility of testing one-tailed null hypotheses described in Chapter 13 exist also for parametric tests that approximate randomization tests.

Having indicated the importance of drawing the same inferences as randomization tests when parametric tests are used to approximate randomization tests, we need now to consider comparison of the assumptions of parametric and randomization tests when they are used to provide the same inferences, i.e., to test the same null hypotheses. It should be recognized that for a comparison of the assumptions underlying parametric or nonparametric P-value determination to be meaningful, the null hypotheses tested must be comparable.

First, consider a comparison within a random sampling model. It is commonplace for advocates of nonparametric tests to state that those tests free the users from assumptions of normality and homogeneity of variance associated with parametric tests. This statement is not true. It is true that the rank or raw-data permutation tests do not require those assumptions but it is not true that parametric tests require both of those assumptions to test the same null hypothesis as the nonparametric tests. It has been noted (Edgington, 1965) that although the assumptions of normality and homogeneity of variance are both necessary for parametric tests of the null hypothesis of identity of population means, only the assumption of normality is necessary for using parametric tables to test the null hypothesis of identity of populations, the null hypothesis tested by nonparametric (permutation) tests. Of course, random sampling is necessary for both types of tests to justify any statistical inferences about populations.

Next, consider a comparison within a random assignment model. Although valid tests of hypotheses about populations are possible for either permutation tests or parametric tests, only randomization tests can test hypotheses on the basis of random assignment. Parametric tests assume normality, which means random sampling of a normal population.

14.4 Randomization Test Theory

Randomization tests frequently are described in their simplest form, as in Example 1.1 and Example 1.2, which can be quite useful for an introduction to the randomization test procedure. However, as the subsequent chapters showed making effective use of the potential of randomization tests necessitates dealing with designs that are not so simple that intuition alone is likely to provide sufficient guidance. An understanding of the fundamentals of randomization tests, i.e., P-value determination by the randomization test procedure, requires recognizing the logical links among the null hypothesis, the experimental randomization, and the permuting of data to determine the P-value, which is not always easy in the absence of theory.

Theoretical contributions by Chung and Fraser (1958) and Kempthorne and Doerfler (1969) have profound consequences for the use of randomization tests. This chapter relies heavily on the permutation group concept presented by Chung and Fraser and the randomization-referral concept of Kempthorne and Doerfler. The present chapter also is based on an earlier examination of the implications of the Chung-Fraser view of randomization tests based on *reference subsets* (Edgington, 1983). Reference subsets are subsets of *primary reference sets*, which are reference sets consisting of data permutations associated with all possible randomizations (random assignments).

14.5 Systematically Closed Reference Sets

The reference set is the core of a randomization test; it is the "significance table" to which the obtained test statistic is referred to determine the P-value associated with the experimental results. The proportion of the data permutations in the reference set with test statistic values as large as (or in some cases, as small as) the value for the experimental results is the P-value. Associated with each data permutation in a reference set is a randomization, and so the reference set is intended to be representative of outcomes under alternative randomizations (random assignments) within a set of randomizations under the null hypothesis. Because there are many types of randomization tests that have been and could be developed, it will be helpful for an experimenter to have general rules of ensuring validity rather than to deal with each test separately as it is produced.

In specifying how to ensure the validity of a systematic randomization test, it is helpful to think of two steps:

1. Randomly select the experimental randomization (random assignment) from a set of randomizations.

2. On the basis of the obtained data, transform every randomization in the set of randomizations into a data permutation showing the experimental results and test statistic that would, under the null hypothesis, have been provided if that randomization (random assignment) had resulted from the random assignment procedure. This procedure provides a reference set of data permutations for determining the P-value of the obtained results.

Following the above two steps ensures validity because it guarantees systematic closure of reference sets, a reference set being closed for systematic data permutation if, when the null hypothesis is true: the same reference set would be produced, no matter which data permutation represented the obtained randomization.

A systematically closed reference set will therefore show the outcomes (data permutations) associated with every randomization represented in it, but additionally the rank of a test statistic value within the set is independent of whether it is the obtained test statistic value or one that could have been obtained. The independence of the reference set from the random assignment performed and from the experimental results is a theoretically significant property of a closed reference set.

14.6 Permutation Groups

Clearly, from what has been said the validity of a reference set is not a function of its composition. A reference set is valid because of the validity of the procedure that produced it. To ensure that a reference set based on systematic data permutation is valid, we should follow a procedure of generating reference sets that are closed. If the reference set is not closed, the rank of a data permutation in a reference set can vary according to which data permutation is the obtained data permutation because of the shifting of reference sets, and there is no way to determine the probability under the null hypothesis of getting an obtained data permutation that has one of the k largest of the n test statistic values associated with its reference set.

After indicating the impracticality of randomization tests based on all randomizations for experiments with m subjects for one treatment and n for another when m and n are large, Chung and Fraser (1958, p. 733) suggested that when samples are large we could consider the data permutations for a reference subset:

> There is no essential reason why all the permutations or combinations need be used. Suppose we have some rules for permuting the sequence of $m + n$ observations. Successive application of these permutations may produce new permutations but eventually there will be no new permutations produced. The resulting collection of different permutations of the sequence of $m + n$ observations is then a group.

Note that the group property refers to the procedure for permuting the data, not to the set of data permutations resulting from application of the procedure. The essence of a permutation group is contained in the following definition:

> *A permutation group is a collection of permuting (rearranging) operations that produce the same set when applied to elements in the set, regardless of the initial element to which the operations are applied.*

Permuting or manipulation of experimental results is the standard procedure for determining a reference set of data permutations, and for such a procedure to provide a closed reference set requires that it be a permutation group, but a set produced by a permutation group is not necessarily closed. It is closed only if, when the null hypothesis is true, the data permutations in the reference set represent outcomes for the set of randomizations from which the random assignment was selected.

14.7 Data-Permuting and Randomization-Referral Procedures

The typical method to construct a reference set for a randomization test, which will be called the *data-permuting* procedure, is the procedure discussed in this book. This is the same procedure as that employed for permutation tests for nonexperimental data, although the justification for manipulating data in a given manner for randomization tests is quite different from the justification of data permutation for permutation tests on nonexperimental data.

With the standard data-permuting procedure, the experimental results are manipulated to produce a reference set without explicit consideration of which assignment (randomization) is associated with the experimental results. When a measurement is moved from one treatment designation to another, it represents an alternative assignment of the subject associated with that measurement, but what particular subject provided the measurement is irrelevant. It is the usual way of performing a randomization test and the basic idea is that under the null hypothesis, random assignment of a subject is in effect random assignment of the measurement that subject would provide, no matter what the assignment. (This is in stark contrast to the rationale for permuting nonexperimental data, which is that the various data permutations in the reference set represent divisions or configurations of the same data but not of the same subjects.)

An alternative way to construct a reference set is the *randomization-referral* approach. This approach is of little practical value but it is useful in revealing the rationale underlying the data-permuting procedure for a randomization test. To construct a reference set for a randomization test based on the set of all possible randomizations, that set — which of course must contain the obtained randomization — is transformed into data permutations comprising the reference set by noting the subject (or other experimental unit) that provided each measurement in the experimental results. For a randomization-referral test based on a subset of all randomizations, the set of all randomizations is divided into subsets in advance and the randomization subset containing the obtained randomization is transformed into a reference subset.

In linking the randomization test procedure to random assignment, Kempthorne has consistently described the construction of a reference set in terms of the randomization-referral approach. For example, Kempthorne and Doerfler (1969, p. 237) indicated that if an experimenter wants to perform a randomization test,

> he will superimpose on the data each of the possible plans and will obtain $C_1, C_2, \ldots C_M$, one of which will be C_{obs}. He will then count the proportion of M values, $C_1, C_2, \ldots C_M$, which equal or exceed C_{obs}.

In this quotation, the "plans" are the assignment possibilities, or randomizations, and the "proportion" that is determined will be designated as the P-value. Inasmuch as there are several randomizations and only one set of data, we will consider the randomization-referral approach to involve superimposing the data on the randomization rather than vice versa. More specifically, each experimental unit has a "name" (an identifying label), and the randomizations showing the distribution of those experimental units are transformed into data permutations by substituting for each experimental unit name or label the measurement that experimental unit contributed to the experimental results. This is what was meant earlier by the statement that the random assignment of a subject would (under the null hypothesis) be random assignment of the subject's measurement to a treatment. Although simply stated, the ramifications of the statement are far-reaching and well worth examining in detail, which is the essence of the randomization-referral approach.

Example 14.1. Consider the random assignment of four subjects, designated as *a*, *b*, *c*, and *d*, to treatments A and B with two subjects for each treatment The following is the set of six possible "plans" or randomizations, which can be constructed prior to random assignment to treatments:

A	B	A	B	A	B	A	B	A	B	A	B
a	c	a	b	a	b	b	a	b	a	c	a
b	d	c	d	d	c	c	d	d	c	d	b

The random assignment performed by randomly selecting two subjects for A and having the remaining two subjects take treatment B is the same as randomly selecting one of the six randomizations in the above set of randomizations. Step 1 in Section 14.5 then was carried out in the random selection of a randomization. We will now consider performance of the second step in Section 14.5. Suppose the actual randomization was the second, and that the results were A: *a*(8), *c*(6) and B: *b*(4), *d*(5). The above set of randomizations is then transformed into a reference set of data permutations by substituting for each letter designating a subject the obtained measurement for that subject to provide the following closed reference set of data permutations:

	A	B	A	B	A	B	A	B	A	B	A	B
	8	6	8	4	8	4	4	8	4	8	6	8
	4	5	6	5	5	6	6	5	5	6	5	4
T_A	12		14		13		10		9		11	

Whatever test statistic is to be used is computed at this stage. (However, the test statistic to be computed must be decided upon in advance or be chosen in some other manner that is independent of which data permutation represents the obtained results.) For a one-tailed test where A was expected to provide the larger measurements, we could use T_A as the one-tailed test statistic and the P-value — which is the proportion of T_As as large as 14, the value for the obtained (second) data permutation — is 1/6, as it is the only data permutation in the set with such a large value of T_A.

The validity criterion given in Section 3.3 was that the probability of a P-value of *p* must be no greater than *p* when the null hypothesis is true. It can readily be shown that the randomization-referral procedure has that property. Although illustrated above for only four subjects randomly assigned to two treatments with two subjects for each treatment, the principle of deriving a data permutation for each randomization on the basis of the obtained results is applicable to experiments with any number of subjects and any number of treatment levels. For each randomization, there is a corresponding data permutation derivable by the randomization-referral procedure, and random assignment is random selection of one of the equally probable randomizations and thus random selection of one of the equally probable data permutations. With no ties in the test statistic values over n data permutations, the probability of random selection of the data permutation with the largest value is $1/n$, of selecting one of the two largest values is $2/n$, or in general, one of the k largest values is k/n. Allowing for the possibility of ties, the probability of getting one of the k largest of n test statistic values is no greater than k/n. As the P-value calculated by the randomization-referral procedure is k/n, the procedure is valid.

That the probability of selecting one of the k largest of n test statistic values is no greater than k/n is true for any set of numerical values, each of which has the same probability of being selected. This property of randomly sampled sets is extremely useful in assessing the validity of randomization tests with systematic or random data permutation.

Example 14.2. Prior to random assignment, suppose the experimenter in Example 14.1 had represented the six possible randomizations in the order given and decided to use only the first three or the last three as the set for deriving three data permutations, the choice to be based on whether the actual randomization falls in the first or second half of the row of possible randomizations. Now if the second randomization was the randomization performed in the experiment, the P-value for T_A of 14 for the corresponding data permutation — which would be 1/6 if the reference set for all six randomizations was used — would be 1/3 with the use of the reference set consisting of only the first three data permutations. Or if the sixth randomization was performed in the experiment, the P-value for T_A of 11 for the corresponding data permutation — which would be 4/6 if the reference set of all six randomizations was used — would be 1/3 with the use of the reference set consisting of the last three data permutations. Despite the discrepancy in P-values for the reference set for all randomizations and the subset of that reference set, either method is valid. (Of course, it would not be valid to get a P-value by both methods and then use the smaller P-value as if such "data-snooping" had not occurred.)

The random assignment procedure performed by the experimenter is equivalent to randomly selecting the first or second half of the randomizations, then randomly selecting one of the randomizations from the half that was selected. To say that a randomization that determined the subset of randomizations is randomly selected from that subset is to say that all members of the subset have the same probability of being selected. Whether a subset of the set of all equally probable randomizations was determined by an experimenter prior to the random assignment or was determined by the resulting randomization is irrelevant: the probability of getting a P-value of k/n (where n is the number of data permutations in the subset) is no greater than k/n.

A tactic that will be employed in subsequent sections of this chapter to demonstrate the validity of a reference set for a data-permuting test will be to show that the same reference set can be generated by a valid randomization-referral counterpart to the data-permuting test.

14.8 Invariance of Measurements under the Null Hypothesis

The typical randomization test is a test of the null hypothesis of invariance of the measurements of experimental units over treatment conditions. Tests of one-tailed null hypotheses are only apparent exceptions because, as was shown in Chapter 13, when P-values for those null hypotheses can be derived they are P-values borrowed from two-tailed null hypotheses of invariance of measurements. The null hypotheses tested in Chapter 12 and Chapter 13 seemed to specify a change in measurements over treatments, not invariance, but in every case the null hypothesis was a relationship between treatment and response magnitudes that was hypothesized to be invariant.

Quantitative laws tested in Chapter 12 specified a proportional relationship between treatments and responses to be invariant over levels of treatments. The null hypothesis therefore was that the relationship (response/ treatment or treatment/response) was invariant over treatment levels, and that relationship provided a numerical value ("measurement") for every subject that was permuted over treatment levels.

The tests of magnitude of effect in Chapter 13 hardly seem to be testing a null hypothesis of invariance of measurements over treatment levels, and the testing of null hypotheses about additive effects in Section 14.7 seem to involve the fluctuation of measurements from one data permutation to another. However, the transformation of those measurements by subtracting Δ from the A measurements before permuting the A and B measurements allowed for invariance of the measurements over data permutations for the simpler procedure. This transformation is not fundamentally different from those in Chapter 12. To test the null hypothesis A − B = 5, we could assign codes of 5 and 0 to A and B treatments, respectively. Then the "measurement" or numerical relationship to be permuted for each subject is (*raw measurement − treatment code*), which of course gives the same values to be permuted as subtracting 5 from the A measurements before permuting them and the original B measurements. Looked at in terms of expressing each measurement for a subject as (*raw measurement − treatment code*) makes the extension to more than two treatments obvious. For the null hypothesis that for treatments A, B, and C there will be a 5-point drop in magnitude of effect from A to B and B to C, we can assign codes of 10, 5, and 0 to A, B, and C, respectively, and subtract the codes from the raw measurements to obtain the values to be permuted.

14.9 General and Restricted Null Hypotheses

The reference set for testing a null hypothesis depends on two factors: the randomization performed in the experiment and the null hypothesis. The *general null hypothesis* is the hypothesis of no effect of any aspect of the treatment manipulation on the "measurement" of any experimental unit, where "measurement" can refer to the original measurement provided by some measuring device or any transformation of it, such as the ratio of the original measurement to the treatment magnitude for testing quantitative laws, as in Chapter 12. Alternatively, a *restricted null hypothesis* is the hypothesis of no effect of certain aspects of the treatment manipulation on the measurement of any experimental unit.

Two types of restricted null hypotheses that have been discussed in earlier chapters are: hypotheses restricted to certain levels of a treatment manipulation, such as for a planned comparison of treatments A and B after an experiment involving treatments A, B, and C; and hypotheses restricted to certain factors, as in a test of factor A in a factorial experiment. Reference sets consisting of data permutations for all possible random assignments, although appropriate for testing general null hypotheses, cannot be used to test restricted null hypotheses because data permutations for certain randomizations are indeterminate. Those data permutations are indeterminate that are associated with randomizations differing from the obtained randomization in the assignment of subjects with respect to a treatment manipulation not specified in the null hypothesis.

14.10   Reference Sets for General Null Hypotheses

In the following example, we will consider the relationship between the data-permuting and the randomization-referral approaches in generating a systematic reference set for testing a general null hypothesis. The link between the random assignment procedure and the permuting of the data for testing a general null hypothesis is so direct that it is the common way of providing a rationale for randomization tests: random assignment of subjects to treatments, under the null hypothesis, is the random assignment of measurements of subjects (or other experimental units) to treatments. The randomization test rationale for testing general null hypotheses is easiest to understand and is presented here prior to examining the rationale of more complex tests.

Example 14.3. Consider an experiment to determine the effect of drug dosage on reaction time. It is predicted that larger dosages will lead to longer reaction times. Four subjects are randomly assigned to 15, 20, 25, and 30 units of a certain drug, one subject per dosage level. The reaction time measurements associated with the dosages are 20, 37, 39, and 41, respectively. A randomization test using product-moment r as a test statistic could be conducted by the use of Program 8.1. There would be 24 data permutations in the reference set, and the highest possible correlation is associated with the obtained results, so the one-tailed P-value is 1/24, or about 0.04.

We will now consider how the test would be carried out, using a reference set derived by a randomization-referral procedure. The entire randomization set of 24 randomizations is constructed before the experiment, using the letters *a*, *b*, *c*, and *d* to designate particular subjects. Every possible randomization can be represented by a sequence of the four letters. Suppose the actual assignment is *acbd*, which means that subject *a* was assigned to 15, *c* to 20, *b* to 25, and *d* to 30 mg. This is the randomization randomly selected from the set of randomizations by the random assignment procedure. In conjunction with the experimental results, the obtained randomization provides the basis for transforming the set of possible randomizations into a set of data permutations comprising the reference set. Under the null hypothesis, the measurement associated with each subject in the experiment is the measurement that subject would have provided under any of the alternative randomizations. As the measurements for subjects *a*, *c*, *b*, and *d* were 20, 37, 39, and 41, respectively, those numerical values can be substituted for the respective letters in each of the randomizations to transform a randomization into a data permutation. (Again, it must be emphasized that the measurements in the data permutations could be any transformation whatsoever of the measurements from a measuring device, including those in Chapter 12 and Chapter 13.) The proportion of data permutations in the resulting reference set that have correlation coefficients as large as 0.87, the value for the experimental results, is the P-value. The obtained data permutation is the only one of the 24 with such a large correlation; thus, the P-value is 1/24, or about 0.04, the same as for the data-permuting procedure.

The reference set for the randomization-referral procedure is valid, and the validity of the data-permuting test can be shown by demonstrating that it has the same reference set and thus necessarily gives the same P-value. The randomizations that could result from the random assignment procedure are the 24 randomizations represented by permutations of the letters *a*, *b*, *c*, and *d*, and the randomization-referral procedure associates 24 permutations of the measurements 20, 37, 39, and 41 with those randomizations. A data-permuting procedure that arranges the four measurements in all 24 possible sequences produces the same reference set by directly — rather than indirectly — generating the data permutations as a result of applying the same procedure to permute the letters as was used to generate the set of randomizations.

14.11   Reference Subsets for General Null Hypotheses

Next, consider how to generate a reference subset to test the general null hypothesis of no effect of drug dosage on reaction time for the experimental results in Example 14.3. First, we will consider the conduct of a data-permuting test and then justify it by comparing it to the corresponding randomization-referral test.

Example 14.4. The rule here for permuting the data to get a reference subset is very similar to the one used as an example of a permutation group by

TABLE 14.1

Data Permutations and Associated Correlation Coefficients

	Drug Dosage			
15	20	25	30	r
20	37	39	41	0.87
37	39	41	20	−0.66
39	41	20	37	−0.36
41	20	37	39	0.15

Chung and Fraser (1958) in the article cited in Section 14.6: Move the first measurement of the experimental results (i.e., 20, 37, 39, 41) to the last position to generate a second data permutation, transform that data permutation in the same way to generate a third data permutation, and so on until no new data permutations are produced. Application of these operations results in the data permutations shown in Table 14.1. There are only four data permutations in the reference subset shown in Table 14.1 because transforming the fourth data permutation by moving "41" to fourth place in the sequence gives the initial data permutation. The collection of operations providing this subset of the primary reference set thus constitutes a permutation group. This permutation group produces only a subset of the data permutations associated with all randomizations, consisting of 4 of the 24 data permutations in the primary reference set that could be used for testing the same null hypothesis.

The validity of the reference subset of four data permutations generated by the data-permuting procedure can be demonstrated by considering how the same reference subset could be generated by a randomization-referral procedure. Before the experiment, we consider the set of all randomizations to determine whether it can be partitioned in such a way that the members of each subset could be generated by applying to any member the same permutation group that was applied to the obtained data permutation by the data-permuting procedure. The partition of the randomization set shown in Table 14.2 is such a partition.

Within each randomization subset, all four randomizations can be derived from any randomization in the subset by moving the first letter of a randomization to the fourth position in the sequence to generate a new randomization, transforming that randomization in the same way, and so

TABLE 14.2

Partition of a Randomization Set

Drug Dosage																							
15	20	25	30	15	20	25	30	15	20	25	30	15	20	25	30	15	20	25	30	15	20	25	30
a	b	c	d	a	b	d	c	a	c	b	d	a	c	d	b	a	d	b	c	a	d	c	b
b	c	d	a	b	d	c	a	c	b	d	a	c	d	b	a	d	b	c	a	d	c	b	a
c	d	a	b	d	c	a	b	b	d	a	c	d	b	a	c	b	c	a	d	c	b	a	d
d	a	b	c	c	a	b	d	d	a	c	b	b	a	c	d	c	a	d	b	b	a	d	c

on until no new randomizations are produced. In other words, we apply the same permutation group to the randomizations within subsets as was applied to the data permutations in the data-permuting test to generate the reference subset. The letters designating the subjects could be permuted in the same fashion to produce the subset of randomizations that is randomly sampled as the numbers designating the measurements for those subjects are permuted after the experiment. All 24 randomizations are represented in the partitioned set, and each is contained within a subset; therefore, the randomization that will be associated with the experimental results must be a member of one of the six subsets. No matter which randomization is associated with the experimental results, the data permutations in the reference subset generated by the data-permuting procedure are associated with the randomizations in the subset containing the obtained randomization, and so the same reference set would be used by either the data-permuting or the randomization-referral test. The reference set for the randomization-referral procedure is valid and because the data-permuting procedure provides the same reference set, that set also is valid.

For a test of any general null hypothesis by use of a reference set or subset, a data-permuting procedure is not valid unless the permutation group applied to the obtained data permutation is a group that would be valid to apply to every randomization with a randomization-referral procedure. If the permutation group applied to the obtained data permutation could provide data permutations associated with impossible randomizations, this condition would not be met. Such would be the case if we tossed a coin to assign subjects *a* and *b* to 15 or 20 mg and tossed it again to assign subjects *c* and *d* to 25 or 30 mg and the data were permuted by the procedure described above, whereby the first measurement is moved to the end repeatedly to generate four data permutations. If the obtained data permutation had been associated with randomization *abdc*, the data permutations generated would be those associated with *abdc*, *bdca*, *dcab*, and *cabd*, but the last three randomizations would be impossible under the assignment procedure.

14.12 Reference Subsets for Restricted Null Hypotheses

Reference subsets must be used to test restricted null hypotheses because data permutations for certain randomizations are indeterminate. Those data permutations are indeterminate that are associated with randomizations differing from the obtained randomization in the assignment of subjects with respect to a treatment manipulation not specified in the null hypothesis. Two types of restricted null hypotheses are hypotheses restricted to certain conditions and hypotheses restricted to certain factors. Tests of both types of restricted null hypotheses have been described in earlier chapters, and the following discussion will provide a rationale for those tests, which is fundamentally the same as the rationale in Section 14.11 for using reference subsets in testing general null hypotheses.

14.13 Reference Subsets for Planned and Multiple Comparisons

The reference sets for planned and multiple comparisons do not consist of all data permutations for testing the general null hypothesis of no effect of any aspect of the treatment manipulation because planned and multiple comparison null hypotheses refer only to certain treatment differences.

Example 14.5. Suppose six subjects are assigned randomly to three treatments, with two subjects per treatment, and the following results are obtained: A: 2, 6; B: 3, 8; C: 10, 11. If the null hypothesis is restricted to the difference between the effects of B and C, there is no basis for switching measurements for A with those for B or C in the generation of data permutations because treatment A might have a different effect. If in fact we employed a permutation group for permuting data over all three treatments, the primary reference set generated would not be a closed reference set because there would be no justification for permuting the A measurements. With such data permutation, the reference set would contain data permutations that under the restricted null hypothesis might not be the results that would have been obtained if the randomizations associated with those data permutations had been the actual assignment.

The test of the restricted hypothesis can be carried out by a data-permuting procedure in the following manner. Divide the four measurements for B and C between B and C with two measurements per treatment in all $4!/2!2! = 6$ ways, keeping the assignment of measurements 2 and 6 to A fixed. This permutation group provides the six data permutations shown in Table 14.3, which constitute the reference subset. For each of the data permutations, the one- or two-tailed test statistic for reflecting the expected difference between effects of treatments B and C is computed and the proportion of the test statistic values that are as large as the value for the first (obtained) data permutation is the P-value.

TABLE 14.3

Reference Set of Six Data Permutations

Treatments		
A	B	C
2, 6	3, 8	10, 11
2, 6	3, 10	8, 11
2, 6	2, 11	8, 10
2, 6	8, 10	3, 11
2, 6	8, 11	3, 10
2, 6	10, 11	3, 8

The reference subset provided by this data-permuting procedure is the reference subset that would be produced by a valid randomization-referral approach. The latter would use randomization subsets generated by applying to a randomization subset member the permutation group that was applied to the obtained data permutation to generate the reference subset for the data-permuting procedure. If we knew the obtained randomization — i.e., if we knew the treatment to which each subject was assigned — and we applied the data-permuting permutation group to it, a subset consisting of six randomizations would be generated. We could determine before the experiment all possible randomization subsets that could be produced. Those randomization subsets are the ones resulting from a partitioning of the $6!/2!2!2! = 90$ possible randomizations into 15 subsets. Each subset consists of six randomizations where the two subjects for A are the same and the four subjects for B and C are divided in all six ways between B and C. Whatever the obtained randomization might be, it will belong to one of the 15 subsets, and transforming the randomizations in that subset into data permutations will provide the same subset of randomizations as that given by the data-permuting procedure. To ensure the validity of a reference subset for a restricted null hypothesis, we must use only permutation groups that generate data permutations that, under the null hypothesis, could be derived from the associated randomizations. Holding the measurements for A fixed and permuting the other four measurements over B and C does provide only data permutations that can be determined from their associated randomizations. The permutation group employed in the data-permuting test provides only data permutations that are associated with possible randomizations and are determinate under the restricted null hypothesis. Thus, the reference subset is valid and the data-permuting procedure is valid.

More than one data-permuting procedure can be valid for testing a certain restricted null hypothesis. There may be a number of permutation groups that, in a given situation, could provide a valid reference subset. For example, for a repeated-measures experiment in which each of n subjects takes four treatments, A, B, C, and D, in a random order, a test of no differential effect of treatments A, B, and C could be based on a reference subset consisting of all data permutations under the restricted null hypothesis, of which there are $(3!)^n$, or on a reference subset of 3^n data permutations based on the permutation group that produces only the ABC, BCA, and CAB randomizations of the orders in which any subject would take the treatments.

14.14 Reference Subsets for Factorial Designs

Another type of restricted null hypothesis is one restricted to certain factors in a factorial experimental design. Of course the performance of a randomization test is different from that for planned and multiple comparisons or

TABLE 14.4

Data from a Factorial Experiment

		Levels of Factor A	
		A_1	A_2
Levels of Factor B	B_1	3, 5	8, 7
	B_2	1, 4	10, 12

other tests of restricted null hypotheses, but the principle is the same: the obtained randomization defines the randomization subset that the random assignment procedure randomly samples.

Example 14.6. Suppose eight subjects are assigned randomly to any of four cells in a 2×2 design, with two subjects per cell, and that the results shown in Table 14.4 are obtained. A test dependent on all $8!/2!2!2!2! = 2520$ divisions of the data among the four cells would provide a primary reference set based on a permutation group, but the set would not be valid for testing the effect of only one of the factors because it would include data permutations for randomizations for which, under the restricted null hypothesis, the data conceivably might be different. However, a valid test of the effect of only some of the factors in a completely randomized experiment can be based on a reference subset. Consider a data-permuting procedure to test the effect of factor A. Without assumptions about the effect of factor B, 36 data permutations can be generated on the basis of the experimental results and the null hypothesis of no effect of factor A. The 36 data permutations are those generated by exchanging the B_1 measurements between A_1 and A_2 in the six possible ways, while also exchanging the B_2 measurements between A_1 and A_2 in the six possible ways. These 36 data permutations constitute the reference subset used for determining the P-value of factor A.

Suppose we wish to conduct a randomization-referral test that necessarily will give the same reference subset as that provided by the data-permuting procedure. Prior to the random assignment, we decide to generate the 2520 possible randomizations in two stages. The first stage is to divide the eight subject designations, letters *a* to *h*, into four for B_1 and four for B_2 in all 70 possible ways. In the second stage, within each of the 70 randomization subsets the four subjects for level B_1 and the four subjects for level B_2 are assigned to the same levels of B over all 36 randomizations in the subset. When the members of the subset within which the obtained randomization falls are transformed into data permutations, the resulting reference set is the same as the reference set for the data-permuting test. As the reference set for the randomization-referral test obviously meets the validity requirement, the data-permuting test also must be valid.

14.15 Open Reference Sets: Treatment Intervention and Withdrawal

Justification of randomization test procedures that have closed reference sets has been provided in the last few sections of this chapter. Examples of open reference sets will now be examined. In N-of-1 experiments, one might be interested in testing the effect of either intervention or withdrawal, without assumptions about the effect of the other. When we examine a random assignment procedure that consists of randomly selecting a single pair of treatment times, the earlier of the two being the time for intervention and the latter the time for withdrawal, as in Section 11.16, we see some complications that make it impossible to construct a closed reference set for testing the effect of either intervention or withdrawal alone.

Example 14.7. Consider an experiment where the intervention and withdrawal times are selected from only six pairs of treatment blocks: 8-12, 8-13, 8-14, 9-13, 9-14, and 10-14, where the first number indicates the block of time at the beginning of which the treatment was introduced and the second number the block at the end of which it is withdrawn. The sixth pair of treatment blocks (i.e., 10-14) is selected, and after treatment intervention and withdrawal, the data for all treatment blocks in the study are recorded. To test the effect of treatment intervention without any assumptions about the effect of withdrawal, we decide to compute our test statistic, which is the mean of the treatment blocks minus the mean of the control blocks (the blocks preceding treatment intervention), for each of the six possible randomizations to get our reference set. Under the null hypothesis of no effect of intervention, the test statistic values for the intervals 8-12, 8-13, and so on are just what they would have been if any of those pairs of treatment blocks had been the obtained randomization. Despite faithful representation of alternative outcomes under the null hypothesis, the reference set would not be closed. Only the sixth randomization allows us to infer, under the null hypothesis, the test statistic values for all six randomizations. If the first randomization (i.e., 8-12) had been the obtained randomization, there would have been only the data permutation for that randomization in the reference set because the other five pairs include blocks that would be withdrawal blocks if the first randomization (8-12) was the one performed, and there is no basis for assuming the withdrawal blocks to have no effect when our null hypothesis is restricted to the difference between intervention and control conditions. Depending on the pair of treatment blocks selected for intervention and withdrawal, we would have a reference set of one, two, three, four, five, or six data permutations. The reference set is open with one possible outcome being compared with certain alternatives, whereas those alternatives are not compared with each other. The effect of lack of closure of reference sets in biasing P-values is not always obvious but in the present example, it is easy to show how it could tend to produce small P-values in the absence of treatment effects.

This result can be shown by considering the effect of extremely strong extraneous factors (practice, motivation, and so on) that consistently produced a positively accelerated trend over time in the treatment block measurements even though there was no effect of intervention on the measurements. In this case, within each of the six reference sets the obtained test statistic value — which is compared only with test statistic values for earlier treatment blocks — would have the largest test statistic value. Each of the intervention points would have a probability of 1/6, so there would be a probability of 1/6 of getting a P-value of 1/6 and a probability of 1/6 of getting a P-value of 1/5, so that under the null hypothesis, the probability of getting a P-value as small as 1/5 would be 1/3, rendering the procedure invalid.

Such strong extraneous factors as those that would consistently produce positively accelerated upward trends may be extremely unlikely to be encountered in an experiment, but the validity of a randomization test procedure must not be contingent on the nonexistence of such strong influence. If less consistent upward trends existed in the absence of treatment effects, P-values for the above procedure that produced open reference sets would be affected in the same way only to a lesser degree; the procedure still would be invalid. A similar lack of closure of the reference set would be encountered if we were to try to test the effect of withdrawal alone. The lack of closure of the reference sets for the test of intervention or withdrawal alone means that there is no basis for a valid randomization test of those specific effects with such an experimental design.

14.16 Closed Reference Sets: Dropouts

In Chapter 3, it was noted that lack of independence among experimental subjects invalidates a randomization test. When subjects drop out of experiments without providing measurements for reasons that are related to the particular treatment to which they were assigned, invalidation of a randomization test also occurs. Therefore, independence among subjects or other experimental units and no differential dropping out are assumptions for all randomization tests. (These assumptions are necessary for any statistical test.) The following discussion of closed reference sets when there are dropouts therefore is based on the assumption that subjects who drop out would have dropped out no matter what random assignment to treatments was made.

In Section 4.15, it was indicated that if there are dropouts one can use either the reference set where dropout indicators as well as data are permuted or a subset of it. Although the discussion in that section was in terms of the between-subjects t test, the choice of two or more alternative reference sets in the case of dropouts is available for any experimental design.

The permutation group permutes the data and markers indicating dropouts, so that for some data permutations the dropouts will be associated with certain treatments and for other data permutations with other treatments. As the dropouts to which we are referring provide no measurements, the test statistic should take this fact into consideration in computing means; otherwise, if the sum of the measurements for a treatment is divided by the total number of subjects taking the treatment (including the dropout), the dropout is in effect given a score of 0.

A reference subset is an alternative that can be more readily adapted to the use of conventional randomization test computer programs. The permutation group for generating the reference subset holds the assignment of dropouts to treatments fixed and permutes the remaining data. The two systematic permuting procedures will not necessarily give the same P-value but both are valid procedures.

14.17 Open Reference Sets: Permuting Residuals

In the discussion in Section 4.6, it was stated that F for analysis of covariance should be computed separately for each data permutation. If residuals from regression lines based on the obtained data permutation were permuted, a permutation group could be employed for permuting residuals but the reference sets they produced would not be those that would have been produced if some of the randomizations associated with alternative data permutations had provided the experimental results. The reference sets therefore would be open.

Similarly, if one were to use an interaction test statistic in testing for a treatment effect and permute, as some have proposed, residuals based on deviations from row and column means, there would be lack of closure but closure could be ensured by permuting the raw data and computing the residuals separately for each data permutation before computing the interaction test statistic.

14.18 Sampling a List of Randomizations

The randomization-referral approach uses randomization sets formed prior to an experiment as a means of explaining randomization tests, and the value of that approach has been demonstrated for tests of general and restricted null hypotheses. Actual construction of a randomization set prior to an experiment and transforming the randomizations one at a time into data permutations is not usually a practical alternative to data permuting procedures, but conceiving of randomization-referral methods for performing the same test can help in understanding their data-permuting counterparts.

The set of randomizations can be for any type of assignment procedure, with the random assignment consisting in actual random sampling of a constructed set of randomizations. Brillinger, Jones, and Tukey (1978) recommended such a procedure for random determination of when to "seed" clouds by infiltrating them with chemicals in weather modification experiments because certain temporal patterns of cloud seeding were ruled out as undesirable. In agricultural experimentation, it has also been suggested that when only certain latin square designs or other designs are appropriate for an experiment the actual assignment should be randomly selected from those designs.

Where successive assignment of experimental units (one at a time) would lead to unequally probable randomizations, those randomizations can be listed with one of them to be randomly selected, making all listed randomizations equally probable. This was done in Section 12.16 to permit the use of a randomization test for treatment intervention and withdrawal with equally probable randomizations. Random determination of the point of intervention, followed by random determination of the point of withdrawal of treatment, makes randomizations of intervention-withdrawal combinations unequally probable, and so it was suggested that the intervention-withdrawal combinations be listed and a combination randomly drawn from the list would indicate the assignment to be performed.

14.19 Random Data Permutation: Hypothesis Testing vs. Estimation

The random data permutation procedures described in this book have been for sampling of a systematic reference set with replacement, although sampling without replacement — if practical — could have been employed. Here we will focus on sampling with replacement. The rationale for random data permutation is in connection with hypothesis testing, not estimation, and regarding random data permutation procedures as estimation procedures can cause confusion. Random procedures known as Monte Carlo procedures sometimes are used to determine the P-value based on a random reference set that does not include the obtained data permutation. We will call those procedures Monte Carlo estimation procedures because they estimate valid P-values instead of providing them.

As estimation procedures, Monte Carlo procedures could be and apparently have been regarded as providing P-values that are estimates of the "true" P-values based on systematic tests employing a single data permutation for every possible randomization. As the Monte Carlo P-values only approximate the systematic P-values, they are valid only to the extent that the approximation is close. The random data permutation procedures in this book can be shown to be completely valid but the P-values are not unbiased estimates of the systematic P-values. Alternatively, Monte Carlo estimation procedures that give completely unbiased estimates of the systematic P-values necessarily cannot be valid in their own right.

Let us examine the notion that Monte Carlo estimation procedures using random sampling with replacement, where the obtained data permutation is not included in the reference set, provide unbiased estimates of systematic P-values. To the extent that the null hypothesis is true, the obtained data permutation was randomly selected from a population of data permutations associated with possible randomizations and randomly permuting the data (appropriately) provides a random sample of data permutations from that population. Random sampling of a population with replacement provides unbiased estimates of proportions, and the systematic P-value is the proportion of the systematic reference set with values greater than or equal to a particular value. So if we randomly select n data permutations from the systematic reference set with replacement to estimate from those n data permutations the proportion of data permutations in the "population" with such a large value as the obtained value, we have an unbiased estimation of the P-value.

The random reference set in this book, which is generated in the same way as described but with the obtained data permutation added to the set, does not provide an unbiased estimate of the systematic P-value. The random data permutation P-values are related to the unbiased estimates of systematic P-values in this way: where the unbiased estimates are k/n, the random data permutation P-values are $(k+1)/(n+1)$ as a result of including the obtained data permutation in the reference set. The difference is small for large n (the number of random data permutations), but it can be seen that for any n the random data permutation P-values will be larger than the unbiased estimates of the systematic P-values, and consequently are not unbiased estimates of systematic P-values but if regarded as estimates tend to overestimate, erring on the conservative side.

Next, consider the random data permutation procedure as a valid procedure in its own right, in the sense that it provides P-values by a procedure that, under the null hypothesis, has a probability no greater than p of giving a P-value as small as p. Under the null hypothesis, a correctly programmed random data permutation procedure generates data permutations by the same sampling procedure as generated the obtained data permutation, and so the P-value based on $(k+1)/(n+1)$ is valid. The random data permuting procedure is conservatively biased as an estimator of a systematic P-value but valid in its own right.

The procedure of computing the P-value as the proportion of the randomly generated data permutations with test statistic values as large as the obtained value, using only the n data permutations as a reference set without including the obtained data permutation, can readily be shown not to be valid. If the set of n randomly generated data permutations did not contain any data permutation with test statistic values as large as the obtained value, the P-value would be 0, and with any possibility of a P-value of 0, a procedure cannot be valid. Thus, although the estimation procedure provides an unbiased estimate of the systematic P-value, it is valid only to the extent that it provides a close approximation to the systematic P-value.

It is easy to mistake the intent of the comparison being made between the random data permutation and the Monte Carlo estimation procedures. It is not to propose that in practical application it makes much difference which is used or even that one procedure is always better than the other. The intent is to examine two different points of view that have implications for the interpretation of randomization test results. The estimation viewpoint appears to assign random data permutation procedures a role that can be justified only as providing approximations to the "true" P-values provided by the systematic procedures. Alternatively, the validity viewpoint appears to assign greater importance to having a valid procedure than having a close estimate to the systematic P-value. In this chapter, we are concerned with theory and as the concern throughout the book has been with validity and hypothesis testing rather than estimation (not even estimation of P-values or of magnitude of treatment effects), the random data permutation procedures to be discussed will be those that are valid in their own right, despite the fact that for other purposes the estimation approach may be more appropriate.

14.20 Stochastic Closure When Assignments Are Equally Probable

The validity of a reference set for systematic data permutation is ensured if the reference set is systematically closed, as described in Section 14.5. Such closure is the invariance of the reference set over all assignments represented in the set under the null hypothesis. Employment of a permutation group is necessary but not sufficient to provide a closed reference set. Assessing validity of reference sets by seeing whether they are closed in the sense of meeting the requirement in Section 14.5 is a useful procedure for randomization tests based on systematic data permutation, but a different criterion of closure will be proposed for tests based on random data permutation.

A reference set for random data permutation is randomly produced, so the procedure of generating the reference set cannot ensure that the reference set will be the same as it would have been if some alternative randomization associated with the reference set had been the obtained randomization. However, there is a rule for guaranteeing the validity of either a random or a systematic data permuting procedure. A data-permuting procedure is valid if when H_0 is true, it provides reference sets meeting the following criterion:

> The probability of producing the same reference set would be the same, no matter which data permutation represented the obtained randomization.

A reference set meeting this requirement is a stochastically closed reference set, and the reference set for every test that has been proposed in previous chapters meets this requirement. Note that when the probability of producing the same reference set is 1, the above criterion becomes the criterion of systematically closed reference sets given in Section 14.5.

In Section 3.6, computing the P-value for random data permutation as the proportion of the reference set (which included the obtained data permutation) with a test statistic value as large as the obtained value was shown to be valid whenever each data permutation in the random reference set represents an outcome for the associated randomization under the null hypothesis. Determination that a reference set is stochastically closed validates randomization tests based on systematic or random data permutation. Furthermore, stochastic closure is a useful sign of validity for tests with both systematic and random components, and such tests will be discussed briefly later in the following section.

A generalization of the definition of permutation group in Section 14.6 can be made to accommodate production of stochastically closed reference sets, as follows:

> A stochastic permutation group is a collection of permuting (rearranging) operations that have the same probability of producing the same set when applied to elements in the set, regardless of the initial element to which the operations are applied.

14.21   Systematic Expansion of a Random Reference Set

The concept of a stochastically closed reference set based on a combination of random and systematic procedures can effectively expand the size of a reference set for random data permutation (Edgington, 1989). Such a possibility is of theoretical interest inasmuch as it illustrates an unusual application of the concept of stochastic closure. Its practical value is the result of explicitly or implicitly incorporating systematic permuting procedures, which tend to be faster, into random data permuting procedures.

Example 14.8. An experimenter predicts that treatment A will provide larger measurements than treatment B in an equal-n experiment, so the experimenter decides to use T_A as the test statistic. T_A is computed for the experimental results and for 999 random data permutations. Each of the 1000 data permutations is systematically permuted by transposing the A and B designations for the same partition of data. T_A is calculated for the resulting 1000 data permutations as well, and the proportion of the 2000 data permutations with as large a value of T_A as the obtained results is the P-value. The reference set is stochastically closed.

The practical value of this combination of random and systematic data permuting for equal-n between-subjects t tests is minimal, but the same type of approach can be useful for equal-n correlation trend tests. Given random assignment of the same number of subjects to each of four treatments, A, B, C, and D, with trend coefficients for a predicted linear upward or other specified trend and random data permutation, the reference set of 1000 test statistic values can be enlarged to 24,000 by computing 24 test statistic values for the experimental results and for each of the 999 random permutations of the data. For each of the original 1000 data permutations, the four trend coefficients for treatments A, B, C, and D would be systematically permuted over the four sets of data in all $4! = 24$ ways, and for each of those 24 permutations a trend test statistic would be computed.

In Section 14.11, a permutation group for generating a reference subset for a correlation test of a general null hypothesis consisted in repeatedly moving the first measurement to the last position to produce a new data permutation. Consider a combined systematic-random data permuting procedure for 10 measurements associated with 10 treatment magnitudes with ΣXY computed for each data permutation. Ten data permutations (including the obtained results) are derived from the obtained results using the above systematic procedure. Then 999 random permutations of the data are performed, after each of which the systematic procedure is employed, providing a reference set of 10,000 data permutations.

14.22 Random Ordering of Measurements within Treatments

Example 14.9. In the Chung and Fraser (1958) example of a permutation group, there were several subjects in each of two groups and the test was a test of the difference between the two groups. (Whether the comparison was of two different types of subjects or of two different treatments was not specified.) Suppose there were eight subjects assigned to two groups with four subjects per group. A group of the type Chung and Fraser proposed concerns moving the first two measurements to the end ("rotation of pairs") until no new data permutations are produced, and can be illustrated by the reference set in Table 14.5. In the correlation example in Section 14.9, there was a rule for ordering the obtained data, as the drug dosages were ordered. But the ordering of measurements within A and B for the Chung-Fraser example was not defined. The typical application of a randomization test of a difference between treatments does not need to take into consideration the order of the measurements within the treatments, as the within-treatment ordering does not matter. But the Chung-Fraser group does require a specification of order within treatments.

TABLE 14.5

A Reference Subset Produced by Rotating Pairs of Measurements

Treatment A				Treatment B			
X_1	X_2	X_3	X_4	X_5	X_6	X_7	X_8
X_3	X_4	X_5	X_6	X_7	X_8	X_1	X_2
X_5	X_6	X_7	X_8	X_1	X_2	X_3	X_4
X_7	X_8	X_1	X_2	X_3	X_4	X_5	X_6

Consider why ordering within A and B is necessary for the Chung-Fraser group. The reason is that the reference set will vary according to the initial ordering of data within the A and B treatments. The divisions of experimental units between treatments represented in the reference set depend on the initial ordering. For example, notice that if the sequence of measurements for the first data permutation in Table 14.5 was reversed within treatment A, then the second data permutation would have measurements X_1, X_2, X_5, and X_6 under treatment A, a data permutation that is not included in the subset in Table 14.5.

A specification of the order of measurements within treatments is necessary, but the Chung-Fraser permutation group does not permit systematic ordering within treatments that would make all four data permutations in Table 14.5 representations of possible alternative randomizations. This is the case whether the obtained results are ordered within treatments according to size of measurements, the sequential order of selection of subjects for assignment, the order in which measurements were obtained, or any other information. This problem is easy to demonstrate: for treatment A, observe the intransitivity in the ordering of measurements. For the first data permutation, measurements 1 and 2 precede 3 and 4 on some dimension (not necessarily size of measurements), then for the second data permutation, measurements 3 and 4 precede 5 and 6, then 5 and 6 precede 7 and 8, and finally 7 and 8 precede 1 and 2. Just knowing which four measurements came from A and which four from B is not enough for the Chung-Fraser group approach. It also requires nonarbitrary ordering of measurements within treatments, which cannot be done.

As a systematic data permuting procedure, the permutation group application just considered is inadequate for a randomization test but the addition of a random component to the systematic data permuting can provide a stochastically closed — and therefore valid — reference set. This approach does not require any special information about how the data should be ordered; only the information ordinarily recorded, which is what measurements were obtained under each treatment, is required. The random aspect of the data permuting is introduced first, then followed by the systematic component: randomly order the obtained measurements within A and within B, then rotate the pairs systematically, as shown in Table 14.5. With this approach, the primary set of randomizations would be the $8!/4!4! = 70$ divisions of subjects between treatments, the ordering within treatments within each randomization being irrelevant. The data permutations provided by permuting the data, using the random and systematic procedures jointly, are associated with particular randomizations. Any one of the four randomizations (unordered assignments) associated with the four data permutations would not necessarily generate the same reference set of data permutations. However, the probability is the same, namely $1/(4!)^2$, for each of the four randomizations associated with the data permutations that random ordering of data, followed by rotation of pairs, would generate the same reference set of data permutations. In other words, the procedure would provide a stochastically closed reference set.

A larger stochastically closed reference set can be derived by repeating the procedure in the preceding example in combination with random data permutation in this manner: randomly determine the order of listing of the obtained measurements within A and B, then use the rotation described to produce m systematic data permutations, for each of which there is a test statistic value computed, then randomly order the obtained measurements within A and B again and compute m more systematic data permutations. If n random orderings of the measurements within groups are made, there would be a total of mn data permutations.

14.23 Fixed, Mixed, and Random Models

In experimental design, distinctions are made between fixed, mixed, and random models. A fixed model would be one in which statistical inferences are restricted to the actual treatments that were used in the experiment, whereas a random model would be one in which the treatments are regarded as randomly selected from a population of treatments, and inferences concern the population. Mixed models have components of both fixed and random models, as in some factorial designs in which some factors have fixed levels and others random. For simple ANOVA tests, the difference between a fixed and a random model has to do with the interpretation of results but does not affect the P-value. However, even in simple randomization tests, the P-value can be affected by whether the treatments are or are not randomly selected from a population of treatments.

In Section 10.3, examples were given of randomization tests based on the random selection of treatment levels for matching and proximity tests. It was shown that P-values for some experiments where matching or proximity test statistics are appropriate can be substantially smaller (when there is a treatment effect) if treatments are randomly selected from a set of possible treatments to administer. That is, the random model could be more sensitive than the fixed model in the detection of treatment effects. In the following example, we will illustrate that the random selection of treatments (treatment levels) can increase the sensitivity of randomization tests where a particular type of correlation is predicted.

Example 14.10. An experimenter expects a positive linear correlation between stimulus and response magnitudes for stimulus magnitudes between 10 and 30. The magnitudes to which the subjects can be assigned are restricted to 10, 15, 20, 25, and 30. The null hypothesis to be tested is no differential effect of the treatment magnitudes, and the test statistic to be used in a randomization test is r, the product-moment correlation coefficient. Three subjects are available and random assignment to treatment levels is carried out in this manner: one of the five treatment magnitudes is randomly selected for the first subject, then one of the remaining four for the second subject, and one of the remaining three for the third subject; thus, there are $5 \times 4 \times 3 = 60$ possible randomizations. The randomization that occurred and the experimental results are shown below:

Stimulus magnitude	10	15	20	25	30
Response	3	—	9	11	—

The value of r for the obtained results is 0.996. The three measurements and two dashes, which indicate no measurements, are aligned under the stimulus magnitudes in all 60 possible ways and r is computed for each of the 60 data permutations. Only two other data permutations give a value of r as large as 0.996, the obtained value, namely the following:

Stimulus magnitude	10	15	20	25	30
Response	—	3	—	9	11
Stimulus magnitude	10	15	20	25	30
Response	3	—	—	9	11

The first of the two data permutations gives the same r as the obtained value, 0.996, and the second has a value of r equal to 1.00. The P-value therefore is $3/60 = 0.05$. With a "fixed model," there would have been three stimulus magnitudes to which the three subjects could have been randomly assigned, and the smallest possible P-value would have been $1/6$, or 0.167.

The reason for using r as a test statistic instead of ΣXY is that for this design, ΣXY is not an equivalent test statistic. For instance, the second data permutation shown above has a larger value of ΣXY than the third but a lower r. The nonequivalence of r and ΣXY and the unusual random assignment procedure prevent the use of the correlation programs in Chapter 8 to determine the P-value.

Correlation trend tests could be performed in a similar fashion. If there were five trend coefficients for a predicted linear upward trend, three of the treatment levels could be randomly selected, followed by random assignment of subjects among those three levels. The test statistic could be r, the correlation between the trend coefficients and the individual measurements, for 10 times as many data permutations in the reference set as without random selection of treatments. However, the increase in sensitivity for product-moment correlation or correlation trend tests depends on whether the experimental data tend to show a linear upward trend, not just on the increase in the size of the reference set.

An instance of a mixed model would be a factorial model in which factor A had fixed treatment levels but the levels of factor B were randomly selected, which might be useful in factorial designs in which the effect of factor B was tested by use of a correlation trend test, or in the case of matching or proximity measurement data by use of tests described in Section 10.3.

14.24 Deriving One-Tailed P-Values from Two-Tailed P-Values with Unequal N

With randomization test counterparts of independent t tests that have equal sample sizes, a P-value for a test of the one-tailed null hypothesis when the direction of difference between means is correctly predicted can be computed by halving the P-value for the two-tailed test. This practice is consistent with the relationship between parametric one-tailed and two-tailed P-values for a t test. But for randomization t tests with unequal sample sizes, the one-tailed P-value is not necessarily half of the two-tailed P-value, as was shown in Section 4.10. However, it is sometimes valid to halve two-tailed P-values with unequal sample sizes.

Example 14.11. In Example 4.6, the following data permutations were presented, where the third represented the obtained results:

	A	B	A	B	A	B
	2	3	3	2	5	2
		5		5		3
$|t| =$		1.33		0.04		8.33

The randomization test P-value for the two-tailed t test with $|t|$ or an equivalent test statistic is 1/3 (we used the equivalent $|D|$ test statistic in Example 4.6). The P-value for the one-tailed test, using $(\bar{A} - \bar{B})$ as the test statistic, also is 1/3. Even though the two-tailed P-value for the third (obtained) data permutation is 1/3, the one-tailed test P-value could not be half of 1/3 because with only three data permutations, the smallest possible P-value is 1/3.

There are circumstances where, for a situation like the one in Example 14.10, it would be valid to compute the P-value for a test of the one-tailed null hypothesis as half of the two-tailed P-value. This is the case when there is random determination of the treatment to have the larger number of subjects, as in the following example.

Example 14.12. Suppose the results in the preceding example came from an experiment for which only three subjects were available, making an unequal-n experiment necessary but where the experimenter had no more reason to choose A than to choose B as the treatment administered to only one subject. The experimenter tossed a coin and, getting heads, determined that A would be the treatment with only one subject. By drawing names from a hat, the three subjects then were randomly assigned to treatments A and B, with one subject for A and two for B. There was then a two-stage randomization procedure, and the reference set of all six equally likely data permutations based on both stages of randomization would consist of the three in Example 14.10 plus the following three, where A has two measurements and B only one:

A	B	A	B	A	B
3	2	2	3	2	5
5		5		3	

Knowing ahead of time that of the six data permutations in the reference set three would have one subject for A and three would have two subjects for A, the experimenter decides to use $(\bar{A} - \bar{B})$ instead of T_A as the one-tailed test statistic. The test statistic $(\bar{A} - \bar{B})$ value of 2.5 for the obtained results is the largest of the six values for all of the data permutations and so the P-value is 1/6, necessarily half of the two-tailed P-value for a correctly predicted direction of difference, due to the symmetry of assignment.

If an experimenter requiring unequal-n first randomly determines which of two treatments will have the smaller number of subjects and then randomly assigns subjects to treatments with those sample-size constraints, the one- and two-tailed P-values that the larger reference set would provide can be determined from a reference set half as large using the standard between-subjects t program. The two-tailed P-value computed as for a regular unequal-n design will be the value for the larger reference set based on both stages of randomization, and half of that two-tailed P-value will be the one-tailed P-value that the larger reference set would provide.

When a subject randomly assigned to a treatment in an equal-n between-subjects test design does not provide data that is used in a statistical test for reasons unrelated to the assignment, the derivation of a one-tailed P-value by halving the two-tailed value is valid because the inequality of sample sizes was randomly produced, although not by the intent of the experimenter. The justification is the same as for Example 14.11. This refers to situations where there are dropouts but also to situations in which measurements from some subjects are not used in data analysis, even though the subjects completed an experiment. For example, in Section 14.25 the matter of discarding of extreme measurements is discussed. When discarding of extreme measurements in an equal-n design results in inequality of sample sizes, the halving of the two-tailed P-value is a valid and efficient way of getting a one-tailed P-value. The same is true for experiments where, for example, discarding of an experimental animal's data occurs because conditions discovered during autopsy that are unrelated to experimental treatments disqualify a subject.

14.25 Test Statistics and Adaptive Tests

The discussions of permutation groups and closure of reference sets in this chapter have not been in terms of any particular test statistics because the validity of a randomization test is independent of the test statistic. The test statistic can be chosen to be one that is likely to be sensitive to a certain type of treatment effect.

When a test statistic is chosen prior to the conduct of an experiment, there need be no concern about an improper choice affecting the validity of a randomization test. (It could of course affect the power of the test.) But the validity of an adaptive test, which is one where a test statistic is chosen on the basis of the experimental results, is another matter. Care must be taken to avoid invalidating a randomization test. Hogg (1982, p. 21) indicated that selecting a test statistic on the basis of the research results "and then making the inference from the same data can certainly destroy certain probabilities that are of interest to the statistician." However, he noted that some non-parametric procedures allowed use of adaptive tests without that concern.

Transformations of data on the basis of observed results is in effect altering the test statistic, which would then be computed using the transformed data. A common usage of adaptive tests is to decide whether to discard extreme values, i.e., outliers. In Chapter 4 and Chapter 6, it was shown that the sensitivity of randomization tests is not necessarily adversely affected by the presence of an outlier to the same extent as a parametric test using the same test statistic. Some statisticians (e.g., Spino and Pagano, 1991) nevertheless advocate increasing the sensitivity of a randomization test by detecting and eliminating outliers.

One procedure is to combine the measurements from all treatments into a single distribution and to decide on the basis of the joint distribution whether there are low or high values that should be discarded before per-forming a randomization test. After discarding the data judged to be outliers, the randomization test can be conducted on the remaining data in the usual way. Although the judgment of what data to analyze is made after examining the experimental results, it is made before knowing the treatments associated with the individual measurements. Thus, if the null hypothesis is true, the reference set will be the same no matter what randomization is the one that provided the obtained results. Such decisions, whether the result of subjec-tive or objective criteria, consequently do not affect the validity of a random-ization test.

It is possible to carry out a valid randomization test when data are dis-carded on the basis of their extremeness within treatments, not on the basis of the joint distribution (Spino and Pagano, 1991). Suppose that prior to seeing the obtained results, we decide upon a criterion for determining which measurements to omit or ignore in computing a test statistic as, for example, those measurements more than three standard deviations from a treatment mean. The permuting of the data is done without respect to what measurements are left out of the computation of the test statistic; all mea-surements are permuted but some are not included in the computation of a test statistic. An example would be computing a difference between means for the obtained results on the basis of all data except the measurement value 85 for treatment A, which was over three standard deviations from the A mean, then permuting all of the data (including the measurement value 85) to provide the next data permutation. Whether the 85 or other values would be included in computing the difference between means for that data permutation would be dependent on the distribution of data within treatments. Whether certain measurements are or are not included in the computation of a test statistic thus depends on the data permutation but not on whether that data permutation represents the observed results; consequently, the reference set for such adaptive testing is closed.

14.26 Stochastic Closure When Assignments Are Not Equally Probable

The theory in this chapter — and indeed, the discussion throughout this book — has been directed toward randomization tests for experiments in which all possible assignments are equally probable. This is by far the most common type of randomized experiment but special occasions can call for the treatment of data from experiments where assignments are not equally probable. Even if such occasions are uncommon, they deserve a brief discussion because of their theoretical implications.

Cox (1956) and Kempthorne and Doerfler (1969) considered random assignment when assignments are not equally probable. For randomization tests employing systematic reference sets, the P-value could be the sum of the probabilities of all randomizations associated with data permutations in the reference set having test statistic values greater than or equal to the value for the experimental results. (Of course, when all randomizations are equally probable the P-value for that procedure would be simply the proportion of data permutations with test statistic values greater than or equal to the value for the experimental results.) This is not a data-permuting procedure that would provide a reference set by permuting the data without knowledge of the probability associated with each randomization.

A random data permuting procedure can be carried out when randomizations are not equally probable. The reference sets that are appropriate would not meet the criterion of stochastic closure in Section 14.20. A general criterion of stochastic closure, which reduces to the earlier criterion when randomizations are equally probable, is one that also is satisfactory for situations where randomizations are not equally probable:

> Given a reference set of n data permutations, each of which represents an outcome for the associated randomization under the null hypothesis, the conditional probability that any particular data permutation represents the experimental results is $1/n$ (Edgington, 1989).

An example of unequally probable assignments in an N-of-1 experiment was discussed in Example 11.16. The example concerned random intervention and withdrawal of a treatment over time, where the times at which withdrawal could occur depended on when the intervention took place. This constraint on withdrawal possibilities allowed enumeration of all possible randomizations, each involving a particular pair of times for intervention and withdrawal, but the possible randomizations varied widely in probability of occurrence.

The treatment intervention and withdrawal design allowed for unequally probable randomizations because the withdrawal time was dependent on the treatment intervention time. The inequality of assignment probabilities in general tends to be the result of the assignment probabilities being dependent on assignments that have already been made. Ways of performing random assignment for clinical trials to provide sequential assignment of subjects that has a random component independent of previous assignments but also a component that is dependent on previous assignments have been proposed (Atkinson, 1982; Begg, 1990; Simon, 1979; Smythe and Wei, 1983). For example, a procedure may be to randomly determine for each subject whether to administer treatment A or B but, except for the first subject, to have the probability of administering A be the proportion of B administrations to previous subjects, to render grossly unequal sample sizes improbable without putting undue constraints on the random assignment. The constrained assignment procedure would require a reference set of assignment sequences with unequal probabilities.

In N-of-1 experimentation as well, often there is a reluctance to have the random assignment at a given point in time be independent of previous treatments administered. There are various reasons for not wanting a sequence like AAAABBBB, where treatments are clustered in time or sometimes not wanting rapidly alternating sequences, and the proposal of varying assignment probabilities on the basis of previous assignments in N-of-1 experiments has been made (Onghena and Edgington, 1994).

Example 14.13. Consider the following randomization test application to data from an N-of-1 experiment with adaptive random assignment (Onghena, 1994). The experiment has treatments A and B, each administered three times, with constraints on the random administration to prevent clustering of A or B treatment times. Specifically, the experimenter starts with a "population" of three A and three B treatment designations, which are randomly selected without replacement with the constraint that after the same treatment was selected, say, twice in succession, the other treatment would be given; for example, the sequence ABBBAA would be impossible. Suppose the following randomization and results are obtained when the random assignment is made with the constraint that the same treatment would occur no more than twice in succession:

B	B	A	A	B	A	$p = 0.133$
7	3	8	9	4	6	

With a prediction that A would provide the larger measurements, the test statistic could be T_A, which would be 23 for the obtained results. The probability of the obtained assignment BBAABA is $3/6 \times 2/5 \times 1 \times 2/3 \times 1 \times 1/1 = 0.133$, based on random sampling without replacement from a collection of three As and three Bs with a probability of 1 for the third treatment being A because two Bs preceded it and a probability of 1 for the fifth treatment being B because two As preceded it. The precise two-tailed null hypothesis (upon which the one-tailed test is based) is that the sequence of measurements was not affected by the treatments given at those treatment times. Consequently, the reference set consists of all sequential orders of treatments that could have occurred, paired with the above sequence of measurements. As 6 of 20 possible sequences of three As and three Bs would have too much clustering, there are 14 assignments that the constructive-sequential procedure could have provided, and for each of the 14 associated data permutations the test statistic T_A is calculated. The P-value is the sum of the probabilities of the assignments giving such a large test statistic value. The only other data permutation in the reference set giving such a large T_A value as 23 is the following:

A	B	A	A	B	B	$p = 0.050$
7	3	8	9	4	6	

with a T_A equal to 24. The P-value thus is $0.133 + 0.050 = 0.183$.

In Example 14.13, determining the P-value by determining the proportion of the data permutations in the systematic reference set with as large a test statistic value as the obtained value would not give a valid P-value because the randomizations are not all equally probable. The determination of the P-value by systematic data permutation when the randomizations are not equally probable must be done by adding the probabilities of the randomizations providing test statistic values as large as the obtained value. Alternatively, random data permutation can be performed in such a way that the P-value can validly be computed as the proportion of data permutations with a test statistic value as large as the obtained.

The closure of a reference set ensures its validity. An implication of the criterion of stochastic closure given above is that a reference set will be stochastically closed for a test of the general null hypothesis when the stochastic permutation group employed in the random assignment is the same as the one employed in generating the reference set of data permutations, each data permutation being produced independently of earlier data permutations. To attain such closure, the experimenter "randomly assigns" the data to treatments in the same way the experimental units were randomly assigned.

Example 14.14. To determine the P-value of the results in Example 14.13 by random data permutation, the data permutations are produced with the same constraints as are involved in the random assignment. That is, a computer is programmed to select from three As and three Bs (without replacement) a sequence of six letters with the same treatment designation occurring no more than twice in succession. If 99 sequences were produced, each independently of the preceding sequences, each sequence would provide a T_A value by being paired with the obtained data sequence, and the proportion of those 100 values (consisting of the obtained value plus the other 99) that are as large as the obtained T_A would be the P-value.

The SCRT program included on the accompanying CD permits the user to introduce assignment constraints into random data permuting programs to accommodate various designs with unequal probabilities, as well as more conventional designs. The program can be used for random assignment as well as for permuting data. Generation of the random data permutation reference set must be based on a procedure that "samples" the set of all randomizations with replacement (i.e., independently of preceding samples). Alternatively, random sampling without replacement will not produce a stochastically closed reference set for situations with unequally probable assignments, and a P-value based on the proportion of data permutations in the reference set with a test statistic as large as the obtained value would be invalid. (The sequences or randomizations are sampled with replacement so that the production of one is independent of those produced previously.)

Random data permutation for designs with unequal probabilities of assignment can be performed to test restricted as well as general null hypotheses.

Example 14.15. Suppose an experimenter has randomly assigned treatments A, B, and C to nine treatment times in a N-of-1 experiment, with three administrations of each treatment and with the constraint that the same treatment does not occur more than twice in succession. The following sequence of treatments and results are obtained:

B	A	A	C	B	C	A	B	C
8	5	4	9	6	8	3	7	11

For an overall test of no significant difference among treatments, ΣT^2 could be employed as an equivalent test statistic to F. Random generation of the reference set could be employed to determine the proportion of the reference set of treatment sequences that have as large a value of ΣT^2 as the obtained value. For that test, the data permuting procedure would be a procedure of determining a sequence of treatment administrations and would copy the random assignment procedure.

For a test of the difference between the A and B treatments, those measurements associated with C — namely 9, 8, and 11 — would not be involved. The sequential positions to which the three As and three Bs could be assigned are indicated by dashes in the following diagram:

—	—	—	C	—	C	—	—	C

The constraint that no treatment occur more than twice in succession would only have relevance to the first three positions. With that constraint, the three As and three Bs are randomly assigned to the six positions, and for each resulting sequence T_A is computed for a test of a one-tailed H_0 where A is expected to provide the larger measurements. The P-value is the proportion of the n sequences consisting of the obtained sequence and the randomly generated sequences that have a T_A value as large as the obtained value. This procedure of generating treatment sequences is conditional on the obtained sequential positions of the irrelevant treatment C and thus could not generate certain sequences that would have been possible under the original three-treatment assignment procedure.

With a systematic procedure, the generation of the reference set depends on a knowledge of the probabilities associated with the various assignments, but that is not so for the random permutation procedure. The same group of operations for generating data permutations can be used as was used in the original assignment without knowing what the assignment probabilities are. The procedure of employing the same permutation group in generating data permutations as would provide the relevant set of randomizations renders random data permutation possible when systematic data permutation could not be validly employed.

Example 14.16. Consider a design in which each of 20 subjects takes two treatments, A and B, in an order determined by the toss of a coin where heads means that treatment A will be administered first. If the coin was known to be biased, with a known probability of heads equal to 0.6 rather than 0.5, a valid randomization test with either systematic or random data permutation for unequal probability of assignment could be performed. But only random data permutation could be validly employed if the nature of the bias was unknown. In such a case, the random reference set can be generated by the following procedure:

1. For each of the 20 subjects, examine the experimental results, record the treatment associated with each of the two measurements and which measurement was for the earlier administered treatment;

2. Produce a single data permutation by tossing a coin for each subject and "assigning" the measurement for the earlier treatment to treatment A for heads; otherwise, assign the measurement for the earlier treatment to B;

3. Generate $(n - 1)$ data permutations, which together with the observed results provide a reference set of n data permutations; and

4. Get the P-value by computing the proportion of the n data permutations with test statistic values as large as the value for the experimental results.

Example 14.17. Given three treatments in a repeated-measures design to be given in a random order independently for each subject, a randomization test could be carried out without knowing the probability of various sequences. One of the six sequences ABC, ACB, BAC, BCA, CAB, and CBA could be chosen for a subject depending on whether the roll of a die led to a 1, 2, 3 4, 5, or 6. Even if the die was loaded to make the probabilities of the different faces different, the random data permutation reference set could be derived by rolling that die for each subject and permuting the data accordingly, as for the coin of unknown bias. And of course, one could use a different coin or die for each subject. The random reference set then could be based on using the same coin or die to generate the sequence for a subject as was originally used in the random assignment.

14.27 Questions and Exercises

1. When the experimental results necessarily are in the form of ranks, why would a randomization test be better for determining the P-value than a conventional rank test?

2. Some people are interested in tests of randomness of certain processes but that is not the concern of this book. What are the random processes that experimenters or other researchers can themselves introduce into the research to facilitate statistical inferences?

3. Unit additivity is an assumption for randomization tests that was made by Kempthorne to enable him to draw statistical inferences about differences between means in randomized experiments. What is unit additivity and why is it an unrealistic assumption?

4. An experimenter randomly selects one of three subjects to take treatment A and assigns the remaining two to treatment B. The following results are obtained: A: 12; B: 6, 7. (a) What is the one-tailed P-value for a correct prediction of direction of difference, using the difference between means as a test statistic? (b) What is the two-tailed P-value associated with no prediction of which treatment would have the larger mean?

5. Suppose the experimenter in Question 4 has no basis for determining which treatment should have one subject and which two subjects, so they first randomly make that decision, happening to assign one subject to A and two subjects to B before dividing the subjects between treatments. (a) In this case, for the results in Question 4 what would be the P-value for a one-tailed test? (b) What would be the P-value for a two-tailed test?

6. What is the relationship between a permutation group and a closed reference set?

7. What is a stochastically closed reference set?

8. Give an example of an adaptive random assignment process. In what sense is it "adaptive"?

9. A physician doing a study on losing excess weight does not want to pick from a list of patients which patients to use as subjects. Instead, the physician decides to assign and administer the assigned treatment to patients when they happen to drop into the office over the following month. How could this be done in a manner that would ensure 10 patients assigned to each treatment?

10. In medical experimentation, there is sometimes the desire to have the probability of assignment to one or another treatment vary from subject to subject. Give a simple numerical example to explain what would be necessary to have a valid randomization test in such a situation.

REFERENCES

Atkinson, A.C., Optimum biased coin designs for sequential clinical trials with prognostic factors, *Biometrika*, 69, 61, 1982.

Begg, C.B., On inferences from Wei's biased coin design for clinical trials, *Biometrika*, 77, 467, 1990.

Brillinger, D.R., Jones, L.V., and Tukey, J.W., *The Management of Weather Resources. Volume II. The Role of Statistics in Weather Resources Management*, Report of the Statistical Task Force to the Weather Modification Board, Department of Commerce, Washington, DC, 1978.

Chung, J.H. and Fraser, D.A.S., Randomization tests for a multivariate two-sample problem, *J. Am. Statist. Assn.*, 53, 729, 1958.

Cox, R., A note on weighted randomization, *Ann. Math. Statist.*, 27, 1144, 1956.

Edgington, E.S., The assumption of homogeneity of variance for the *t* test and nonparametric tests, *J. Psychol.*, 59, 177, 1965.

Edgington, E.S., The role of permutation groups in randomization tests, *J. Educ. Statist.*, 8, 121, 1983.

Edgington, E.S., Stochastically closed reference sets, in: *Encyclopedia of Statistical Sciences*, Supplement Volume, Kotz, S. and Johnson, N.L. (Eds.), John Wiley & Sons, New York, 1989.

Hogg, R.V., Adaptive methods, in: *Encyclopedia of Statistical Sciences*, Vol. 1, Kotz, S. and Johnson, N.L. (Eds.), John Wiley & Sons, New York, 1982.

Kempthorne, O., Randomization — II, in: *Encyclopedia of Statistical Sciences*, Vol. 7, Kotz, S. and Johnson, N.L. (Eds.), John Wiley & Sons, New York, 1986.

Kempthorne, O. and Doerfler, T.E., The behaviour of some significance tests under experimental randomization, *Biometrika*, 231, 1969.

Onghena, P., *The Power of Randomization Tests for Single-Case Designs*, Ph.D. thesis, University of Leuven, Belgium, 1994.

Onghena, P. and Edgington, E.S., Randomization tests for restricted alternating treatments designs, *Behav. Res. Therapy*, 32, 783, 1994.

Pitman, E.J.G., Significance tests which may be applied to samples from any populations, *J.R. Statist. Soc. B.*, 4, 119, 1937a.

Pitman, E.J.G., Significance tests which may be applied to samples from any populations. II. The correlation coefficient, *J.R. Statist. Soc. B*, 4, 225, 1937b.

Pitman, E.J.G., Significance tests which may be applied to samples from any populations. III. The analysis of variance test, *Biometrika*, 29, 322, 1938.

Simon, R., Restricted randomization designs in clinical trials, *Biometrics*, 35, 503, 1979.

Smythe, R.T. and Wei, L.J., Significance tests with restricted randomization design, *Biometrika*, 70, 496, 1983.

Spino, C. and Pagano, M., Efficient calculation of the permutation distribution of trimmed means, *J. Am. Statist. Assn.*, 86, 729, 1991.

15

General Guidelines and Software Availability

This final chapter will provide some general guidelines for making effective use of randomization tests and give some references to free and commercial software packages that can be used to perform randomization tests.

15.1 Randomization: Multistage Model

A convenient device for remembering how to permute data in testing a *restricted null hypothesis* is to conceive of the overall random assignment procedure as being made up of consecutive stages (Edgington, 1969). For example, in a completely randomized two-factor experiment, for the purpose of testing the effect of factor B the overall random assignment procedure can be thought of as being composed of two stages: the first stage is the assignment of subjects to levels of factor A and the second stage is the assignment of subjects within levels of A to levels of B. The reference set for testing the restricted null hypothesis of no effect of factor B is produced by permuting the data in the same manner as they would have been permuted to test the hypothesis if the first stage of assignment had been nonrandom, as in blocking the subjects in a randomized block design. To test the effect of factor A, we can regard the assignment of subjects to levels of A as the second stage. The type of restricted null hypothesis that refers to only some of several treatments in a single-factor design would be treated in an analogous manner. With random assignment of subjects to treatments A, B, and C to test the hypothesis of identity of B and C effects, we can regard the first stage of assignment as the assignment of subjects to treatment A or to a treatment that is not A and the second stage as the assignment to B or C of the subjects not assigned to A. To produce the reference subset, we divide the data between B and C in the way the measurements would have been divided to test the hypothesis if the second stage had been the only random assignment performed in the experiment. To test any restricted null hypothesis, all stages of assignment of subjects with respect to treatment variation not specified in the null hypothesis can be regarded as having been completed in a nonrandom manner before the final stage of assignment. The latter is the part of the assignment to treatments that is relevant to the null hypothesis and is treated as the only aspect of the assignment that is random.

15.2 Permuting Data: Data-Exchanging Model

An alternate way of conceiving of the production of a reference set by the data-permuting procedure is in terms of exchanging data. In the following discussion of the data-exchanging model, there will be consideration of the cells among which exchanges of data are permitted by the null hypothesis, a cell being defined as the intersection of a level from each of the factors built into a design. In a completely randomized design, all factors are manipulated, so each cell represents a distinct treatment condition. With randomized block designs, factors used for classifying experimental units are not experimentally manipulated, so two or more cells can represent the same treatment condition for different types of subjects. The null hypothesis determines the cells to which exchanges of data are restricted.

Consider the cells to which subjects can be assigned in an experiment with four levels of factor A and three levels of factor B, as shown in Table 15.1, where the cells are numbered for identification. Table 15.1 will be used to evaluate the permuting of data for single-factor and two-factor experiments and will provide a basis for evaluating more complex designs.

For testing the effect of A within a particular level of B, the data are divided in all possible ways among those four cells. Alternatively, to test the null hypothesis of no differential effect of the first two levels of A, the data are divided between only the first two cells.

For a two-factor, completely randomized experiment, the hypothesis of no effect of either factor is based on a primary reference set resulting from exchanging data in all possible ways among all 12 cells. The hypothesis of no effect of factor A is tested by permuting data within three sets of cells: {1, 2, 3, 4}, {5, 6, 7, 8}, and {9, 10, 11, 12}. This leads to all possible patterns of data over the 12 cells that can be generated by exchanging data within

TABLE 15.1

Data-Exchanging Table

		Levels of Factor A			
		A_1	A_2	A_3	A_4
Levels of Factor B	B_1	1	2	3	4
	B_2	5	6	7	8
	B_3	9	10	11	12

those three sets of cells. The null hypothesis of no effect of either factor at the lowest level of the other factor is the hypothesis of no differential effect of cells 1, 2, 3, 4, 5, and 9, so the permutation group exchanges data among those cells in all possible ways.

Hypotheses for randomized blocks can be expressed in terms of exchanges of data between cells in a similar fashion. For example, in a randomized block design the three levels of B could represent three types of subjects or, for a repeated-measures experiment, the three levels of B could refer to three subjects, each of which was subjected to all four levels of treatment A. The null hypothesis of no effect of factor B is not testable by a randomization test when subjects are not assigned randomly to levels of factor B. However, the main effect of factor A or the effect of A within certain levels of B is testable and the test can be carried out as for a completely randomized design.

15.3 Maximizing Power

The versatility of the randomization test procedure provides the experimenter with so many valid alternatives in conducting experiments and analyzing the data that it is important for him to have a basis for choosing procedures that are most sensitive to treatment effects.

15.3.1 Homogeneity of Subjects

The experimenter should select subjects that are as homogeneous as possible to make sure that between-subject variability does not mask treatment effects. Homogeneous groups of other types can be used to test the generality of a finding for a homogeneous group used in an earlier experiment.

15.3.2 Statistical Control over Between-Subject Variability

Direct control over between-subject variability should be exercised through the selection of homogeneous subjects but when the subjects must be heterogeneous, statistical control may be necessary. Treatments-by-subjects designs can be used to control for heterogeneity of subjects, as can analysis of covariance and partial correlation. Repeated-measures designs and N-of-1 experiments can also be useful for minimizing the unwanted effects of between-subject variability.

15.3.3 Number of Subjects

As the number of subjects increases, the chance of detecting treatment effects also increases, provided additional subjects are as appropriate for the experiment as the early subjects that are available. In N-of-1 experiments, the chance of detecting treatment effects increases as the number of replications of treatment administrations increases, provided, of course, that increasing the number of replications for a subject does not severely affect the quality of responses through boredom, fatigue, or other influences.

15.3.4   Unequal Sample Sizes

Subjects should not be left out of an experiment simply to ensure equal sample sizes, equal cell sizes, or proportionality of cell frequencies. However, when there is a disproportionality of cell frequencies for a factorial design, the test statistic for nondirectional tests should be the special one given in Chapter 6 or else the data should be transformed according to the procedure in that chapter.

15.3.5   Restricted-Alternatives Random Assignment

Sometimes more subjects are available when the assignment possibilities are allowed to vary from subject to subject. However, when restricted-alternatives random assignment is employed the procedures in Chapter 6 for disproportional cell frequencies should be used for nondirectional tests.

15.3.6   Equal Sample Sizes

Although samples should not be restricted in size to ensure equality, equal sample sizes are preferable. With equal sample sizes, there are more potential assignments and, consequently, a smaller P-value is possible. For example, for between-subjects assignment there are $10!/5!5! = 252$ possible assignments of five subjects to one treatment and five to the other, whereas there are only $10!/3!7! = 120$ possible assignments of three subjects to one treatment and seven to the other. Another advantage of equal sample sizes is that statistics and experimental design books frequently give computational formulas for equal sample sizes only. When sample sizes for two treatments must be unequal because of limited numbers of available subjects, randomly determine which treatment will receive the larger number of subjects, thus allowing the two-tailed P-value to be halved for determining the one-tailed P-value as described in Section 14.24.

15.3.7   Counterbalancing

Counterbalanced designs put constraints on the random assignment possibilities and thereby limit the smallness of the P-values that can be attained. Such designs should be avoided unless there is no way to eliminate a strong sequential effect that is expected. Sometimes increasing the time between successive treatments will minimize such effects, serving the function of counterbalancing without restricting the random assignment possibilities.

15.3.8 Treatments-by-Subjects Designs

As discussed in Chapter 6, treatments-by-subjects designs can provide effective control over between-subject variability but there are fewer possible assignments than with one-way ANOVA. Consequently, unless there is interest in the treatment effect on each type of subject separately, the experimenter should strive for homogeneity of subjects to make this design unnecessary.

15.3.9 Precision of Measurements

Measurements should be as precise as possible. There is no necessity for reducing precise measurements to ranks to perform a rank-order test; the same kind of test can be performed with raw data when the P-value is determined by data permutation. Any degrading of precision of measurements should be avoided.

15.3.10 Precision of Test Statistics

To make effective use of the precision of the measurements, the test statistic should take into account the precise measurements, not just the rank order. Means and standard deviations are more precise measures of central tendency and variability and, consequently, they tend to be more appropriate test statistics or test statistic components than are medians and ranges.

15.3.11 Composite Measurements

In a multivariate experiment when measurements are combined to form a single composite measurement, the procedure should be such that the P-value is independent of the units of measurement. One means of accomplishing this is through the use of z scores, as in Section 7.5.

15.3.12 Combining Test Statistics over Dependent Variables

When test statistics are computed separately for each dependent variable in a multivariate experiment and they are added together to get an overall test statistic, the test statistics that are added should be such that the P-value is independent of the units of measurement. For example, in Section 7.6, F is such a test statistic but $\Sigma(T^2/n)$ and SS_B are not.

15.3.13 Directional Test Statistics

For categorical as well as continuous measurements, one-tailed tests should be used when there is a good basis for a directional prediction. A directional test statistic, such as the total frequency in the lower right cell, can be substituted for the nondirectional chi-square test statistic with a 2×2 contingency chi-square test. There is no factorial t counterpart to factorial F for main effects but when there are only two levels of a factor, a one-tailed test for the main effect of that factor over all levels of the other factors can be carried out by using T_L, the total of the measurements for the level predicted to provide the larger measurements, as the test statistic. One-tailed tests are preferable to two-tailed tests when the direction of difference between means or the direction of relationship — as in correlation — can be predicted. However, occasionally, journals require that two-tailed P-values be reported. When there are more than two levels of a treatment and a directional prediction is made, trend tests like those in Chapter 9 can be used. When trend tests are used, they should be designed to accommodate the full specificity of the prediction.

15.3.14   Control over Systematic Trends

When a subject takes a number of treatments in succession in either a N-of-1 or a repeated-measures experiment, and there is a general upward or downward trend over time, a test that compensates for such a trend will be more powerful than one that does not. For instance, one can use a randomization test based on difference scores, analysis of covariance with time as a covariate, or some type of time-series analysis.

15.3.15   Random Sampling of Levels of Treatments

In Chapter 10 and Section 14.23, it was shown that random selection of treatment levels in conjunction with random assignment can substantially increase the power of a test. There are only certain situations where this is possible but careful thought should be given to this possibility when there are only a few experimental units.

15.3.16   Combining of P-values

The P-values for pilot studies should be combined with those for the main study to increase the power of investigations. Similarly, the P-values for different experiments testing the same general H_0 should be combined. P-values also can be combined over replications of N-of-1 experiments with different subjects (or other entities).

15.4   Randomization Test Computer Programs on the CD

As mentioned before, the CD that accompanies this book includes RT4Win and its main source code (an integrated package of the main programs that have been discussed in this book), COMBINE (a program to combine

P-values that was introduced in Chapter 6), and SCRT (a program for randomization tests customized to N-of-1 trials as was mentioned in Chapter 11). The FORTRAN source code on the CD can be used to develop tailor-made randomization tests. Guidelines that follow can be helpful in making effective use of the source code on the CD.

15.4.1 Modifying Programs

The permuting procedures in the programs given on the CD correspond to the common types of random assignment and as a consequence, the programs can be readily modified to serve in many experiments. For instance, a program can be changed by altering the test statistic to make it sensitive to differences in variability instead of differences in size of measurements. Parts of one program can be combined with parts of another to make a new program, as was done in deriving the correlation trend program, Program 9.2. Program 9.2 is appropriate for the assignment associated with one-way ANOVA, so the data-permuting component is that of Program 4.3. However, the test statistics are the same as for the correlation program, Program 8.2, so the component of Program 8.2 that concerns the computation of test statistics was incorporated into Program 9.2. Programs that are not readily derivable from those in this book are ones for special random assignment procedures, like restricted-alternatives random assignment, random assignment for counterbalanced designs, and some of the assignments for N-of-1 operant research designs described in Chapter 11.

15.4.2 Equivalent Test Statistics

For both systematic and random data permutation, simple test statistics that are equivalent to the more conventional test statistics are useful. Not only do they reduce the computer time but they also make it easier to write and check programs.

15.4.3 Random Data Permutation

Random data permutation programs are more practical than those with systematic data permutation because little computer time is required for moderate or large samples. They are perfectly valid in their own right; the closeness of the P-value given by random data permutation to that of systematic data permutation is related to power alone, not to validity. Although the power is less than the power of systematic data permutation, the discrepancy is small with even a few thousand random permutations and can be reduced even further by increasing the number of random permutations employed (see Section 3.11). When a new type of random assignment is used, it is easier to develop a random data-permuting procedure that will match the assignment procedure than to develop an appropriate systematic permuting procedure.

15.4.4   Systematic Listing Procedure

For systematic data permutation programs, a systematic listing procedure is required. For conventional types of random assignment, the data permutation procedures used in the systematic data permutation programs in this book are the required listing procedures. However, for unconventional types of random assignment a new systematic listing procedure must be devised first and then incorporated into the program.

15.4.5   Elimination of Redundant Data Permutations

For systematic data permutation programs, the reduction in the required number of data permutations can be considerable for certain equal sample size designs when redundant data permutations are omitted. When sample sizes are equal, the experimenter using a systematic data permutation program should check to see if the symmetry of the sampling distribution of data permutations allows him to alter the systematic listing procedure so that only a fraction of the total number of data permutations have to be listed. However, this is a refinement that, like the formation of equivalent test statistics, is not essential to a systematic data permutation program and should be carried out only if a program is to be used repeatedly.

15.4.6   Output

A computer program for a randomization test should provide more than a P-value. It should also provide the value of t, F, or other conventional test statistics for the obtained results when the program is for determining the P-value associated with a conventional test statistic by data permutation. Equivalent, simpler test statistics may be used to determine the P-value but the ordinary test statistic should be determined for the obtained results so that it can be reported in the research report. Additionally, other statistics for the obtained results, such as means and standard deviations, may be appropriate to compute sometimes for reporting the results of the analysis.

15.4.7   Testing Programs

Section 3.11 discusses the testing of programs. Test data for which the P-value is known, like the data accompanying the programs in this book, should be used. The systematic programs should give the exact P-value and the random programs should provide a close approximation, given 10,000 random data permutations (Section 3.11). It is best to test the programs with data that do not give the smallest possible P-value because there are several ways for a program to be incorrect and yet give the smallest P-value correctly. In testing a program, it is frequently helpful to have all or some of the data permutations and test statistic values printed out in addition to the P-value for the test data.

15.5 Other Computer Programs

There are many other computer programs that can be put to the service of randomization testing, either because they were developed explicitly following the principles that are at the foundations of this book or because they contain fast routines for permutation or nonparametric tests that match common random assignment procedures, null hypotheses, and test statistics. We first present the programs that are available for free and in a second part the commercial software.

15.5.1 Other Free Computer Programs

The most important other free computer programs are, in alphabetical order, BLOSSOM, NPC-Test, NPFACT, NPSTAT, and RESAMPLING. The names of the developers, the main feature, a reference to a related textbook, and the website for downloading and contact information are in the following list:

15.5.1.1 *BLOSSOM*

Development: Paul W. Mielke, Jr., Brian S. Cade, and Jon D. Richards
Main feature: Distance function approach; multiresponse permutation procedures for one-way, randomized block, paired, and sequential tests
Reference: Mielke and Berry (2001), see also Cai (2006) for an SPSS implementation
Website: *http://www.fort.usgs.gov/Products/Software/Blossom*

15.5.1.2 *NPC-Test*

Development: Fortunato Pesarin and Luigi Salmaso
Main feature: Nonparametric combination method of dependent permutation tests; the website contains a free demo-version, SAS macros, S-Plus code and Matlab code
Reference: Pesarin (2001)
Website: *http://homes.stat.unipd.it/pesarin/index.html*

15.5.1.3 *NPSTAT and NPFACT*

Development: Richard B. May, Michael A. Hunter, and Michael E.J. Masson
Main feature: Randomization test approach; the program accompanies an introductory statistics textbook
Reference: May, Masson, and Hunter (1990)
Website: *http://web.uvic.ca/psyc/software*

15.5.1.4 *RESAMPLING*

Development: David C. Howell
Main feature: Randomization tests are embedded in a resampling approach; the program accompanies an introductory statistics textbook

Reference: Howell (2002)

Website: *http://www.uvm.edu/~dhowell/StatPages/Resampling/Resampling.html*

Free available code for more limited applications is presented in journal articles by Dallal (1988) (Fortran code), Bianchi and Segre (1989) (BASIC code), Baker and Tilbury (1993) (ANSI Standard C; see also *http://lib.stat.cmu.edu/apstat/283*), Chen and Dunlap (1993) (SAS code), and Hayes (1996; 1998) (Macintosh and SPSS code; see also *http://www.comm.ohio-state.edu/ahayes*). The excellent textbooks on resampling by Lunneborg (2000) and Good (2005a; 2005b) contain R code for several randomization tests (also in Resampling Stats and SC; see below for these two software packages). As mentioned in Section 11.23, Microsoft Excel programs for N-of-1 randomization tests are presented in Todman and Dugard (2001) and the stand-alone program RegRand (for regulated randomization tests in multiple-baseline designs) and its availability are described by Koehler and Levin (2000).

15.5.2   Commercial Computer Programs

Many commercial computer programs provide facilities for randomization and permutation tests. Most interesting in this respect are RESAMPLING STATS, RT, SAS, SC, SPSS, StatXact, LogXact, and TESTIMATE. The names of the developers, a one-line description, and the website for information on content, price, and ordering are given in the following list.

15.5.2.1   *RESAMPLING STATS*

Development: Julian Simon and Peter Bruce

Description: Resampling software, available as a stand-alone version or as an add-in for Excel or Matlab

Website: *http://www.resample.com*

15.5.2.2   *RT*

Development: Bryan Manly

Description: Resampling software that can be used together with the textbook by Manly (1997)

Website: *http://www.west-inc.com/computer.php*

15.5.2.3   *SAS*

Development: Jim Goodnight, John Sall, and SAS Insitute Inc.

Description: One of the most popular general statistical software packages that includes many procedures that can provide randomization test P-values

Website: *http://www.sas.com* (see also *http://support.sas.com/faq/003/FAQ00354.html*)

15.5.2.4 SC

Development: Tony Dusoir

Description: Statistical calculator and structured programming language that has many built-in routines for recent distribution-free methods, robust methods, and methods based on permutation

Website: *http://www.mole-soft.demon.co.uk*

15.5.2.5 SPSS

Development: Norman H. Nie, C. Hadlai Hull, Dale H. Bent, and SPSS Inc.

Description: Among the most widely used programs for statistical analysis in social science research; the Exact Tests module includes many procedures that can provide randomization test P-values

Website: *http://www.spss.com*

15.5.2.6 StatXact and LogXact

Development: Cyrus Mehta and Nitin Patel

Description: Comprehensive package for exact nonparametric inference and logistic regression, available as a stand-alone version or as an add-on to SAS

Website: *http://www.cytel.com/home/default.asp*

15.5.2.7 Testimate

Development: Volker W. Rahlfs

Description: Statistical software package for exact tests, effect size measures, confidence intervals, test for equivalence, and tests for noninferiority

Website: *http://www.idvgauting.de/eng/frame.htm*

More details on some of these computer programs and additional algorithms can be found in the Files Area of the EXACT-STATS mailing list of the National Academic Mailing List Service JISCmail (UK) (see *http://www.jiscmail.ac.uk/lists/exact-stats.html*).

REFERENCES

Baker, R.D. and Tilbury, J.B., Rapid computation of the permutation paired and grouped *t* tests, *Appl. Stat.*, 42, 432, 1993.

Bianchi, E. and Segre, G., A program in BASIC for calculating randomization tests, *Comput. Meth. Prog. Bio.*, 30, 305, 1989.

Cai, L., Multi-response permutation procedure as an alternative to the analysis of variance: An SPSS implementation, *Behav. Res. Meth.*, 38, 51, 2006.

Chen, R.S. and Dunlap, W.P., SAS procedures for approximate randomization tests, *Behav. Res. Meth. Ins. C.*, 25, 406, 1993.

Dallal, G.E., PITMAN: A FORTRAN program for exact randomization tests, *Comput. Biomed. Res.*, 21, 9, 1988.

Edgington, E.S., *Statistical Inference: The Distribution-Free Approach*, McGraw-Hill, New York, 1969.

Good, P., *Introduction to Statistics through Resampling Methods and R/S-PLUS*, John Wiley & Sons, London, 2005a.

Good, P., *Permutation, Parametric and Bootstrap Tests of Hypotheses* (3rd ed.), Springer, New York, 2005b.

Hayes, A.F., PERMUSTAT: Randomization tests for the Macintosh, *Behav. Res. Meth. Ins. C.*, 28, 473, 1996.

Hayes, A.F., SPSS procedures for approximate randomization tests, *Behav. Res. Meth. Ins. C.*, 30, 536, 1998.

Howell, D.C., *Statistical Methods for Psychology* (5th ed.), Duxbury, Belmont, CA, 2002.

Koehler, M.J. and Levin, J.R., RegRand: Statistical software for the multiple-baseline design, *Behav. Res. Meth. Ins. C.*, 32, 367, 2000.

Lunneborg, C.E., *Data Analysis by Resampling: Concepts and Applications*, Duxbury, Belmont, CA, 2000.

Manly, B.F.J., *Randomization, Bootstrap, and Monte Carlo Methods in Biology* (2nd ed.), Chapman & Hall, London, 1997.

May, R.B., Masson, M.E.J., and Hunter, M.A., *Application of Statistics in Behavioral Research*, Harper & Row, New York, 1990.

Mielke, P.W. and Berry, K.J., *Permutation Methods: A Distance Function Approach*, Springer, Wien, 2001.

Pesarin, F., *Multivariate Permutation Tests with Applications in Biostatistics*, John Wiley & Sons, New York, 2001.

Todman, J.B. and Dugard, P., *Single-Case and Small-n Experimental Designs: A Practical Guide to Randomization Tests*, Erlbaum, Mahwah, NJ, 2001.

Index